T0291570

HELIOPHYSICS

Active Stars, their Astrospheres, and Impacts on Planetary Environments

Heliophysics is a fast-developing scientific discipline that integrates studies of the Sun's variability, the surrounding heliosphere, and the environment and climate of planets. It encompasses all the variability that we know as space weather, but also enables us to understand the conditions leading to the relatively stable environment that supports life on Earth: our planet is in an orbit that has kept it within the habitable zone throughout the Sun's evolution, without overly detrimental effects of flares and coronal mass ejections, shielded well enough from galactic cosmic rays, and with the right amount of dynamo action.

This volume, the fourth in this series, explores what makes the conditions on Earth "just right" to sustain life, by comparing Earth to other solar-system planets, by comparing solar magnetic activity to that of other stars, and by looking at the properties of evolving exoplanet systems. By taking an interdisciplinary approach and using comparative heliophysics, the authors illustrate how we can learn about our local cosmos by looking beyond it, and, in doing so, also enable the converse.

Supplementary online resources are provided, including lecture presentations, problem sets, and exercise labs, making this ideal as a textbook for advanced undergraduate and graduate-level courses, as well as a foundational reference for researchers in the many subdisciplines of helio- and astrophysics.

The four volumes in the Heliophysics series are:

1. *Heliophysics – Plasma Physics of the Local Cosmos*
2. *Heliophysics – Space Storms and Radiation: Causes and Effects*
3. *Heliophysics – Evolving Solar Activity and the Climates of Space and Earth*
4. *Heliophysics – Active Stars, their Astrospheres, and Impacts on Planetary Environments*

CAROLUS J. SCHRIJVER is a Senior Fellow at Lockheed Martin Advanced Technology Center, California, where his work focuses primarily on the magnetic field in the solar atmosphere. His research includes the dynamics of magnetic field at and above the solar surface, how that shapes the heliosphere, and how it powers solar eruptions that are the drivers of space storms around Earth. Dr. Schrijver is involved in several NASA missions to better understand the Sun and the heliosphere. He is co-editor of the first three books in this series, and has published a popular science book entitled *Living with the Stars* on the multitude of connections between the human body, the Earth, the planets, and the stars.

FRANCES BAGENAL is Professor of Astrophysical and Planetary Sciences at the University of Colorado, Boulder. Her research comprises the synthesis of data analysis and theory in the study of space plasmas, and she has specialized in the fields of planetary magnetospheres and the solar corona. Professor Bagenal has been involved in NASA missions to planetary objects including Voyager, Galileo, and Deep Space 1, and she is an investigator on the New Horizons mission to Pluto, and the Juno mission to Jupiter.

JAN J. SOJKA is Department Head in the Department of Physics at Utah State University. His research explores how our upper atmosphere and its ionosphere are coupled with the magnetosphere and driven by the Sun and solar wind in an effort to unravel impacts on our technologies, focusing particularly on the modeling of the ionosphere and its couplings and drivers, to enable assessments of how well our theories conform to reality. For over four decades, Professor Sojka has been involved with sounding rockets, magnetospheric and ionospheric satellite missions, as well as numerous groundbased facilities. He is currently Co-Investigator on the NASA SDOs EVE instrument observing the X-EUV solar irradiance.

HELIOPHYSICS

Active Stars, their Astrospheres, and Impacts on Planetary Environments

Edited by
CAROLUS J. SCHRIJVER
Lockheed Martin Advanced Technology Center

FRANCES BAGENAL
University of Colorado

and

JAN J. SOJKA
Utah State University

CAMBRIDGE
UNIVERSITY PRESS

University Printing House, Cambridge CB2 8BS, United Kingdom

One Liberty Plaza, 20th Floor, New York, NY 10006, USA

477 Williamstown Road, Port Melbourne, VIC 3207, Australia

4843/24, 2nd Floor, Ansari Road, Daryaganj, Delhi - 110002, India

79 Anson Road, #06-04/06, Singapore 079906

Cambridge University Press is part of the University of Cambridge.

It furthers the University's mission by disseminating knowledge in the pursuit of education, learning and research at the highest international levels of excellence.

www.cambridge.org
Information on this title: www.cambridge.org/9781107462397

First published 2016
First paperback edition 2017

A catalogue record for this publication is available from the British Library

ISBN 978-1-107-09047-7 Hardback
ISBN 978-1-107-46239-7 Paperback

Contents

The plates are to be found between pages 116 and 117

Preface

Anyone who has ever seen a picture of Earth taken from deep space can be forgiven for thinking of these two words: "splendid isolation." Surrounded by millions of miles of uninterrupted black, the fragile blue globe seems profoundly alone, disconnected from anything else.

Nothing could be further from the truth: Earth is profoundly connected to our star.

The bright blue disk is just the most obvious evidence. A non-stop flood of sunlight warms the planet, simultaneously allowing us to live and to see. Invisible connections are equally profound. Solar radiation puffs up our atmosphere, altering its structure and chemistry. Solar winds buffet our magnetosphere, lighting up polar skies with curtains of light, and driving currents of electricity through the soil below. Solar magnetism deflects cosmic rays, moderating the effect of the Galaxy on our tiny home in space.

Years ago, the study of the Sun–Earth connection was edgy stuff. Big Thinkers held the planet and the star to be a system. From this synthesis emerged many new ideas and a new discipline called "heliophysics."

Now we know that they weren't thinking big enough. Like Earth, every world in the solar system is connected to its star. From the surface chemistry of Mercury, to the tattered atmosphere of Mars, to the flowing ices of Pluto, the fingerprints of solar activity may be found in all corners of the heliosphere.

In pop culture, people trace the "seven degrees of separation" between themselves and actor Kevin Bacon. Earth is connected much more closely to alien worlds. The central role of the Sun puts us just *one* degree of separation away from scores of planets, dwarf planets, moons, asteroids, and comets throughout the solar system. This proximity tells us something important: what we learn about those strange places, we also learn about ourselves.

The connectedness of things is the subject of this book: *Active Stars, their Astrospheres, and Impacts on Planetary Environments*. In 13 graduate-level chapters, experts lay out new ideas about how stars carve out a place in the galaxy to shape their own solar systems. The chapters touch on subjects ranging from magnetic reconnection and magnetohydrodynamics to climate and aeronomy. It may be one of the most interdisciplinary textbooks ever written – at least in the physical sciences.

Indeed, the themes laid out in this text are so interdisciplinary that their proper synthesis requires more than one book. This fourth volume of the Heliophysics series implicitly makes the case for a new research discipline: comparative heliophysics. As humans and their robots spread throughout the solar system, we will need this kind of interdisciplinary approach to understand the places we visit and to anticipate the dangers. What is the weather like on Titan today? How will a solar storm affect the ices of Europe? Is it safe to land on that comet?

These questions cannot be answered in "splendid isolation." Indeed, there really is no such thing . . . under the Sun.

Madhulika Guhathakurta, NASA/LWS program scientist,
Heliophysics Division,
Science Mission Directorate

Editors' notes

The Heliophysics series focuses on the physics of space weather events that start at the Sun and influence the Earth's environment and society's susceptibility to these processes. The Sun's variability affects not only Earth, but also the atmospheres, ionospheres, and magnetospheres of all other bodies throughout the solar system. The solar system offers a wider variety of conditions under which the interaction of bodies with a plasma environment can be studied than are encountered around Earth alone: there are planets with and without large-scale magnetic fields and associated magnetospheres; planetary atmospheres display a variety of thicknesses and compositions; satellites of the giant planets reveal how interactions occur with subsonic and sub-Alfvénic flows whereas the solar wind interacts with supersonic and super-Alfvénic impacts.

Analogous to the use of other environments in our own planetary system to learn more about Earth's environmental conditions we can look at the variety of other stars to learn more about our own Sun and its magnetism. With the realization that most stars support some form of planetary system, the variety of star–planet interactions to be considered also reaches far beyond our own present-day solar system: we can learn about the history and future of heliophysical processes throughout the life of the solar system by considering exoplanet systems, and we can envision environments elsewhere in the Galaxy by studying our solar system. Hence, the theme of this volume: comparative heliophysics.

The chapters in this volume, as in the others in this series, generally adhere to common practices in the disciplines that they primarily cover in their uses of symbols and units. We intentionally did not attempt to homogenize these across the contents to make it easier for the reader to connect the various chapters to the more detailed professional literature.

Additional resources

The texts were developed during a summer school series for heliophysics, taught at the facilities of the University Corporation for Atmospheric Research, in Boulder, Colorado, funded by the NASA Living With a Star program. Additional information, including text updates, lecture materials, (color) figures and movies, and teaching materials developed for the school can be found at http://www.vsp.ucar.edu/Heliophysics.

An online volume (Heliophysics V) describing the impacts of space weather on society in chapters looking into economic impacts, business opportunities, impacts on the electric power grid, and consequences of space weather for radio waves used in, e.g., navigation, can be accessed freely via the web.[1]

Definitions of many solar–terrestrial terms can be found via the index of each volume; a comprehensive list can be found at a web page maintained by NOAA's Space Weather Prediction Center.[2]

A note on star names and spectral designations

In this volume, several chapters refer to stars other than the Sun, using a variety of different naming conventions.[3] Here is a selective summary that covers most of what the reader will encounter in this volume. Stars that are relatively bright in the sky are often named after the constellation in which they occur, of which the Latin name is commonly abbreviated using three letters taken from the beginning of the word or words in their name, preceded by a Greek letter, starting with α for the brightest (e.g., α Cen for the brightest star in the constellation Centaurus) and continuing through the Greek alphabet, and continuing with numbers after the letters are all used. Alternative names include (1) old Arabic names (modified over time, such as Betelgeuse), (2) catalog numbers (such as the Henry Draper catalog numbers, e.g., HD 114762), (3) a letter combination preceding the constellation abbreviation which often designates a member of a variable star catalog (e.g., EK Dra), or (4) a rather unimaginative but straightforward combination of right ascension and declination that specifies the location in the sky, sometimes preceded

[1] http://www.vsp.ucar.edu/Heliophysics/science-resources-textbooks.shtml
[2] http://www.swpc.noaa.gov/content/space-weather-glossary
[3] See a summary at http://en.wikipedia.org/wiki/Star_designation

by a group of letters that specifies the type of star (e.g., the pulsar PSR 1257+12). Stars are often known by multiple names (users of, e.g., the Astrophysics Data System, ADS, can hunt through the literature by using databases of synonyms to search for studies on stars using multiple names simultaneously). In case of multiple stars, a capital following the name indicates which component is meant, most commonly differentiating between the brightest and next-brightest one as "A" and "B". When a lower-case letter follows a star name, the designation refers to a planet orbiting the star.

Spectral types of stars are a measure of their spectral properties, long since resequenced in order of decreasing effective temperature (and thus in order of color from blue to red): O, B, A, F, G, K, M, L, T, Y, followed by a number from 1 to 9 as a subdivision. The Roman numeral that follows it is a measure for stellar radius or surface gravity, and indirectly of relative age: V for a mature main-sequence star (such as our Sun), IV for a slightly evolved star, and then III, II, and I for evolved giant, bright giant, and supergiant stars.

Stellar astronomers characterize spectral lines by the abbreviated Latin name of the element followed by a Roman numeral that designates the ionization stage of the element, e.g., the designation "Mg II" refers to a spectral line (or, if no particular name or wavelength follows, the entire set of spectral lines) of singly ionized magnesium. Sometimes, authors may use the spectral designation to characterize the population of the emitting atoms (e.g., "D I atoms" for neutral deuterium). Particularly strong spectral lines may have names followed by a Greek letter (such as "Lyman α", the first in the Lyman series), or a capital letter which refers to a strong line identified in the solar spectrum (such as the Ca II H and K lines) or their equivalent transition in another element when indicated in lower case (e.g., for the H and K lines: the Mg II h and k lines).

Heliophysics

helio-, pref., on the Sun and environs, from the Greek helios.
physics, n., the science of matter and energy and their interactions.

Heliophysics is the

- *comprehensive new term for the science of the Sun–Solar System Connection.*
- *exploration, discovery, and understanding of our space environment.*
- *system science that unites all of the linked phenomena in the region of the cosmos influenced by a star like our Sun.*

Heliophysics concentrates on the Sun and its effects on Earth, the other planets of the solar system, and the changing conditions in space. Heliophysics studies the magnetosphere, ionosphere, thermosphere, mesosphere, and upper atmosphere of the Earth and other planets. Heliophysics combines the science of the Sun, corona, heliosphere and geospace. Heliophysics encompasses cosmic rays and particle acceleration, space weather and radiation, dust and magnetic reconnection, solar activity and stellar cycles, aeronomy and space plasmas, magnetic fields and global change, and the interactions of the solar system with our galaxy.

From NASA's "Heliophysics. The New Science of the Sun–Solar System Connection: Recommended Roadmap for Science and Technology 2005–2035."

1

Introduction

CAROLUS J. SCHRIJVER, FRANCES BAGENAL, AND JAN J. SOJKA

1.1 Comparative heliophysics

Our knowledge of conditions – past, present, and future – throughout the solar system is rapidly advancing. We can specify the details of the evolution of the Sun, and can represent the conditions of its magnetic activity over time with fair certainty. We have explored – both remotely and in situ – many of the "worlds" throughout the solar system, from Mercury to Pluto, including asteroids and comets. Along with an appreciation of the diversity of conditions in space around all these environments comes the realization that the conditions here on Earth are remarkably just right for the only world that we know to sustain life. Specifically, Earth is in an orbit that has kept it within the range in orbital distances within which the planet has maintained liquid water at its surface throughout the past several billions of years. It orbits far enough from the Sun that tidal forces could not lock Earth's rotation to its orbital motion, thus allowing solar irradiance to be effectively distributed over the sphere on the relatively short time scale of the terrestrial day. Moreover, since the Late Heavy Bombardment some 4 billion years ago, Earth has received limited environmental debris (asteroids, comets, ...). The impacts of the Sun's magnetic activity on the Earth have also been relatively benign, without overly detrimental effects of flares and coronal mass ejections. Moreover, the Earth is shielded well enough from galactic cosmic rays by both solar wind and terrestrial magnetism. And then there are the added conditions related to the internal properties of the planet that contribute to its ability to sustain life, including plate tectonics and dynamo action. In short, the Earth's characteristics are "just right" (see, e.g., Ch. 4 in Vol. III, and Ward and Brownlee, 2000, and also Vidotto *et al.*, 2014, and references therein, for select topics on cool stars and space weather and planetary habitability).

All of these circumstances conspire to a relatively stable environment, both in terms of the terrestrial climate and in terms of what we nowadays call "space weather". In order to understand the processes that create these environmental conditions, and to appreciate the influences of space weather on the planet that is our only known compatible habitat, we need to advance a field of science called "heliophysics". Heliophysics is the field that encompasses the Sun–solar system connections, and that therefore concentrates on the Sun and on its effects on Earth, the other planets of the solar system, and the changing conditions in space. As such, heliophysics has to cover an extraordinary range of topics, essentially being equivalent to ecology in attempting to describe a complex web of often non-linear interactions that certainly do not stop at the boundaries of the traditional scientific (sub-)disciplines or at the perimeters of domains maintained by funding agencies.

Heliophysics tackles the workings of an interconnected system of magnetic field and plasma that couples the Sun with her entourage that forms the solar system. Indeed, it is the commonality of plasma physics of this solar system family that has enabled the amalgamation of solar physics, cosmic ray physics, solar wind physics, magnetospheric physics, ionospheric physics, and more into heliophysics. What plasma physics does on the Sun to generate huge bursts of energy as photons, solar wind structures, and energetic particles, plasma physics also does on the planets, in effect protecting them from these energetic outbursts via magnetospheres and ionospheres, whose activity at times interferes with our technological tools on and near Earth.

The "habitable zone" of a planetary system is commonly defined loosely as the range of distances of planets relative to their parent star within which water can exist at the planet surface as a liquid. But from a heliophysics point of view, the definition of "habitable" also includes how well the planet is shielded from what the nearby star or the surrounding galaxy throws at it. In that sense, a habitable planet is one that provides a magnetic shield that stalls and deflects the energetic solar wind structures and cosmic rays respectively. The upper atmosphere of a habitable planet similarly plays the role of absorbing harmful radiations, in the process creating the planet's ionosphere. The ionospheric shield also safely dissipates the energy carried by huge currents generated in the magnetosphere as the planet's magnetic field stalls the solar wind structures. Heliophysics can thus be viewed as the integrating science that will determine which exoplanets are habitable to a technologically advanced life form. In fact, the aspects of this definition of "habitable planet" require understanding of space weather.

This is the fourth volume in a series that introduces the rich spectrum of topics covered in heliophysics (see Table 1.1 for the volume titles and Table 1.2 for thematically sorted chapter titles). Where the preceding three volumes focus on describing and understanding the conditions in our local cosmos, particularly where the Sun and its planets are concerned, here we take a somewhat different perspective. The focus of this volume is what we can learn from other stars, from

Table 1.1 *Titles of the volumes in the Heliophysics series. References in this volume to chapters in other volumes use the numbering as in this table.*

Volume	Title and focus
I	Plasma physics of the local cosmos
II	Space storms and radiation: causes and effects
III	Evolving solar activity and the climates of space and Earth
IV	Active stars, their astrospheres, and impacts on planetary environments
V[a]	Space weather and society

[a] Available online at http://www.vsp.ucar.edu/Heliophysics/science-resources-textbooks.shtml

Table 1.2 *Chapters and their authors in the Heliophysics series sorted by theme (continued on the next page), not showing introductory chapters.*

Universal and fundamental processes, diagnostics, and methods	
I.2. Introduction to heliophysics	*T. Bogdan*
I.3. Creation and destruction of magnetic field	*M. Rempel*
I.4. Magnetic field topology	*D. Longcope*
I.5. Magnetic reconnection	*T. Forbes*
I.6. Structures of the magnetic field	*M. Moldwin* et al.
II.3 In-situ detection of energetic particles	*G. Gloeckler*
II.4 Radiative signatures of energetic particles	*T. Bastian*
II.7 Shocks in heliophysics	*M. Opher*
II.8 Particle acceleration in shocks	*D. Krauss-Varban*
II.9 Energetic particle transport	*J. Giacalone*
II.11 Energization of trapped particles	*J. Green*
IV.11 Dusty plasmas	*M. Horányi*
IV.12 Energetic-particle environments in the solar system	*N. Krupp*
IV.13 Heliophysics with radio scintillation and occultation	*M. Bisi*
Stars, their planetary systems, planetary habitability, and climates	
III.3 Formation and early evol. of stars and proto-planetary disks	*L. Hartmann*
III.4 Planetary habitability on astronomical time scales	*D. Brownlee*
III.11 Astrophysical influences on planetary climate systems	*J. Beer*
III.12 Assessing the Sun-climate relationship in paleoclimate records	*T. Crowley*
III.14 Long-term evolution of the geospace climate	*J. Sojka*
III.15 Waves and transport processes in atmosph. and oceans	*R. Walterscheid*
IV.5 Characteristics of planetary systems	*D. Fischer & J. Wang*
IV.7 Climates of terrestrial planets	*D. Brain*

Table 1.2 *(Continued from the previous page)*

The Sun, its dynamo, and its magnetic activity; past, present, and future	
I.8. The solar atmosphere	*V. Hansteen*
II.5 Observations of solar and stellar eruptions, flares, and jets	*H. Hudson*
II.6 Models of coronal mass ejections and flares	*T. Forbes*
III.2 Long-term evolution of magnetic activity of Sun-like stars	*C. Schrijver*
III.5 Solar internal flows and dynamo action	*M. Miesch*
III.6 Modeling solar and stellar dynamos	*P. Charbonneau*
III.10 Solar irradiance: measurements and models	*J. Lean & T. Woods*
IV.2 Solar explosive activity throughout the evol. of the solar system	*R. Osten*
Astro-/heliospheres, the interstellar environment, and galactic cosmic rays	
I.7. Turbulence in space plasmas	*C. Smith*
I.9. Stellar winds and magnetic fields	*V. Hansteen*
III.8 The structure and evolution of the 3D solar wind	*J. Gosling*
III.9 The heliosphere and cosmic rays	*J. Jokipii*
IV.3 Astrospheres, stellar winds, and the interst. medium	*B. Wood & J. Linsky*
IV.4 Effects of stellar eruptions throughout astrospheres	*O. Cohen*
Dynamos and environments of planets, moons, asteroids, and comets	
I.10. Fundamentals of planetary magnetospheres	*V. Vasyliūnas*
I.11. Solar-wind magnetosphere coupling	*F. Toffoletto & G. Siscoe*
I.13. Comparative planetary environments	*F. Bagenal*
II.10 Energy conversion in planetary magnetospheres	*V. Vasyliūnas*
III.7 Planetary fields and dynamos	*U. Christensen*
IV.6 Planetary dynamos: updates and new frontiers	*S. Stanley*
IV.10 Moons, asteroids, and comets interact. with their surround.	*M. Kivelson*
Planetary upper atmospheres	
I.12. On the ionosphere and chromosphere	*T. Fuller-Rowell & C. Schrijver*
II.12 Flares, CMEs, and atmospheric responses	*T. Fuller-Rowell & S. Solomon*
III.13 Ionospheres of the terrestrial planets	*S. Solomon*
III.16 Solar variability, climate, and atmosph. photochemistry	*G. Brasseur* et al.
IV.8 Upper atmospheres of the giant planets	*L. Moore* et al.
IV.9 Aeronomy of terrestrial upper atmospheres	*D. Siskind & S. Bougher*
Technological and societal impacts of space weather phenomena	
II.2 Introduction to space storms and radiation	*S. Odenwald*
II.13 Energetic particles and manned spaceflight	*S. Guetersloh & N. Zapp*
II.14 Energetic particles and technology	*A. Tribble*
V.2 Space weather: impacts, mitigation, forecasting	*S. Odenwald*
V.3 Commercial space weather in response to societal needs	*W. Tobiska*
V.4 The impact of space weather on the electric power grid	*D. Boteler*
V.5 Radio waves for communication and ionospheric probing	*N. Jakowski*

other planetary systems, and from non-planetary bodies within the solar system. With this, we attempt to open up a view of what could be termed "comparative heliophysics" that aims to learn about our local cosmos by looking beyond it, and in doing so also enables the converse.

Where experimental physics offers researchers the possibility to set key parameters to desired values, albeit to a range of values generally limited by the instrumental setup, heliophysics does not offer that valuable option to study the dependence of processes on any one of the environmental parameters. We cannot change the mass, age, spin rate, or chemical composition of the Sun or of the Earth. Nor can we change the distance at which our home planet orbits its parent star. This is a major complication not only in quantifying how these and other parameters affect the Sun–Earth connections, but even in determining which parameters are of critical importance in shaping the climate of space throughout the solar system.

As our observational technologies advance, and as our observational archives grow, however, even heliophysics can approximate an experimental science by looking not only at the present-day Sun, Earth, and other planets, but by studying conditions of a multitude of Sun-like stars, of the other bodies within the solar system, and increasingly of other planetary (or exoplanet) systems.

1.2 Exoplanets

Until the mid 1990s, no planets were known to orbit Sun-like stars other than those orbiting the Sun. Within the following two decades, our capabilities and knowledge advanced to the estimate – using statistical arguments based on the observed sample – that most stars that we see in the sky have a planetary system with one or more planets; even binary stars show evidence of circumbinary planets in at least 1 in 10 cases (note that roughly 1 in 2 "stars" in the sky is actually a binary system composed of two, often quite dissimilar, stars). Chapter 5 outlines the observational instrumentation and techniques that enabled this dramatic shift in our knowledge.

Instrumental limitations and the properties and orbital parameters of exoplanets have thus far kept the number of detected exoplanet systems at several thousand. But that number suffices to enable us to learn much about the formation and evolution of planetary systems and their planets, as discussed in Ch. 5.

For example, the combination of observations and numerical experiments suggests that gas giants accumulate up to a few hundred Earth masses of material – first the solids and then increasingly rapidly gases – within a matter of a few million years. This process is aided in its efficiency by the migration of growing planets within the young planetary system: planets are not bound to their initial orbits, but can migrate either inward or outward, subject to gravitational interactions, thus having access to a large volume of the primordial disk from which to collect material. Interestingly, it appears that it is the very collection process of matter onto the growing planet that causes mass redistributions within the disk so that their tidal effects can make planets migrate, particularly if other planets are forming elsewhere in the system, while the gravitational coupling between multiple young planets in eccentric orbits can scatter bodies around (both in distance from their central star and in orbital inclination).

Planets may form into three distinct categories: terrestrial planets with radii up to about $1.7R_{\oplus}$, dwarf gas planets up to about $4R_{\oplus}$, and giant gas planets beyond that. Their radii and their atmospheric constituents appear to reflect their migration history within the primordial disk, but the details of these dependencies continue to be uncovered. The lower-radius group looks to be mostly rocky, while the intermediate-radius group shows evidence of an increasingly large gaseous envelope with increasing radius.

1.3 Cool stars and their space weather

Many stars exhibit signatures of stellar magnetism resembling what we see on the Sun. Even if spectroscopy does not provide direct evidence for a complex, evolving surface magnetic field, many stars do display one or more of a variety of tell-tale signatures: cool starspots that modulate the brightness as the stars rotate and the spots evolve, a warm chromosphere enveloping the star, a hot corona glowing in the EUV or in X-rays, or flaring. All this activity is rooted in the stellar dynamo process (see, e.g., Chs. 3 and 4 in Vol. I, and Chs. 2, 5, and 6 in Vol. III) that converts mechanical energy in 3D convective plasma motions of sufficient complexity into electromagnetic energy, directed into large-scale organization by the Coriolis forces acting within the rotating star.

Magnetically dominated phenomena in the atmospheres of stars bear similarities to what we observe on the Sun in a class of stars that we refer to as "cool stars". Whether old or young, more or less massive than the Sun, these stars have two characteristics that set them apart from other stars. The first, as can be inferred from their name, is that they all have surface temperatures that are relatively low, making them yellow, orange, or red in appearance. It is another, related common characteristic that is directly responsible for making all cool stars exhibit solar-like magnetic activity: they all have a zone that extends up to their surfaces – i.e., an envelope – in which convective energy transport (through overturning bulk plasma motions) carries most of the energy outward.

This convection zone may be a shallow layer (as in mid-F type main-sequence stars of 1.2 solar masses) around a "radiative interior" (in which energy is transported primarily by a random walk of photons) or it may reach all the way to the very center (for main-sequence stars much cooler than the Sun, of spectral type late-M or cooler, around 0.3 solar masses or less). In the long-lived phase of stellar adulthood – when we characterize them as "main-sequence stars" – stellar mass may be used as differentiator: stars with masses below roughly 1.2 solar masses are cool stars. But stars in their infancy (prior to reaching hydrostatic stability that follows initiation of nuclear fusion of hydrogen) or in their old age (after running out of most of the fusible hydrogen) have convective envelopes regardless of their mass, making it unambiguously clear that it is the convection that powers the stellar dynamo. We know from stellar population studies that the characteristics of the convective motions and their response to stellar rotation (through the Coriolis force) are the controlling parameters that set stellar magnetic activity levels.

The evolution of stars like the Sun and of their magnetic activity was summarized in Chs. 2 and 3 of Vol. III (for other introductory discussions of solar and stellar magnetism, see, e.g., Schrijver and Zwaan, 2000, and Reiners, 2012). In this volume, Ch. 2 uses observations of stellar flaring to piece together the history of impulsive activity of the Sun from its pre-main-sequence phase to old age. Observations of many stars and with different instruments sensitive to different energy ranges along the electromagnetic spectrum suggest that observable stellar flares (which are typically much more energetic than even the largest solar flares simply because smaller ones hide below present-day detection thresholds or within the quiescent background glow of the corona) have many properties in common, supporting the assumption that stellar and solar flares share the basic picture that was developed for magnetically driven, reconnection-powered solar impulsive events. Although that general picture appears to be widely applicable, we note that observations of stellar flares, and indeed also of solar flares, generally occur in narrow spectral bands that can range from hard X-rays to radio, or in broad spectral windows without spectral information. The result is that it remains a challenge to put solar and stellar flare observations on a common energy scale (preferably the bolometric scale that measures all of the temporary, sudden increase in photon output from stars that is the definition of flaring; see, e.g., Schrijver *et al.*, 2012).

The young Sun, in the first hundreds of millions of years of its main-sequence life, would most likely have exhibited flaring with energies exceeding 100 up to perhaps 1000 times the total energy involved in the largest recent flares, and would have displayed even these major events as frequently as roughly once a week. Over time, the frequency of flaring dropped sharply as the Sun lost angular momentum through the magnetized solar wind, a process that continues today albeit at a much reduced rate. This decrease in flare frequency would have been rapid over the first billion years or so of solar history, likely having slowed considerably over the most recent few Gyr. Whether there is an upper cutoff energy to solar flares that might also have dropped over time remains to be established: we do not know if the present-day Sun can still generate such enormous flares, albeit at vastly reduced frequency. On the one hand, we have no evidence that it did so in any available record, be it written or geological. On the other hand, there are observations of stars that look to be quite like the Sun in terms of internal structure and rotation rate that have been observed to produce flares with energies hundreds of times larger than the largest flare recorded for the Sun in the past half century (see Sect. 2.2.3, and Schrijver and Beer, 2014).

Binary stars (in particular the tidally interacting compact ones) provide yet another experimental environment, both on dynamo activity and on flaring: in such stars, flare energies can reach up to 100 000 times those seen for the most energetic solar flares, possibly through tidal effects on dynamo action or through magnetic coupling between the stars. Although such large flares attract observers for the obvious reason that they stand out from the background emission of the stellar system and that they occur so often that they are readily captured within the limits of observing windows, it remains to be seen whether such enormous energies could ever have been released from flaring from the single star that we call the Sun at any phase of its evolution.

Although energetic flares can be observed on other stars throughout the electromagnetic spectrum, there are no unambiguous indicators of coronal mass ejections or energetic particle populations in the astrospheres of those stars; for those properties, we need to rely on solar and solar-system observations combined with numerical experimentation at least until detection techniques are developed that are far more sensitive than we have access to today.

1.4 Astrospheres, stellar winds, and cosmic rays

The interaction of the solar wind and the interstellar medium is discussed in Ch. 3. One of the effects of the solar wind is that Galactic Cosmic Rays (GCRs) are held at bay fairly effectively (e.g., Ch. 9 in Vol. II, and Ch. 9 in Vol. III). Changes in the solar wind and its embedded magnetic field over evolutionary to short-term time scales are associated with changes in the spectrum of the GCRs that can penetrate into the solar system to reach Earth orbit, because they affect the propagation of the GCRs into the inner solar system and also because these changes move the very boundary of the solar wind from where GCRs start their random-walk diffusive penetration of the solar system.

Over the billions of years in the history of the Sun, the solar system will have encountered very different interstellar medium (ISM) environments. ISM density contrasts reach up to a factor of a million, with densities ranging from some $0.005\,\mathrm{cm}^{-3}$ in tenuous, warm clouds to $10^4\,\mathrm{cm}^{-3}$ in dense, cool molecular clouds. Increasingly dense ISM environments would decrease the size of the heliosphere by pushing the heliopause – that separates the solar wind from the ISM with its GCRs – inward. Chapter 3 discusses how the heliosphere would vary with the properties of the solar wind and of the interstellar medium. Models suggest that the Earth would be shielded from the full effect of GCRs under almost all circumstances. Even at an ISM density 100 times higher than in the local interstellar cloud (LIC), the termination shock (TS) would still be at about 10 Sun–Earth distances (astronomical units, or AU). Only at a density of multiple hundreds of times that in the LIC would the TS move inside of the Sun's present-day habitable zone (cf., Ch. 4 in Vol. III), exposing Earth to the full intensity of GCRs as well as to interstellar dust. A nearby supernova would do the same, plus expose Earth to the remnants of the stellar explosion themselves; this may have happened a mere 3 Myr ago, leaving a signature in the ^{60}Fe radioisotope in the Earth's crust (see Sect. 3.5).

When the Sun was younger, it most likely had a larger mass loss. How much larger remains to be established, but observations of Sun-like stars (Sect. 3.7.2) suggest the mass-loss rate may have been at least some 50 times higher than at present. For the present ISM conditions, that would have meant a heliosphere some seven time larger, with correspondingly much lower GCR intensities (cf., Ch. 11 in Vol. III). In the distant future, with weakening solar activity, the Sun's mass-loss rate is expected to decrease by a factor somewhere between 4 and 7 prior to the beginning of the subgiant and giant phases (Eq. (3.2)), and the heliosphere will correspondingly gradually shrink to about half its present size (to roughly Pluto's orbital radius; assuming present-day ISM conditions), and the GCR population in the inner solar system will thereby gradually increase.

Chapter 4 shows how models suggest that the GCR intensity around 1 GeV has likely increased by more than a factor of 100 since the Archean era (ending some 2.5 billion years ago) after the Late Heavy Bombardment when evidence for single-celled life is first seen on Earth (cf., Ch. 4 in Vol. III). Whether energetic particles originating from the more active young Sun would have compensated for the low GCR intensities around the young Earth remains to be seen: although young stars flare much more intensely and frequently (Ch. 2), we simply do not know enough about any associated coronal mass ejection (CME) activity to usefully constrain "SEP" events for the young Sun: whereas radio observations of stars may give some information on energetic particles trapped within coronae, no techniques are currently available to reveal energetic particles moving outward through astrospheres.

1.5 Astrophysical dynamos and space weather

Computer simulations provide us with glimpses of what space-weather phenomena may be like if the orbit of the planet is changed, or if the level of activity of the star is changed. Chapter 4 discusses, among other things, the effect of a variable stellar wind on a hypothetical magnetized planet at about a fraction of only a few tenths of the Sun–Earth distance. First, the rapid orbital motion would cause the planet's magnetotail to persistently be driven away from the essentially radial direction relative to the Sun–planet line that it has for Earth, by many tens of degrees. When coronal mass ejections (CMEs) would hit this planet, the magnetotail would be forced into the near-radial direction, thus shifting the associated auroral zones and the range of impacts of magnetospheric storms.

Planetary magnetism appears to be an important ingredient in the evolution of a planetary atmosphere, in part by setting the stripping effects of stellar winds. It also keeps energetic particles largely away from these atmospheres. As in the case of stellar dynamos, planetary dynamos require the combination of overturning motions in a conducting fluid and a deep-seated driver of such motions (see Ch. 6). The fluid in this case is indeed a liquid, of a viscosity that is considerable compared to that of the stellar gaseous plasma. The driver of the flows can be heat from impacts onto the planet (or smaller body), or released from nuclear fission of radio-active elements deep inside the planet, or from tidal forces. Or the driver of the flow may be a gradient in chemical composition maintained by a chemically differentiating phase transition from fluid to solid as the planet cools.

Astrophysical dynamos persist for a long time in bodies of sufficient size. Earth's magnetic field, for example, although variable in its detailed pattern and dominant directionality (i.e., polarity), appears to have maintained a roughly constant net strength over at least the past 3 billion years (Ch. 6) despite the fact that the dominant driver of the internal convective motions switched from thermal (nuclear decay) to compositional (chemical differentiation driven by Ni and Fe condensation onto the core, releasing lighter elements including Si, S, and O) about a billion years ago.

Astrophysical dynamos appear to generate structures in the magnetic field on a wide range of scales. This is immediately evident on the Sun, where we see bipolar regions emerge and evolve that span a range from the observational resolution limit (of order 100 km) up the full solar diameter (1.4 million km). For Earth, we cannot directly measure the small-scale structures, because the top of the dynamo region is deep inside the Earth, so that only the lower orders are readily measurable at the surface (see, e.g., Ch. 7 in Vol. III). If we look away from the solar surface by the same relative distance, we see a similarly "simple" structure, dominated by the dipole and quadrupole terms that shape the heliospheric magnetic field.

In smaller bodies, one key problem for exciting and maintaining a magnetic field is simply that their size lowers the magnetic Reynolds number by making the diffusion of magnetic field more effective relative to the motions shaping that field, all other things being equal (cf., Eq. (6.3)). In those smaller bodies with limited available primordial thermal energy or nuclear fission energy, and larger surface-to-volume ratios that speed up cooling, dynamos that do manage to operate should shut down sooner than in larger bodies (a process that may take only millions of years in planetesimals with radii above some 100 km, see Ch. 6). This is likely to be what happened on Mars in which no current global dynamo is functioning. But chemical composition may lead to exceptions: Ch. 5 describes the possibility that a sulfur-rich interior of Ganymede might lead to iron snow or FeS "hail" to maintain the only dynamo known to work in a satellite. Earth's Moon has no current dynamo, but it appears it sustained one longer than would be expected. Planetary dynamos, despite fairly successful modeling of Earth's dynamo (e.g., Ch. 7 in Vol. III) and Jupiter's dynamo (Ch. 6), have quite some secrets to disclose. There, study of solar system objects will help us figure out what happens in exoplanets where (as discussed in Sect. 6.3.2) there is a great deal of variability to be anticipated related to composition, size, and orbit.

1.6 Heliophysics of planetary atmospheres

The evolution of planetary systems, the structure and dynamics of the formed planets and the impacts of asteroids, comets, and dust on them, and even the properties of planetary dynamos, all contribute to the conditions in planetary atmospheres. Chapter 7 describes this for the terrestrial planets Venus, Earth, and Mars. The atmospheric conditions at the surfaces of these three planets differ greatly in composition, pressure, temperature, relative humidity, and types of (or even absence of) precipitation (see the summary in Table 7.1). Even the general patterns of atmospheric circulation differ greatly, leading to very different "seasons" and latitudinal climate zones on the terrestrial planets. The slow rotation of Venus enables effective exchange of thermal energy between poles and equator, greatly weakening the effects of different insolation at different latitudes relative to the pole–equator climate contrasts we see at Earth. The small tilt angle of Venus' rotation axis means that the small pole–equator differences are weakened even more in any seasonal activity. On Mars, despite a current tilt of its rotation axis relative to its orbital plane that is similar to that of Earth that contributes to seasonal changes, the main seasonal variations (including the occasional planet-encompassing dust storms that form during the austral summer season when Mars is near perihelion) are associated with Mars' orbital eccentricity that modulates the solar power input by some 44% from perihelion to aphelion.

Climates are set by a combination of stellar irradiance (i.e., by the properties of the central star and of the planet's orbit), planetary albedo, and atmospheric greenhouse gases (Ch. 7 and Ch. 11 in Vol. III). The latter two combine to elevate the surface temperatures by some 500 K, 30 K, and 5 K for Venus, Earth, and Mars, respectively, relative to the equilibrium mean surface temperatures set by orbital radius and solar irradiance in the absence of planetary atmospheres.

On Earth, the global mean temperature has changed by 10–15 °C and the surface pressure by a factor of two over its history. This has to do with geochemical changes, evolving life, and gravitational interactions with the other planets. Evidence for climate change on Venus is largely masked by a makeover of its surface within the past several hundred million years, but the chemistry of its atmosphere (notably the D/H ratio) suggests that it has lost much of at least the water content of its atmosphere. Mars' largely transparent atmosphere provides easier access to that planet's climate history. It appears that Mars has had liquid water and a much thicker atmosphere with a surface pressure exceeding 0.5 bar to account for many of its surface features; by now, it may have lost somewhere between half to 90% of its original atmosphere. Chapter 7 discusses the variety of processes involved in the erosion of atmospheres: thermal escape, photochemical escape, and sputtering for neutral atoms and molecules, and ion outflow, ion pickup, and bulk plasma escape for ions.

On geological time scales, climates are influenced by the processes that power planetary dynamos, both through bulk chemical evolution driven by internal processes and atmospheric losses driven from outside the planet: internal fluid motions are associated with volcanic activity (which releases various gases into the atmosphere – for Earth's life forms, most importantly CO_2) and continental drift, that either change or maintain greenhouse gases and albedo. And they are coupled to the evolution of the stellar dynamo: both the output of high-energy X-ray and EUV photons that can dissociate molecules and energize or ionize atoms, and the efficacy of the stripping effects of the stellar wind, decrease over time as the star loses its angular momentum causing its dynamo action to weaken.

Estimates for present-day atmospheric escape rates are uncertain to about two orders of magnitude. Chapter 7 points out that the uncertainty range is rather intriguing from the perspective of heliophysics: at the lower end of the range of estimates, heliophysical drivers are unimportant for each of the three terrestrial planets considered, while at the upper end, heliophysical drivers of the atmospheric losses are dominant. Estimates for past conditions are even more uncertain, being subject to changes in solar and solar-wind conditions, atmospheric composition, and planetary dynamo action. It is fascinating to see that the atmospheric escape rates for Venus, Mars, and Earth are comparable within their uncertainties, despite the differences in their atmospheres, surface gravity, and dynamo action. The latter distinction in the magnitude of the planetary magnetic fields has caused speculation that whereas these fields may protect a planetary atmosphere by keeping the solar wind at a distance, it may also add to atmospheric losses by harnessing some of the solar-wind energy in heating and ionization effects within its topmost atmospheric regions, making it ambiguous at present what planetary magnetism does to planetary atmospheric losses as driven by solar activity. There is clearly a lot to be learned in this arena.

1.7 Aeronomy and magnetospheres

The study of comparative aeronomy involves understanding differences in the aeronomy processes occurring in a planet's upper atmosphere where the lower neutral atmosphere transitions to a mixture of ions plus neutrals in which plasma processes contribute significantly to the energetics (Vol. I, Ch. 12 provides an introduction to ionospheres). In our solar system not only are there examples of terrestrial planets as well as of gas giants, but within each group there are quite different aeronomy processes at work. These differences are partly due to significantly different chemistry of the neutral atmospheres. A planet with an intrinsic magnetic field will also provide additional plasma processes that couple the planet's ionosphere, via the magnetosphere, to the solar wind in dramatically different ways to that of an unmagnetized ionosphere.

The distance of a planet from the Sun provides yet another aeronomy difference concerning the energy deposition and effectiveness of ionization in generating the ionosphere. Where the terrestrial thermosphere is predominantly heated by solar irradiance, the temperatures in the thermospheres of the gas giants appear to require more: particle precipitation from their magnetospheres (by Joule heating) provides likely at least an order of magnitude more energy than solar input, while an as yet unknown contribution by dissipation of wind shear and waves is also being considered (Ch. 8).

Aeronomy focuses on the energetics that drive neutral–ion processes. On a micro-scale, these processes involve chemical reactions of both neutral and charged particles. On a macro-scale, the dynamics of the neutral and ion populations is driven from the atmosphere below and the mostly ionized region above. In this volume, Chs. 8 and 9 describe comparative aeronomy for the gas giants and terrestrial planets respectively. How the terrestrial atmosphere evolves and hence impacts the planet's aeronomy is presented in Ch. 7. Chapter 12 discusses the energetic particle environment of the solar system that, together with solar irradiance, creates the ionosphere and is an aeronomy driver. The gap between our aeronomy knowledge based on our solar system versus what we currently know about exoplanets can be inferred from how we detect exoplanets, as discussed in Ch. 5.

The major commonality of the terrestrial upper atmospheres is the dominance of atomic oxygen, O, at the expense of molecular species (N_2, O_2 for Earth, CO_2 for Mars and Venus) above $\sim$150–200 km for all three planets (cf., Chs. 7 and 9 in this volume, and Chs. 13 and 14 in Vol. III). These are the altitudes where the densities are low enough for orbiting satellites (hence the importance of aeronomy for spaceflight, see Vol. III, Ch. 14) but also where solar photons are absorbed, heating and ionizing the upper atmosphere. Chapter 9 shows the steep temperature gradients in all three terrestrial atmospheres above 120–150 km with factors of $\sim$2 differences between minimum and maximum solar activity. But the true value in comparative aeronomy of the terrestrial planets comes less from their similarities than from their differences. The very different magnetic field structures, tides, and planet-wide atmospheric circulations of the terrestrial planets provide opportunities to test the detailed models developed for the Earth under contrasting conditions at Mars and Venus.

The giant planets of our solar system, Jupiter, Saturn, Uranus, Neptune as well as Jovian exoplanets of other systems, all have upper atmospheres dominated by hydrogen chemistry. One might easily expect the ionization of molecular H_2 to produce H_2^+ and H^+, but infrared (IR) observations of Jupiter in 1988 gave the first detection of H_3^+ outside the lab (Drossart *et al.*, 1989), produced by dissociative charge exchange of H_2^+ and H_2. Chapter 8 discusses the chemistry of neutral and ion species in the upper atmospheres of the giant planets and compares the relative importance of solar UV, dissipation of magnetospheric currents, and auroral precipitation for heating and ionization at the different planets. While the upper atmospheres of Jupiter and Saturn are fairly well described (as measured using stellar occultations in the UV, or at radio wavelengths by spacecraft communicating with Earth as they move behind the planet), the upper atmospheres of Uranus and Neptune are very poorly understood (as are their magnetospheres, see Vol. I, Ch. 13). Chapter 8 describes how observations of auroral emission (in the UV from excited H_2, from thermal IR emissions of H_3^+, X-rays from precipitating energetic ions, or auroral radio emissions from non-thermal electrons) provide an important diagnostic of magnetospheric processes. While the UV and radio emissions indicate immediate responses to changes in auroral precipitation, the thermal IR from H_3^+ has a slower response and indicates long-term deposition of energy into the ionosphere-thermosphere via Joule heating. Thus, observations at different parts of the spectrum provide diagnostics of different aspects of ionosphere–magnetosphere coupling, a valuable tool for understanding the giant planets of our solar system and may prove to be useful for exploring processes at exoplanets.

1.8 Dimensionless heliophysics: from heliosphere to dust

The variety of conditions under which bodies in the solar system interact with the surrounding plasma (either the solar wind or magnetospheric plasma) is described in Ch. 10, focusing in particular on the interactions with moons, asteroids, and comets. That chapter presents a wonderful example of what we can learn by looking at "universal processes" that occur throughout the solar system, and indeed the universe. Here, we see how to partition conditions depending on the relative strengths of effects through the use of dimensionless numbers that express ratios of flow or wave speeds: the Alfvénic Mach number, M_{A} (which measures flow speeds in terms of the Alfvén speed, v_{A}); M_{ms} (which measures flow speeds in terms of the magnetosonic fast mode speed $(c_{\mathrm{s}}^2 + v_{\mathrm{A}}^2)^{1/2}$, with c_{s} the sound speed); and the plasma β (which compares gas pressure to magnetic pressure; note that $\beta^{1/2}$ is the ratio of sound to Alfvén speeds), are particularly useful when grouping interactions of magnetized plasma flows into similar classes (cf., Table 10.1; see also Ch. 13 in Vol. I). Other important factors include whether the object has an atmosphere (either a gravitationally bound gas or one that may be sustained only as a result of the interaction with the solar wind, or may be maintained against losses by geysers or volcanoes). Does the body have an intrinsic magnetic field or not? Does it allow electrical currents to flow through it or does it not, i.e., does it behave as a conductor or as a non-conducting object? Is the body larger or smaller than the gyration radius of the solar-wind particles, or – alternatively phrased – is magnetohydrodynamics a valid approximation?

With just these few dimensionless numbers and properties of the solar-system bodies we are already looking into a six-dimensional space in terms of interactions, creating $6! = 720$ potentially distinct types of interactions to explore (leaving geometrical distinctions aside). These include the non-magnetized, non-conducting Moon when outside of the terrestrial magnetosphere, where the solar wind slams onto the lunar surface with little to stop it. Another example is the magnetized Earth with its conducting upper atmosphere and interior where the magnetic fields of solar wind and Earth field conspire to form a bowshock that deflects essentially all of the solar wind to flow around an extended, highly asymmetric volume that we call the magnetosphere. Beyond these six dimensions lie factors that differentiate by temporal behavior and by the directions of the flows and fields. Here, we see contrasts between the steady sub-Alfvénic plasma flow with field opposite to that of the body around which it flows as in the case of Ganymede moving through Jupiter's magnetosphere, or Earth embedded in the solar wind with its variable strength, direction, and speed. And yet (as discussed in Sect. 10.6.1) we see that the reconnection between the body's field and that in the inflowing solar wind or the Jovian magnetospheric "headwind" in both cases is intermittent, despite the near-perfectly steady and symmetric conditions around Ganymede.

When we extend the size spectrum of bodies under consideration downward well beyond that of asteroids and comets, at some point another transition occurs: when we reach the scale of "dust" (at roughly a micron, or 10^{-11} g, and below), the dust particles interact in such a way with the surrounding medium that in effect they become part of the plasma or are at least strongly affected by it and the magnetic field that it carries (see Ch. 11).

One example in which dust is used as a probe of heliospheric conditions is the effect of the solar cycle, through the heliospheric magnetic field, on the population of interstellar dust grains moving through the heliosphere. Their electrical charge causes them to respond as a population depending on the direction of the heliospheric magnetic field, as do galactic cosmic rays. But their low velocity (of some 26 km/s rather than near light speed) means that they probe the state of the heliospheric field (which they spend up to two decades traversing) with a very different response time, resulting in a phase lag of order half a sunspot cycle. Another population of dust has its origin in the destruction of comets in the inner solar system, after which radiation pressure pushes this dust outwards, on near-hyperbolic pathways away from the Sun, even as it serves as a source of neutrals and ions (through sputtering, sublimation, or collisional vaporization) that join the solar wind as pickup (atomic and molecular) ions. Within planetary magnetospheres, where volcanism of moons is another source of dust, a population of small, positively charged dust grains may even migrate outwards, to be expelled by a combination of their electrical charge subject to the apparent electric field associated with the rotating planetary field. Such evicted particles likely contribute to dust streams observed about Jupiter and Saturn (Sect. 11.3.2).

1.9 Energetic particles as diagnostic tools for heliophysics

Charged particles interact with magnetic fields in a wide variety of ways, some being accelerated to higher energies in the process. Chapter 12 discusses how detection of energetic particles in different regions of the heliosphere can act as diagnostic for such acceleration processes. For example, reconfiguration of the Sun's strong magnetic field produces expulsion of coronal plasma and entwined magnetic flux as a coronal mass ejection (CME). Particles are accelerated both during the initial reconnection and at the shock front as the CME travels through the heliosphere. Within the roughly dipolar magnetic fields of planetary magnetospheres charged particles exhibit gyro, bounce, and drift motions on increasing time scales. Perturbations of the local fields on these time scales can accelerate particles and populate the radiation belts within planetary magnetospheres (see also Vol. II, Ch. 11). Further acceleration processes within planetary magnetospheres result from large-scale motions (e.g., the Dungey cycle at Earth driven by reconnection at the day-side magnetopause and within the magnetotail) or via more diffusive processes during centrifugally driven flux tube interchange motions in rotation-driven magnetospheres of the outer planets (Vol. II, Ch. 10). The most energetic particles are generated well outside the heliosphere as galactic cosmic rays (GCRs) penetrating first across the heliopause and then into magnetospheres where they interact both with atmospheres (where further energetic particles are produced via cosmic-ray albedo neutron decay) and solid surfaces of moons (where radiolysis of ices can produce hydrocarbon compounds, perhaps pre-biotic materials). The tracking of energetic particles is important, therefore, not only as tracers of magnetic topology (i.e., open versus closed magnetic flux) and as an indication of acceleration processes, but also key for habitability perhaps below icy surfaces (e.g., Europa, Enceladus) as well as for human activity in space (Vol. II, Chs. 13 and 14).

1.10 Radio signals as diagnostic tools for heliophysics

The vast area of heliophysics is probed by analyzing the properties of light and particles: the glow of the various domains of the solar atmosphere help us understand its properties, and the reflection of sunlight or the intrinsic glow from planets and other bodies in the heliosphere enable us to diagnose conditions without sending probes to explore these conditions in situ. Where observations of light invariably integrate along the line of sight, in-situ measurements enable determination of local properties of chemical compositions, ionic and energetic particle populations, dust, and electromagnetic fields, although providing only a local snapshot along their migration pathways. In-situ measurements of energetic particles ultimately inform us about their origin and their interactions with the medium through which they propagated to wherever they are measured. In-situ probes are sampling conditions around Earth, other planets and their moons, near or on comets and asteroids, throughout the heliosphere and – with the Voyager spacecraft – even beyond that cavity carved out of the interstellar medium by the solar wind. Chapter 3 in Vol. II describes how particle measurements can be made, and the chapters that describe what they tell us include Ch. 12 in this volume, and Chs. 8–11 in Vol. II, and Ch. 9 in Vol. III.

Similarly, photons can tell us about the source regions (Ch. 4 in Vol. II) and the regions through which they propagate. Even almost entirely transparent media can be probed by photons by using the way that light is diffracted and scattered by density irregularities of the constituent particle populations. Chapter 13 focuses on a particular wavelength range, namely radio waves. It tells us what we can learn by looking at distant radio sources, both within the galaxy and far beyond its bounds, about the properties of the solar corona and the solar wind. Frequent or even continuous measurements of radio sources such as quasars and pulsars across the background sky enable three-dimensional (3D) tomography of coronal mass ejections throughout the inner heliosphere.

These same observations also contain information about the properties of the interstellar medium through which the signals propagated, and about the terrestrial ionosphere as the final refractive pathway through which extraterrestrial radio waves propagate prior to reaching the radio telescopes around the globe. Observing distant radio sources or using man-made radio transmitters during occultations can also tell us about planetary ionospheres and neutral atmospheres (e.g., Ch. 8 in this volume; as do occultations at UV and IR wavelengths) as well as about the Sun's corona (for which Faraday rotation of polarized signals offer the means to also sense the properties of the magnetic field, in addition to the scintillation signal that is sensitive to density contrasts).

1.11 Chapter outlines

Chapter 2 discusses explosive events on the Sun and on stars like the Sun in temperature and internal structure. These involve the changing of magnetic configurations and subsequent liberation of energy. The chapter reviews how flares manifest themselves differentially across the electromagnetic spectrum – from hard X-rays to radio – and the difficulty in observing any associated energetic particles and coronal mass ejections that propagate into astrospheres. The sections in this chapter discuss key stellar and wavelength-dependent parameters important to a discussion of explosive events. The examination of explosive events proceeds as a function of the star's age, from birth to beyond the current age of the Sun, with energies of over a thousand times that of the largest solar flares down to flares that disappear in the background atmospheric emission of the Sun-like "cool stars".

Chapter 3 discusses interactions of stars with their galactic environment, in the context of how that might affect orbiting planets. The characteristics of astrospheres depend on the properties of the stellar winds and on those of the interstellar medium (ISM) into which the winds are expanding. The chapter therefore includes a discussion of what is known about stellar winds and how they evolve with time, and a discussion of different sorts of ISM within the galaxy. Particular attention is given to our own Sun and to the surrounding heliosphere as it is carved out of the ISM by the solar wind. Observational studies of global heliospheric structure are reviewed, and speculation is offered as to how this heliospheric structure might change in response to different ISM environments and long-term solar-wind evolution.

Chapter 4 reviews how changes in stellar magnetic activity of Sun-like stars throughout their evolution translate into changes in their stellar winds, the structure of their interplanetary space, as well as of their astrospheres, the transport of particles, and the propagation and evolution of CMEs. Young, rapidly spinning stars have strong high-latitude fields and substantial effects of rotational forces in their winds. Old, slowly spinning stars have winds with fields wound up into the Parker-spiral pattern characteristic of the heliospheric structure. All stars likely exhibit coronal mass ejections (CMEs) that contribute in different degrees to stellar angular momentum loss. CMEs and the background stellar winds affect the evolution of planetary atmospheres, and the activity of planetary magnetospheres, depending also on the planet's orbital radius. The strength, pattern, and dynamics of astrospheric fields changes the GCR population that penetrates deep into astrospheres; Earth has been subjected to the associated changes in GCR intensities and energy spectra over its 4.5-billion year history.

Chapter 5 focuses on exoplanet systems. Search techniques for exoplanets include Doppler measurements, transit photometry, microlensing, direct imaging, and astrometry. Each detection technique has some type of observational incompleteness that imposes a biased view of the underlying population of exoplanets. With such partial information, we must piece together an understanding of exoplanet architectures by counting planets in the regimes where techniques are robust and then estimating correction factors when possible. This chapter therefore begins with a review of the exoplanet detection techniques with particular consideration of the observational biases. Only then can we discuss the implications of the rapidly growing multitude of exoplanet systems for planet formation with an eye toward how our solar system compares. We can begin to address questions about how planetary systems form and evolve. How do planetary systems – including gas and debris disks, and planetary migration by mutual gravitational interactions – evolve over time? In what ways do exoplanetary systems mirror our solar system? How are they different? Does the presence of a binary star affect planet formation? What are the formation histories of "terrestrial" planets and of gas giants? Are Earth analogs common?

Chapter 6 reviews planetary dynamos, which produce the magnetospheres that are a crucial component of the study of heliophysics and the interactions of solar and stellar winds with planets. This chapter reviews the current understanding of planetary dynamos, including theoretical foundations, and our knowledge of planetary magnetic fields from spacecraft data. It discusses planetary interior structure and processes responsible for the different magnetic fields seen in the solar system, focusing on recent findings. It also discusses the possibilities of extrasolar planet magnetic fields and what we can learn about exoplanets from them.

Chapter 7 addresses how a planet's atmosphere and surface habitability are inextricably linked. Life at Earth's surface has been made possible by atmospheric conditions, and has modified them. Planetary climates are not static, but instead change as energy or particles are added or removed from the atmosphere. The Sun plays a large role in shaping climate in a variety of ways. Of particular interest for this volume are heliophysical processes that remove particles from the upper layers of terrestrial planet atmospheres. This chapter describes the present-day climates of Venus, Earth, and Mars, and evidence for how their atmospheres have changed over time. The present understanding of the role of heliophysical processes in driving these changes is summarized. Special attention is paid to how atmospheric escape varies with changes in solar photon fluxes, solar wind, and the interplanetary magnetic field, and the role of global-scale planetary magnetic fields in inhibiting escape. We discuss the importance of applying lessons from the terrestrial planets to exoplanetary atmospheres.

Chapter 8 reviews how the upper atmospheres of planets represent a key transition region between a dense atmosphere below and a tenuous space environment above. An array of complex processes from below and from above lead to a highly coupled system, with neutrals, plasmas, and electromagnetic processes linking surfaces to magnetospheres and to the solar wind, and ultimately to the Sun itself. Evidence of these coupling processes includes various upper-atmospheric emissions, such as dayglow and nightglow, resulting from the absorption of solar photons, and aurorae, produced from the energy deposition of energetic particles from the space environment. The chapter gives an overview of the current state of knowledge of giant-planet atmospheres, first focusing on thermospheres and ionospheres, next on the processes coupling planetary atmospheres and magnetospheres, and finally on the auroral emissions resulting from those coupling processes. Giant-planet aurorae are spectacular displays of magnetosphere–atmosphere coupling, the most powerful in the solar system. Furthermore, by studying the giant planets in our own neighborhood we can lay a solid foundation for understanding the rapidly accumulating zoo of exoplanets.

Chapter 9 reviews how upper atmospheres of terrestrial planets are affected by processes from above and from below. One of the new paradigms in solar–terrestrial research is that meteorological forcing from the lower atmospheres of planets can affect planetary upper atmospheres and drive space weather. Concurrently, general circulation models of planetary lower atmospheres are being coupled with planetary thermospheres and ionospheres. We are thus witnessing the beginning of the era of whole-atmosphere modeling of space weather. Ultimately, these whole-atmosphere models will be linked to heliophysics models to produce true solar–terrestrial simulation systems. This chapter focuses on the specific phenomena that have been recently identified to couple the lower atmospheres of terrestrial planets, mainly Earth and Mars, with their respective upper atmospheres and ionospheres. It describes the basic state of terrestrial planetary upper atmospheres and ionospheres and how they depend upon atmospheric composition. It also discusses specific differences between the two worlds such as Martian dust storms or effects from the differing magnetic-field environment of both planets.

Chapter 10 explores couplings between smaller bodies in the solar system and their space environment. A body such as a planet, a moon, an asteroid, or a comet, typically enveloped in a tenuous neutral gas, perturbs its surroundings in a flowing, magnetized plasma. The structure of field and plasma resulting from the interaction depends on properties of the ambient plasma, of the atmosphere and internal structure of the body onto which the plasma flows, and on the relative speeds expressed in terms of propagation velocities of acoustic and magnetohydrodynamic waves. This chapter addresses the interaction of solar-system plasmas with a number of small bodies of the solar system by exploring magnetohydrodynamic couplings as a function of several dimensionless numbers. The interaction regions surrounding the small bodies of interest vary in global geometric configuration, in spatial extent relative to the size of the central body, and in the nature of the plasma disturbances.

Chapter 11 starts with an introduction of selected heliospheric observations that indicate dusty plasma effects on the flow of interstellar dust through the heliosphere, the effects of dust impacts on electric-field measurements on spacecraft, and on the ion composition of the solar wind. After describing the basic dust-charging processes, the chapter describes the unusual dynamics of charged dust particles in planetary magnetospheres, and the possible role dusty plasma waves might play in cometary environments. Dust particles immersed in plasmas and UV radiation collect electrostatic charges and respond to electromagnetic forces in addition to all the other forces acting on uncharged grains. Simultaneously, dust can alter its plasma environment as it can act both as a sink and a source of ions, and electrons. Dust particles in plasmas are unusual charge carriers. They are many orders of magnitude heavier than any other plasma particles, and they can have time-dependent (positive or negative) charges that are many orders of magnitude larger. Their presence can influence the collective plasma behavior, for example, by altering the traditional plasma wave modes and by triggering new types of waves and instabilities. Dusty plasmas represent the most general form of space, laboratory, and industrial plasmas.

Chapter 12 argues that energetic-particle environments in the solar system are fundamental to understand the plasma-physics processes in our universe. Energetic ions and electrons are found everywhere in the heliosphere and beyond. The knowledge of their composition, energy, and spatial distributions in different plasma environments provides an enormous source of information to investigate the origin, evolution, and current state of our solar system. Originating outside and inside the heliosphere the distribution of energetic particles is widely used to study acceleration phenomena in space near shocks, to study the configuration and dynamics of planetary magnetospheres, atmospheres, rings, neutral gas and dust clouds in the vicinity of planets and moons. They are also very important to study the interaction processes between planets/moons and their local plasma environment and can help to discover formerly unknown objects.

Chapter 13 explores how remote sensing can tell us about tenuous environments, primarily by focusing on radio waves as diagnostics. The density irregularities in the solar wind can be observed by remote sensing techniques using ground-based or space-based coronagraphs for the inner solar wind as it emerges from the Sun corona, by space-based white-light heliospheric imagers, and by radio observations of distant compact radio sources to indirectly observe the solar wind through scintillation and Faraday rotation in the radio signal received from distant astronomical or artificial radio sources. This chapter focuses on the development of radio scintillation methods since the middle of the twentieth century, from localized observations of irregularities along particular source directions with a single antenna, to large-scale, three-dimensional, evolving tomographic modeling of outflow velocities and density structures throughout large parts of the inner heliosphere including the inclusion and determination of magnetic fields, that can utilize multi-antenna, multi-wavelength observational methods.

2

Solar explosive activity throughout the evolution of the solar system

RACHEL OSTEN

The time scales on which the Sun varies range from seconds to billions of years. Explosive events on the Sun involve the changing of magnetic configurations and subsequent liberation of energy, and processes associated with this energy transformation occur on the shortest of these time scales. However, the nature of the explosive events can change over the course of the Sun's evolutionary lifetime. This chapter covers the history of explosive events throughout the evolution of the solar system, with a focus on the observed manifestations of explosive events and what we can learn by studying them. Chapters in other volumes of this series (see Table 1.2) address parts of this topic. Chapter 2 of Vol. III describes the long-term evolution of magnetic activity of Sun-like stars, and Ch. 5 of Vol. II discusses observations of solar and stellar eruptions, flares, and jets. A description of radiative signatures of accelerated particles (primarily with application to solar flares) can be found in Ch. 4 of Vol. II, while Ch. 6 of that volume tackles models of coronal mass ejections and flares.

While the main theme of this chapter is the changing nature of these explosive events over evolutionary time scales, the influence of other important stellar parameters – such as rotation rate, convection zone depth, and the influence of binarity on rotation – are also discussed where appropriate. The focus of the chapter is to use an event-driven discussion of explosive events, keeping in mind how the parameters change with some key stellar parameters. This treatment is meant to illustrate key properties of explosive events that are known in stars of these age ranges.

An explosive event is made up of several distinct components, from energetic particles to coronal mass ejections to flares. While on the Sun these three components can be studied separately, on stars other than the Sun the identification and study of explosive events is limited to the radiative manifestation of the event, namely stellar flares. As diagnosed from the Sun, flares involve particle acceleration and plasma heating. They occur as a consequence of magnetic reconnection somewhere in the outer atmosphere, yet involve all stellar atmospheric layers. Owing to a diversity of physical processes involved, they produce emissions across the electromagnetic spectrum, that for the Sun have been recorded from km-wavelength radio waves to the highest-energy gamma-rays.

A fundamental part of the solar–stellar connection (and the rationale for this chapter) is the assumption that processes studied in detail on the Sun are also operating on other stars, and that the differences can be related to fundamental physical differences between the stars.

The major sections in this chapter discuss events according to the ages of the stars. We begin with a review of processes likely to have consequences for planetary climates and space weather, referencing chapters in Vols. I–III as appropriate, and supplementing those chapters with information specific to diagnosing time-variable emissions. We begin with a brief mention of processes on evolutionary time scales, hearkening to what has been discussed in other chapters. Flares have been discussed in other reviews: see Benz and Güdel (2010) for an in-depth characterization of flares on the Sun and stars; Güdel (2007) provides a review of especially magnetic activity characteristics of the Sun in time; Shibata and Magara (2011) reviewed magnetohydrodynamic processes occurring during solar (and by extension, stellar) flares. This chapter's focus is unique, being driven by a discussion of the events themselves and how key stellar parameters, notably stellar age, control their characteristics.

2.1 Key parameters important to a discussion of explosive events

There are stellar parameters whose influence is important to understand. In addition, understanding the information that can be extracted from studies of flares at multiple wavelengths is necessary to studying these events. Table 2.1 gives an idea of how flares in some of these kinds of stars differ from those on the Sun and from each other, in terms of energetics, duration, and increases in intensity at visible wavelengths and X-ray wavelengths.

Table 2.1 *Comparing large solar and stellar flares*

	Energy (erg)	Max. duration	Intensity increase (visible)	Relative intensity increase in X-rays
Sun	10^{32}	$\sim$5 h	1.000 270	6000$\times$
M dwarfs	10^{35}	several days	1000$\times$	1000$\times$
Young stars	10^{36}	$\sim$1 day	small	50$\times$
Close binaries	10^{38}	$\sim$week	1.2	120$\times$

2.1.1 Stellar parameters

What are the key stellar parameters to consider in this discussion of explosive events on solar-type stars? First, let us define what we mean by "solar-type stars" (see also Ch. 2 in Vol. III). Stars are characterized by their spectral type, which essentially is a measure of their surface temperature or their color, going from hot to cool: for a long time, the spectral types were O, B, A, F, G, K, and M, but recently types L, T, and Y were added to describe even cooler stellar objects. Another characterization differentiates by surface gravity; those we refer to here are main-sequence stars (or dwarf stars) and evolved or giant stars, with a class of moderately evolved subgiants in between. What makes stars of solar type is that they have an outer convective envelope that starts from early-F to mid-F (depending on gravity) and ranges ever deeper as we move towards cooler (or "later") spectral type, until the stars become fully convective by about mid-M-type dwarfs. Because all such stars lie on one side of a brightness–temperature (or magnitude–color) diagram (known as the Hertzsprung–Russell or HR diagram (such as Fig. 2.8 in Vol. III), they are generally referred to as "cool stars".

Explosive events on cool stars at their heart are powered by changing magnetic configurations in the stellar outer atmosphere and consequent liberation of energy. They are thus a manifestation of magnetic activity, and ultimately relate to dynamo processes in the stellar interior generating magnetic fields. Age and rotation are coupled because of the well-known stellar spin-down with time, often referred to as the "Skumanich $t^{-1/2}$ law" (see also Ch. 2 in Vol. III). In co-eval stellar clusters there is a spread of rotation at a fixed age, and the magnitude of this spread decreases with advancing stellar age (Ayres, 1997, and references therein). Stars are born rotating rapidly because of the conservation of angular momentum in the compression of the cloud out of which the star is formed (e.g., Chs. 2 and 3 in Vol. III). Rapid rotation engenders enhanced magnetic activity. Coronal structures produced as a result of this magnetic activity provide a torque that can spin the star down, as well as mass loss through a stellar wind in open magnetic field structures which can remove angular momentum from the star. Rotation and activity are fundamentally coupled, and an age–activity relation arises from the rotation and age relationship.

Rotation is not the only factor controlling magnetic activity. Convection zone depth, or its overturn time, also influences the ability of the star to generate vigorous and efficient magnetic fields (Noyes *et al.*, 1984). Because of the internal structure of stars in the cool half of the Hertzsprung–Russell (HR) diagram, outer convection zone depth traces mass, at least for stars less than about 1.2 solar masses, and down to the limit at which stars become fully convective, about 0.3 $M_{\odot}$ (e.g., Ch. 2 in Vol. III). The Rossby number, $R_0 = P_{rot}/\tau_{conv}$, which gives the dimensionless ratio of stellar rotation period to convective turnover time, connects rotation, convection, and magnetic activity in cool stars on the main sequence. The dynamo generation of magnetic fields by dynamo action is driven by an interplay between rotation and convection (Charbonneau, 2010).

Explosive events occur as the result of liberation of energy from a changing magnetic field configuration, and characteristic magnetic flux densities in stars, where they can be measured independently, correlate with coronal emission. This is not really an independent variable, because age, rotation, and convection zone depth all drive the ability of the star to generate sufficiently strong magnetic field that covers a large fraction of the star.

Rotation rates may be affected by the presence of a stellar companion: for cool stellar binaries with separations less than about 10 stellar radii the time scales for synchronization and circularization are short enough that tidal locking effects dominate (Zahn, 1989), and a binary can maintain fast rotation for longer ages than a single star can.

We focus on stars on or near the main sequence in this discussion, but it is interesting to note that the radiative component of explosive events, with characteristics similar to those seen on the Sun, have been found on evolved stars. The usual explanations involved enhanced rotation due to the presence of a binary companion (the so-called RS CVn and Algol systems, with the latter having an additional complication of mass transfer taking place), or to engulfment of a companion (the single active evolved stars known as FK Com systems). Stars that start out life as intermediate-mass, non-convective stars can develop a thin outer convection zone as they evolve into the cool half of the HR diagram during their giant phase; they retain the fast rotation from their time on the main sequence and have exhibited flares (Ayres *et al.*, 1999). Flares have also been detected on intermediate mass stars after their ascent up and return down the giant branch (Ayres *et al.*, 2001), although the means by which the return of magnetic activity happens is less clear in this case.

2.1.2 Wavelength-dependent parameters

This chapter is event focused, so a discussion of the key parameters that can be extracted from observations of explosive events at a variety of wavelength regions where stellar explosive events are most studied is appropriate.

Table 2.2 summarizes the commonalities in manifestations of solar and stellar eruptive events. Stellar flares are not observed panchromatically (because there are no stellar spectral irradiance monitors), so these commonalities are built on a wavelength region by wavelength region basis, and on a star by star basis, with a sampling bias because some types of stars are more often observed for their

Table 2.2 *Solar/stellar explosive event commonalities*

Manifestation	Solar events	Stellar events[a]	Emission seen outside of flares?
	Impulsive phase		
Non-thermal hard X-ray emission	✔	?	no
Radio gyrosynchrotron/synchrotron	✔	✔	yes
Coherent radio emission	✔	✔	no
Transition region far-ultraviolet lines	✔	✔	yes
Hot black body optical/UV continuum	✔	✔	no
Coronal mass ejection	✔	?	no
Solar energetic particles	✔	?	no
	Gradual phase		
Coronal emission	✔	✔	yes
Optical chromospheric lines	✔	✔	yes

[a] Across different types of stars.

flares (e.g., M-type dwarfs). The picture that emerges, however, is generally one of agreement between the basic physical processes at work in solar and stellar explosive events.

Many flare tracers also exist outside times of flares, with a variation in degree. As an example, emission lines and continuum from the corona are seen in quiescence and in flares, with an increase in both line and continuum intensity during the flare (the amount of increase of lines and continuum depends on the flare temperature). Other flare tracers are unique to the flare event and point to a variation in kind. An example of this is the blue–optical continuum which forms during a white-light flare, and is not seen outside of any flare-like increase in visible light. The nature of these flare diagnostics, as to a change of degree or kind, are noted in Table 2.2.

An individual flare can be divided into two main phases: impulsive and gradual. This generally refers to the timing of emissions relative to the processes thought to be occurring in the flare. In the standard picture, the initial energy conversion caused by magnetic reconnection powers particle acceleration and possibly – depending on the energetics and on the magnetic configurations – a mass ejection. The downward-directed particles become trapped in loops and emit non-thermal incoherent radio emission (see Ch. 4 of Vol. III). Coherently emitting particles can be traveling either upwards out of the atmosphere or downwards into the atmosphere. Once the trapped particles precipitate from the magnetic trap, they deposit their energy in or just above the photosphere, producing thick-target non-thermal bremsstrahlung emission. This energy deposition results in the

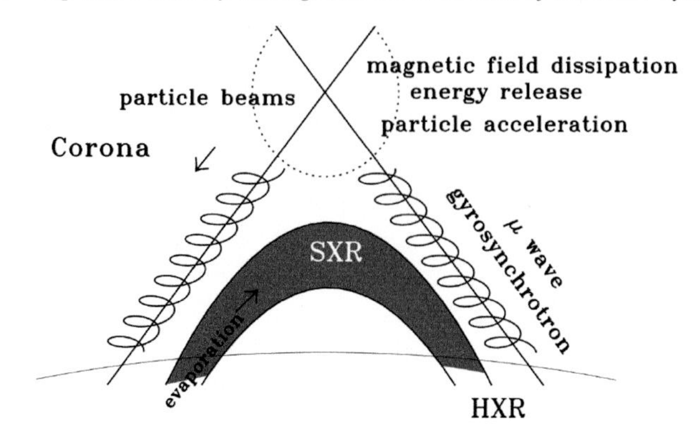

Fig. 2.1 Schematic arrangement in the outer atmosphere of the Sun or a comparable cool star indicating the flow of energy during a flare: a flare starts with magnetic reconnection high in the atmosphere that accelerates particles, leading to motion along field lines upward away from or downward towards the visible surface. Resulting emissions include hard X-rays (HXR), soft X-rays (SXR), and microwave emission. (Figure from Osten, 2002.)

heating of the photospheric material to temperatures near 10^4 K, and emissions from far-ultraviolet (FUV) lines. All of this is associated with the impulsive phase of the flare. The flow of energy at this point proceeds back into the upper atmosphere, with line emission from the lower chromosphere. Thermal X-ray emission occurs as well. As the energy input into the system decreases, emissions of all flare components return to the pre-flare level (see Fig. 2.1).

2.1.2.1 Correlations among multi-wavelength flare emissions

The processes involved in an explosive event occur across different layers of the stellar atmosphere, which respond with differing time scales to the time-dependent flow of energy in the explosive event. This leads to a number of expected correlations between facets of the event. Seeing these correlations in stellar flares gives support to the interpretation of the stellar flares using a solar flare scenario.

The Neupert Effect relationship is often used in stellar multi-wavelength flare studies to establish concordance with or likely disagreement with solar-flare models. It was formulated originally to describe the integral relationship between markers in a solar flare corresponding to the action of non-thermal particles, and the response from the atmosphere to the deposition of energy from these particles as it appears in coronal radiation. Written more generally,

$$L_{\mathrm{gradual}}(t) \propto \int_{t_0}^{t} L_{\mathrm{impulsive}}(t')\ dt', \tag{2.1}$$

where $L_{\text{impulsive}}(t')$ is the time variation of an impulsive phase process which diagnoses the presence and action of particles accelerated in the explosive event (for stellar studies usually radio gyrosynchrotron, transition region FUV emission lines, or photospheric UV–optical continuum emissions), and $L_{\text{gradual}}(t)$ is the intensity corresponding to the gradual phase (usually coronal emission, but some chromospheric emission lines display the Neupert Effect as well). The interpretation is that the gradual phase emission is responding to the buildup of energy that occurs as a result of the energy deposition being diagnosed by the impulsive phase emission. Correspondences in this "expected" manner are often used to bolster support for an interpretation of a stellar explosive event following the solar model. It is important to note, though, that there are also instances where the *opposite* behavior is observed in stellar flares (gradual phase emission leading the impulsive phase; see, for example, Bower *et al.*, 2003; Osten *et al.*, 2000). And not all solar flares follow the standard flare scenario.

Scalings between flare emissions occurring in different parts of the stellar atmosphere are also used to establish the validity of the solar flare interpretation. Butler *et al.* (1988) showed a linear relationship between the integrated energy emitted from the corona and from the chromosphere, both delineating the gradual phase of a flare. The interpretation into the tight correlation between two emission diagnostics with a large difference in formation temperature ($\sim 10^4$ K for chromospheric Hγ to 10^7 K for coronal X-ray emission) may originate from heating by X-ray back warming (see discussion in Hawley *et al.*, 2003). Kowalski *et al.* (2013) showed that the chromospheric Ca II K line showed a delayed response compared to the Balmer lines, and exhibited a Neupert-like effect with the impulsive flare markers.

There is often a close association between individual impulsive phase markers. FUV emission lines and optical continuum emissions have a similar temporal relationship (Hawley *et al.*, 2003). In principle the radio and optical components of a flare should behave similarly, as it is thought that they have a common origin in the action of accelerated particles. The radio emission is formed directly as the result of trapped electrons accelerated in the presence of a magnetic field, whereas the optical emission and other impulsive tracers are formed after energetic electrons precipitate from the magnetic trap and interact with material lower in the atmosphere. Figure 2.2 shows an example of a large radio flare occurring at the same time as a moderately large optical flare. In practice, because there are two emission mechanisms which can produce radiation in the cm-wavelength range, the interpretation of the radio light curve is not straightforward. As described in Sect. 2.1.2.3, circular polarization can be used as a discriminant between incoherent flare gyrosynchrotron emission (which should have little circular polarization) and coherent emission (which is often 100% circularly polarized). Gagne *et al.* (1998) and Osten *et al.* (2004) found that unpolarized radio flares had a higher rate

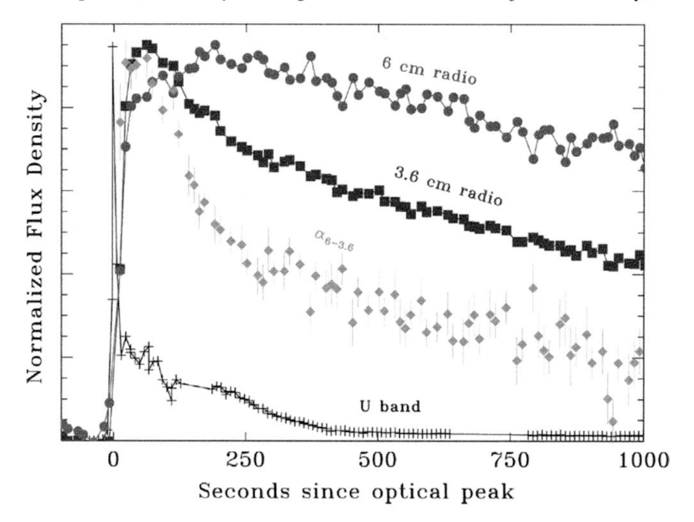

Fig. 2.2 Example radio-optical flare seen on the nearby M dwarf flare star EV Lac (from Osten *et al.*, 2005). Circles and squares show time evolution of radio flux density at wavelengths of 6 and 3.6 cm, respectively. Diamonds show the time variation of the radio spectral index, a measure of the relative strength of emission at the two wavelengths. The plus symbols show the temporal variability of optical emission in the U filter (cf. Fig. 2.5). Because the radio flux arises due to gyrosynchrotron emission from a population of accelerated particles, and the optical emission originates from a black body formed as the result of deposition of accelerated particles in the lower stellar atmosphere, one expects to see a temporal correlation between the two, which is observed for this flare.

of association with optical flares or other impulsive phase flare markers than the polarized radio flares.

2.1.2.2 Non-thermal hard X-ray emission

Chapter 4 of Vol. II discusses non-thermal bremsstrahlung hard X-ray emission from solar flares, and that same mechanism is believed to operate in stellar flares as well. This emission component originates due to the acceleration of electrons in the flare, and is not detectable outside of flares (on the Sun).

Optically thin non-thermal hard X-ray emission is important because it allows for the determination of the properties of the accelerated electrons (their energy distribution, and thus the total kinetic energy of accelerated electrons) in the flare. Yet, with the exception of one exceptionally bright stellar flare detected out to photon energies of $\approx$200 keV (see Fig. 5.11 of Vol. II), this flare signature remains elusive in stellar flares. Part of this stems from the inefficiency of non-thermal bremsstrahlung emission compared with thermal emission, being a factor of $\sim 10^5$ weaker. Another factor that has affected detections of non-thermal hard X-ray emission from stellar flares is the tendency for these flares to show evidence of very hot, *thermal* plasma at temperatures up to and exceeding 100 MK (compared with peak temperatures of $\sim$20 MK for solar flares). The sensitivity of astronomical hard X-ray detectors typically falls off above a few tens of keV, meaning that the tail of thermal emission from the hot plasma can easily overwhelm any small non-thermal continuum signature which might be present.

2.1.2.3 Radio bursts

In general the description of radiation from energetic particles reviewed in Ch. 4 of Vol. II suffices for the case of stellar flares as well as solar flares. Both incoherent (mostly gyrosynchrotron) as well as coherent processes are observed in stellar flares.

Gyrosynchrotron emission is seen both during stellar flares and outside of them, leading to the suggestion that the "quiescent" emission is composed of decaying flares (White and Franciosini, 1995). Multi-frequency observations of gyrosynchrotron emission can be used to infer the spectral distribution of the accelerated particles if the emission is optically thin: the observed spectral index α (determined assuming the radio flux S_ν varies with frequency ν as $S_\nu \propto \nu^\alpha$) relates to the power-law index of the distribution of energetic particles. If the emission is optically thick, the variation of flux density with frequency can be used to determine the effective temperature of the emitting electrons, the size of the radio-emitting source, the magnetic field strength in the radio-emitting source, and the total number of accelerated electrons contributing to the emission, as well as their changes with time. See Osten *et al.* (2005) for an example analysis in a large radio flare on a nearby M dwarf.

Some of the largest stellar radio flares even show a rising component at high frequencies (above a few tens of GHz), attributable to synchrotron emission from highly relativistic particles (Massi *et al.*, 2006). Similar types of "Terahertz" flares have been observed on the Sun, but current explanations require extreme parameters to explain them using known emission mechanisms (Krucker *et al.*, 2013). This population of accelerated electrons is trapped within the magnetic loop and may represent electrons with MeV energies, far in excess of the electron energies usually producing incoherent radio flares.

Coherent emissions are only inferred when bursty (extremely time-variable) emission is observed. There are two main coherent mechanisms seen in solar flares and inferred for stellar flares: either plasma radiation or emissions deriving from a cyclotron maser process. Section 4.3.1.2 in Vol. II covers plasma radiation. Unlike most solar coherent emissions, those on stars are typically observed to be almost 100% circularly polarized. This and their short time scales – as small as 1 ms, which implies a high brightness temperature – are what lead to the classification as coherent emission. They can have brightness temperatures far in excess of the upper limit on brightness temperature of 10^{12} K for incoherent emission, set by the balance between inverse Compton scattering and synchrotron radiative losses (Kellermann and Pauliny-Toth, 1969).

Information can be extracted from coherent radio bursts constraining some fundamental quantities by associating the observed radio frequency with one of two frequencies associated with the plasma: the electron gyrofrequency $\nu_B = 2.8 \times 10^6 B$ MHz, or the plasma frequency $\nu_p \approx 9000\sqrt{n_e}$ MHz. The relative location of the emission in the stellar atmosphere can then be found by using a spatial relationship for this emission; e.g., assuming that the density obeys a barometric law, and the base density and scale height can also be calculated. Some coherent radio bursts exhibit a simple structure in a plot of intensity versus frequency and time, wherein the burst drifts with time and frequency. In these cases, the observed drift rate can be related to the speed with which the exciter is traveling through the stellar atmosphere by using the relation

$$\frac{d\nu}{dt} = \frac{\partial \nu}{\partial n_e}\frac{\partial n_e}{\partial h}\frac{\partial h}{\partial s}\frac{\partial s}{\partial t}, \tag{2.2}$$

or

$$\frac{d\nu}{dt} = \frac{\partial \nu}{\partial B}\frac{\partial B}{\partial s}\frac{\partial s}{\partial t}, \tag{2.3}$$

where h is a radial distance from the stellar surface, and s is the path length traveled through the atmosphere. For a barometric atmosphere the first equation above reduces to $\dot{\nu} = \nu \cos\theta v_B/(2H_n)$, where v_B is the exciter speed whose motions cause the plasma radiation, H_n is the density scale height, and θ the angle between the propagation direction and the radial direction from the surface. If the coherent radio burst is due to cyclotron maser instability, then the drift rate of the radio bursts can be analyzed using the second equation above, which reduces to $\dot{\nu} \approx v_B \nu^{4/3}/(47 B_0^{1/3} \lambda_B)$, where the magnetic field geometry is assumed to be dipolar, λ_B is the length scale of the dipole field, and B_0 is the magnetic field strength in the radio-emitting source. The duration of individual bursts, if small enough, can be related to the collisional damping time scale and the thermal temperature of the plasma constrained. See discussion in Osten and Bastian (2008) for an example.

2.1.2.4 Transition-region far-ultraviolet lines

The magnetic transition region of a cool stellar atmosphere is the location where the plasma β, defined as

$$\beta = \frac{P_g}{P_{\text{mag}}}, \tag{2.4}$$

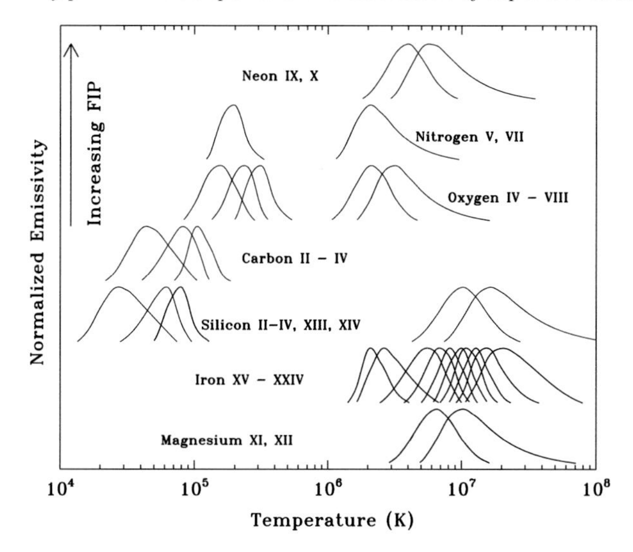

Fig. 2.3 Temperature coverage of chromospheric, transition region, and coronal lines from different elements and ionic stages obtained by combining spectra from ultraviolet through X-ray wavelengths; for the stellar case these instruments are the Space Telescope Imaging Spectrograph on the Hubble Space Telescope (HST/STIS), the Far-Ultraviolet Spectroscopic Explorer (FUSE), and the Chandra X-ray Observatory High Energy Transmission Grating Spectrometer (HETGS). The elements are ordered in terms of increasing first ionization potential (FIP), a quantity that gives the amount of energy required to remove the outermost electron from an atom of this element. These lines are formed under the conditions of collisional ionization equilibrium, and are generally optically thin, two conditions that permit inversion of their intensities to determine the temperature structure from the upper chromosphere to corona. (Figure from Osten, 2002.)

where P_g is the gas pressure and P_{mag} is the magnetic pressure, transitions from a value of $\gg$1 in the lower atmosphere to $\ll$ 1 in the upper stellar atmosphere. It separates the region in the lower atmosphere where gas pressures dominate dynamics from the upper atmosphere where magnetic pressure dominates. Transition region emission largely shows up in the far-ultraviolet (FUV) spectrum, and is dominated by emission lines from ionized species. The formation temperatures span a range, from about $10^{4.4}$ K to $10^{5.2}$ K, and represent stages from lithium-like ions to doubly ionized species. Figure 2.3 shows the temperature dependence of some transition region lines from ions of silicon, carbon, oxygen, and nitrogen. The plasma producing these transitions is considered to be in collisional ionization equilibrium, with the atomic process of collisional excitations being balanced by radiative de-excitations.

Transition region emission is seen during and outside of flares. During stellar flares these lines show an increase in the total intensity, as well as an increase in line broadening. The transitions are for the most part optically thin, which together with being in collisional ionization equilibrium, means that the observed line intensity can be used to determine the volume emission measure of the emitting plasma. The volume emission measure, or VEM, is

$$\mathrm{VEM} = \int n_e n_H dV, \tag{2.5}$$

where n_e is the electron density, n_H the hydrogen density, and the integral is over the emitting volume V; the units of VEM are cm^{-3}. Examining the behavior of emission lines tracing different temperatures determines the multithermal nature of the flare. While often multiple temperatures are present, and responding to the flare energy input in this part of the atmosphere (see Fig. 2.4 for a flare in multiple chromospheric and transition region lines), there are examples of energy releases apparently confined to a small part of the atmosphere, as in the flare observed on the young solar analog star EK Dra (Ayres and France, 2010).

Time-resolved spectra with high enough spectral resolution reveal the presence of line broadening and line shifts during stellar flares. These measurements can be used to quantify the magnitude of turbulent motions in the atmosphere, and bulk flows, respectively. The increase in line broadening is parametrized as

$$\left(\frac{\Delta\lambda}{\lambda}\right)^2 = 3.07 \times 10^{-11} \left(\frac{2k_B T_{\max}}{m_i} + \xi^2 + v_{\mathrm{inst}}^2\right) \tag{2.6}$$

(after Wood *et al.*, 1997; Osten *et al.*, 2006), where $\Delta\lambda$ is the line profile width, λ is the wavelength of line center, k_B is Boltzmann's constant, $T_{\max}$ is the formation temperature of the line under conditions of ionization equilibrium, m_i is the ion mass, ξ is the most probable non-thermal velocity, and v_{inst} is the instrumental broadening of the line profile. Doppler velocity shifts to the line profile, if any, return information on the bulk flows using the standard Doppler shift formula. Blue shifts would be associated with material leaving the stellar surface (evaporation), and red shifts associated with material traveling from above the stellar surface back towards it (condensation).

2.1.2.5 *Hot black-body UV/optical continuum*

The impulsive (seconds–minutes) and often dramatic (factors up to 1000) increase in the white-light emission from M-type dwarf stars during flares was historically the first indication that stars were capable of producing short-time-scale flares as observed on the Sun. Because the spectral energy distribution of cool M-type dwarf stars (typically T_{eff} ~3000 K) peaks towards longer visual wavelengths, the sudden increase in the blue–optical wavelengths seen during flares is easy to recognize. Figure 2.5 displays a representative optical spectrum of an M dwarf during and

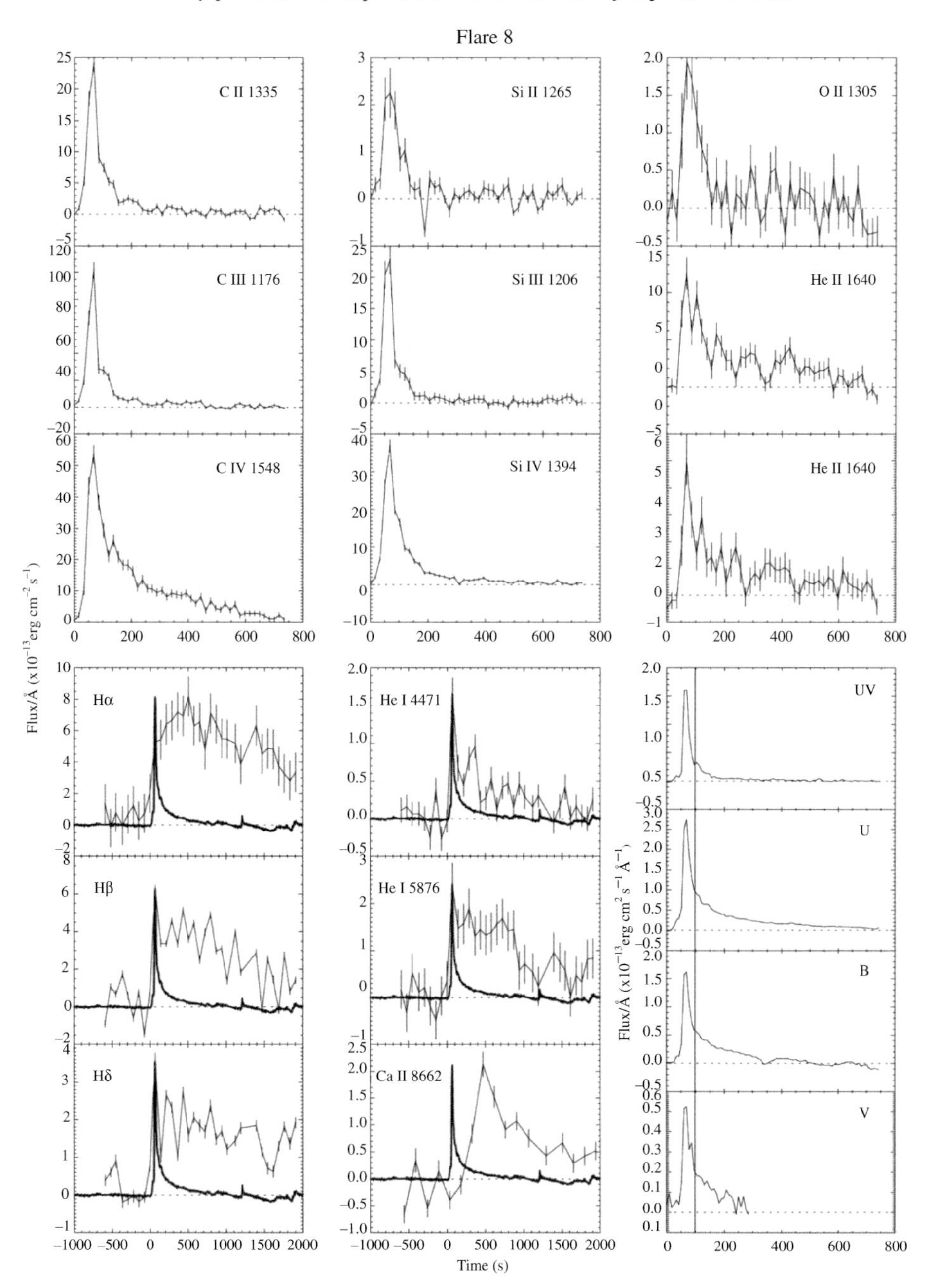

Fig. 2.4 Behavior of emission lines in the optical and ultraviolet during a well-studied flare on the nearby flare star AD Leo (from Hawley *et al.*, 2003). Ultraviolet emission lines (C II, Si II, O I, C III, Si III, He II, C IV, Si IV, N V) have time scales comparable to that seen in the broadband filters (UV, U, B, V) that measure primarily impulsive flare continuum emission. In contrast, the variation of emission lines originating from lower in the stellar atmosphere (Hα, Hβ, Hδ, He I, and Ca II) have a delayed response to the flare energy input seen from the hot black body, as well as a longer decay time scale.

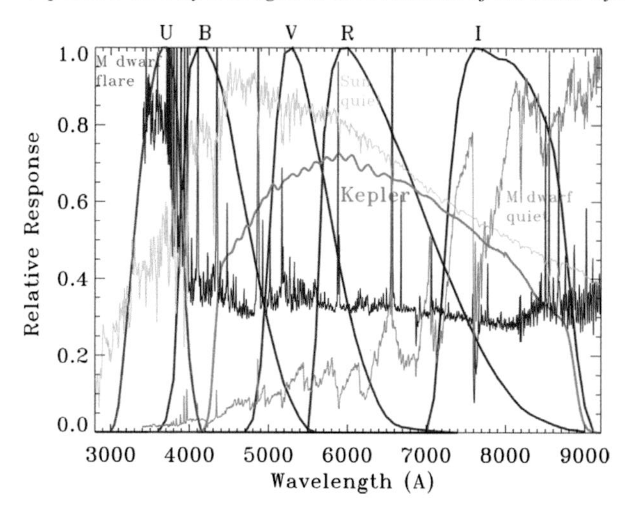

Fig. 2.5 Wavelength coverage of several standard filters used in optical astronomy, along with the wavelength coverage of the filter used in the Kepler mission. Overplotted are also spectral energy distributions of a quiescent and flaring M-dwarf atmosphere, taken from Kowalski *et al.* (2013). The solar spectrum is the 1985 Wehrli Standard Extraterrestrial Solar Irradiance Spectrum from http://rredc.nrel.gov/solar/spectra/am0/. A black and white version of this figure will appear in some formats. For the color version, please refer to the plate section.

outside of a flare; the rise in blue–optical continuum flux during the flare is evident. For reference, the wavelength coverage of some optical broadband filters are also shown: Johnson UVBRI filters, as well as the bandpass response used in the Kepler mission.

The spectral energy distribution of these white-light flares shows a dramatic change from that of the quiescent star. Broadband filters show a roughly black-body behavior, with the temperature of the black body near 10^4 K (Hawley *et al.*, 1995). The flare continuum can be diagnosed by relating the flare continuum flux to the black-body temperature and the area covered by the flare using the relation

$$F_\lambda = \frac{x R_\star^2}{d^2} \pi B_\lambda(T_{\mathrm{BB}}), \tag{2.7}$$

where F_λ, with units of erg cm^{-2} s^{-1} Å^{-1}, is the measured flare flux at effective wavelength λ, $R_\star$ is the stellar radius, d is the distance to the star, T_{BB} is the black-body temperature, and $B_\lambda(T_{\mathrm{BB}})$ is the Planck function evaluated at wavelength λ and temperature T_{BB}. The factor x then gives the fractional area of the star covered by the flaring material, which for the most energetic flares approaches 10^{-3} of the total stellar surface (Hawley *et al.*, 2003).

Observationally, there is a close association between other impulsive phase flare signatures and optical/UV continuum increases. The black-body emission is thus thought to have its origins in the action of the accelerated electrons which dominate the impulsive phase. However, detailed modeling of the radiative hydrodynamic response of an M-dwarf atmosphere to the flare energy input specified by a solar-like prescription for electron-beam heating cannot reproduce the magnitude of observed flare enhancements in stellar flares because the electrons do not penetrate deeply enough into the stellar atmosphere (Allred *et al.*, 2006). The close observed correspondence between the white-light flare and signatures associated with accelerated particles (see discussion in Sect. 2.1.2.1) lends credence to the idea that accelerated particles are involved in the white-light flare. Then the key difference between solar and stellar white-light flares may be the applicability of a solar-type electron beam.

Although this flare signature has been studied most often from flares on M dwarfs, the black-body blue–optical continuum shape may be a common feature of white-light flares from more types of stars. Kretzschmar (2011) performed a superposed epoch analysis for solar white-light flares and detected signals corresponding to small enhancements at the times of even moderate (C-class) solar flares. The spectral signature, taken from a few broadband optical filters, appears to be consistent with continuum emission of black-body shape characterized by a temperature of about 9000 K.

Traditionally, the cool K and M dwarfs have been studied the most for their optical flare signatures, due to the enhanced contrast with the underlying stellar emission. Other studies have expanded the stellar types where impulsive optical flares have been observed, from solar-type stars (Maehara *et al.*, 2012; Osten *et al.*, 2012) to the coolest M dwarfs (Stelzer *et al.*, 2006). There have even been reports of white-light flares on more massive A-type stars (Balona, 2012). The ability to make sensitive relative photometric measurements with a long observing time has enabled these advances. Future astronomical telescopes will use the time domain to find transiting exoplanets, enabling an exploitation of these data for finding new classes of flaring stars.

2.1.2.6 Stellar coronal mass ejections

Coronal mass ejections (CMEs) are an important component of many solar explosive events. Chapters 6 and 12 in Vol. II discuss models of solar CMEs and their relation to flares and the entirety of the eruptive event. Additionally, they are an important component of space weather. Solar studies have demonstrated that the kinetic energy of CMEs is of the same order as the bolometric radiated flare energy from a sample of solar eruptive flares (Emslie *et al.*, 2012).

Based on the close association between flares and CMEs on the Sun, and the apparent similarity in behavior between solar and stellar flares (albeit scaled up by orders of magnitude), it is natural to extrapolate scalings between solar flares and CMEs to investigate the impact of stellar CMEs. This is a speculative affair due to the fact there are no confirmed detections of stellar CMEs (Leitzinger *et al.*, 2011). The workhorse of solar CME observations is the coronagraph, but the requirements on sensitivity, time, and inner working angle so far have been prohibitive for astronomical coronagraphs. Other methods have been proposed, such as interpretation of varying hydrogen column densities in X-ray spectra obtained during flares (Franciosini *et al.*, 2001) or evidence for weak high-velocity features in the wings of optical emission lines during flares (Houdebine, 1996). Osten *et al.* (2013) speculated that a large drop in mid-IR flux from a star with a debris disk could have been caused by a massive CME from the star.

Several studies have established scaling relations between solar flares and CMEs and extrapolated to the case of stellar flares (see Sect. 4.2.2 for such scalings; also Aarnio *et al.*, 2012; Drake *et al.*, 2013). The implied total CME-driven stellar-mass loss for young active stars is high, at levels from 10^{-12}–10^{-9} $M_\odot$ yr^{-1} (compared to about 3×10^{-14} $M_\odot$ yr^{-1} for the total mass loss of the present-day Sun; see Ch. 2 in Vol. III). These high levels are at odds with constraints on stellar mass loss established for young solar-like stars (Matt and Pudritz, 2007). An apparent resolution to this conflict lies in a breakdown in the solar–stellar connection as regards stellar eruptive events, but this has not been demonstrated conclusively. Other signatures of CMEs in the electromagnetic spectrum that can be exploited are low-frequency type-II radio bursts, which have a distinctive signature of a burst which drifts in frequency and time as the MHD shock produced by the CME travels through the tenuous outer stellar atmosphere. The discussions in Chs. 4 and 5 of Vol. II provide more description of solar CMEs.

2.1.2.7 Stellar energetic particles

Solar eruptive events involve the production of energetic particles, which are observed throughout the solar system. Chapter 9 of Vol. II discusses energetic particles produced by solar eruptive events, which generally have kinetic energies in excess of 0.1 MeV/nucleon. The acceleration mechanisms in the solar case have their origin in either the flare process or in a shock associated with the coronal mass ejection, and the energies can extend up to 7 GeV/nucleon. In contrast to the accelerated particles observed in the solar atmosphere during the flare, which are generally electrons trapped in magnetic loops or ions observed at low heights in the solar corona, these energetic particles are outwardly directed from the star, and detected 1 AU or more from the Sun. We have no constraint on the presence of stellar energetic particles in astrospheres associated with stellar explosive events. Application of scaling laws between solar flare X-ray photons and proton fluences are often used to explore parameter space for the influence of stellar explosive events from young stars in affecting material in the early solar nebula (the meteoritic isotopic abundance anomalies; Feigelson *et al.*, 2002), as well as the influence of eruptive events on habitability of extrasolar planets (Segura *et al.*, 2010). It is not clear how applicable these scaling laws are, and they do not (as yet) provide testable constraints for stellar cases.

2.1.2.8 X-ray flares

Stars produce X-ray emission during and outside of flares. The emission is well-described by a plasma in collisional ionization equilibrium, where ionic transitions with collisional excitations balanced by radiative de-excitations give rise to numerous emission lines in the extreme ultraviolet to X-ray spectral bandpass. Continuum emission contributes as well, predominantly from free-free emission but including free-bound and 2-photon processes (Dere *et al.*, 1997). Figure 2.3 shows the temperature coverage probed by the highly ionized species in the X-ray regime, at temperatures above 10^6 K. More description of the atomic physics involved in the production of this emission can be found in Mewe (1999). Spectral analysis of X-ray flares returns key quantities describing the spectral energy distribution of the flaring plasma: the plasma temperature and volume emission measure, and the abundances of the elements contributing significantly to the emission (C, N, O, Ne, Mg, Fe, Si, S for the most part). In situations where the spectral resolution is high enough, changes in the electron density of coronal plasma can be discerned during flares. A much more exhaustive review of stellar X-ray emission and flares in particular can be found in Güdel and Nazé (2009).

The major change in the emission during flares is the sudden creation of denser, hotter plasma, produced as the result of chromospheric evaporation lower in the atmosphere. Temperature and volume emission measure increase, with a characteristic increase during the rise phase of the flare and a return to quiescent conditions in the decay phase. Figure 2.6 shows an example X-ray flare seen on the tidally locked active binary system named TZ CrB (composed of two nearly solar-type stars with 1.1 day orbital/rotational periods) in the energy range 0.8–10 keV (1.2–16 Å). Outside of the flares plasma temperatures are already elevated compared to the Sun – quiescent temperatures of 20 MK are par for the course, while these temperatures are usually only achieved on the Sun during solar flares – and during stellar flares the temperature increases beyond that value, with maximum values around 10^8 K.

The abundance of the flaring coronal material can exhibit a change when compared against the abundance of non-flaring coronal material. Outside of flares, the X-ray spectrum of magnetically active stars typically shows abundances less than

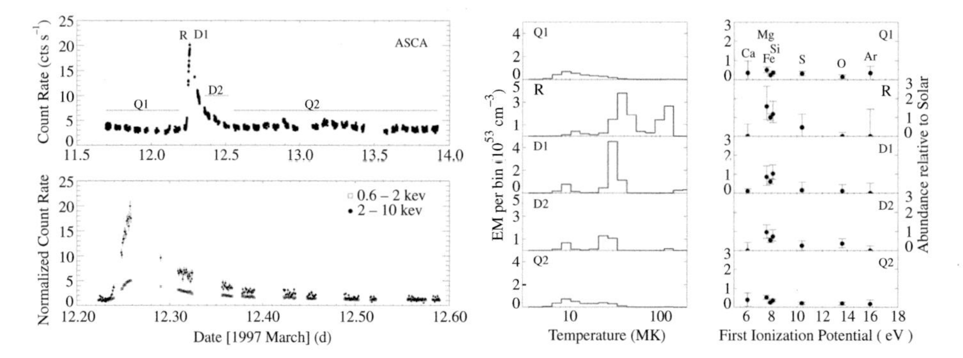

Fig. 2.6 Example stellar flare X-ray light curve (left panel), with the temporal segments analyzed spectroscopically identified. The panel below this total X-ray intensity light curve shows the time variation in two bandpasses: the harder energy bandpass is much more impulsive, reflecting the creation of very hot plasma during the flare rise phase. The right panel shows corresponding change of temperatures and abundances in the phases of the flare. This flare was observed on the active binary system TZ CrB, which is composed of two solar-like stars rotating at a period of 1.1 days (from Osten *et al.*, 2000).

the solar photospheric values (Brinkman *et al.*, 2001). During the rise phase of flares, the coronal abundances *increase*, signaling a temporary increase in material in the corona, presumably brought up from lower in the stellar atmosphere where the abundance is higher. This is in line with expectations from chromospheric evaporation and abundance patterns in coronal emission seen outside of flares, which in the most active stars show a lowered amount of low first ionization potential (FIP) elements compared to the solar photosphere. There are flares for which no change in abundance pattern occurs relative to the quiescent emission and the interpretation here is more complex (Osten *et al.*, 2010).

Indirect constraints on the length scales of stellar flaring coronal plasmas generally suggest length scales longer than those measured during solar coronal flares. The number of assumptions required to employ such techniques may affect the outcome. The VEM, a measurable parameter from spectral fitting, depends on electron density, ion density, and emitting volume, so with a constraint on VEM and an assumption about density and coronal geometry length scales can be estimated to first order. Other methods rely on applying a balance between radiative and conductive losses in the decay phase of the flare to derive a constraint on loop length (van den Oord *et al.*, 1988), or describe the change in stored magnetic energy with time using a simple prescription for the magnetic configuration (Kopp and Poletto, 1984). The most common application to describe length scales in stellar X-ray flares has been the work of Reale *et al.* (1997). The technique uses information on *e*-folding time scale in the decay phase of the flare light curve, as well as time-resolved spectral analysis of the change of temperature and volume emission measure with time. Assuming that the flare geometry and hence volume do not change appreciably with time, the volume-emission measure changes can be related to changes of density. Hydrodynamic modeling of decaying flaring loops with a range of dimensions and time scales provides a quantification of the amount of heating occurring in the flare decay, which can be assessed observationally by the slope of the points in the temperature–density plane during the flare decay. As with most parametrized efforts, a single semi-circular loop with a constant cross-section is assumed for loop geometry. Hydrodynamic modeling of decaying flaring loops with a range of dimensions and time scales is folded into instrument response, allowing for a refinement of the estimates and application to specific instruments and bandpasses. The hydrodynamic method applied to solar flares returns values of flaring loop semi-lengths of $\sim 10^9$ cm, or $L/R_\odot$ ~0.01, typical of solar active region flaring loops (Reale *et al.*, 1997). The method applied to stellar flares returns loop semi-lengths which are larger, in both an absolute sense and when compared against the radius of the star: Stelzer *et al.* (2006) analyzed a flare on a very low mass dwarf, finding a length scale of 1.4×10^{10} cm, or $L/R_\star$ ~2, while Osten *et al.* (2010) dissected a flare on a higher-mass M dwarf, finding a slightly smaller length in absolute units, but a relative size scale of $L/R_\star$ ~0.4 when considering the factor of ≈ three difference in radius of the two stars. The study of Favata *et al.* (2005) examined flares on young (~1 Myr-old) stars, and found loop semi-lengths as large as $\sim 10^{12}$ cm, exceeding the stellar radius by large values (factors of tens). Because of its simplicity and dependence on relatively easily derived observable parameters, this approach has been used most widely to diagnose length scales in stellar flaring coronal plasmas.

2.1.2.9 Chromospheric lines

Chromospheric emission lines originate in the gradual phase of the flare. As demonstrated in Fig. 2.4, these lines (like Hα) show a much longer decay time scale before returning to quiescent conditions than for markers of the impulsive phase. These lines also exhibit a delay in responding to the energy input: their intensity maxima occur significantly after that of the impulsive phase flare emission. These chromospheric lines are optically thick (in contrast to the transition region and coronal emission lines originating higher up in the stellar atmosphere) and so inferring physical conditions in the flaring atmosphere requires the use of radiative transfer models to describe how the radiation interacts with the material. Consideration of the changing conditions during a flare additionally needs a dynamical treatment, such as that in Allred *et al.* (2006).

2.1.2.10 A note about energy partition within flares

There are very few multi-wavelength observations of stellar flares where we can get an entire picture of an explosive event, including the energetics of the various components described in the above sections. Even on the Sun this is only possible for the largest events, and even then some crucial pieces of information are missing and must be extrapolated (Emslie *et al.*, 2012). This requires stellar astronomers to rely on solar scalings that can be applied to stellar flares, or application of results from a detailed study of one stellar flare to flares on different types of stars. So when considering flare measurements made in one wavelength region, this energy release needs to be placed in context with what might be appearing in other regions of the electromagnetic spectrum. Nevertheless, a consistent picture does appear to emerge when considering the radiative output of solar and stellar flares. Kretzschmar (2011) showed that optical continuum flare radiation carries about 70% of the total radiative energy, as measured by total solar irradiance. Emslie *et al.* (2012) showed that X-ray radiated energy of solar flares accounts for about 20% of the total bolometric radiated energy. Multi-wavelength studies of M dwarf stellar flares (Hawley *et al.*, 1995) show similar ratios between the hot continuum radiation in the optical bandpass and soft X-ray radiated energy. The contribution to the flare energy from accelerated particles is less constrained in both solar and stellar flares due to lack of constraint of key parameters, but lower limits suggest that it may well exceed that in the radiated energy (Emslie *et al.*, 2012; Smith *et al.*, 2005).

2.1.3 Flares in aggregate

As examples of dissipative processes, stellar flares are expected to follow a power-law size distribution. Any wavelength region where flares exhibit emission can be used thus; however, in the coronal regime the size distribution of flares takes on added importance. The flare-size distribution is characterized by a power-law α such that

$$\frac{dN}{dE} = kE^{-\alpha}, \tag{2.8}$$

where dN/dE gives the number of flares occurring per unit time per unit energy, k is a constant, and E is the flare energy in the bandpass being considered. As noted by Parker (1988), for values of $\alpha > 2$ it is possible to integrate the contribution of smaller and smaller flares until the quiescent X-ray luminosity of the star is achieved. This is termed the "nanoflare heating hypothesis". This ensemble approach determines the relative importance of flares of different sizes. While many authors (notably X-ray astronomers) give a lot of importance to the value of α, the overall flare rate is also important. As Hilton (2011) demonstrated, M dwarfs with differing activity levels (from inactive to active to the most active M dwarfs) still produce flares, and with similar values of α, yet the overall flare rate can vary by orders of magnitude.

2.2 Time scales: explosive events on stars, young to old

Although age is one of a star's fundamental parameters it is also the most difficult to measure accurately. Soderblom (2010) provides a summary of the available techniques for age-dating stars and ensembles of stars. Each has its own advantages and drawbacks, and the uncertainty associated with a particular age measurement can vary widely. For the present purposes broad age categories are used to delineate some key times within the life of a star, and relate what is known about explosive events on stars within this range. It is important to remember that within each broad age category other parameters like stellar mass and rotation are important to consider, and for a variety of reasons the observational constraints that populate this parameter space suffer heavily from bias (in stellar type and wavelength regions) and observational limitations.

2.2.1 Stellar infancy: birth to the zero-age main sequence

Chapter 3 of Vol. III describes the formation and early evolution of stars and protoplanetary disks, and Fig. 1.1 of Ch. 1 in Vol. III gives key events in the history of the Sun. Although the age range in this first age category is only a percent or so of the total main-sequence lifetime of the star, there are several important steps to the life of the star that occur during this time. Before discussing what is known about explosive events on stars of this age range, it is necessary to set the stage of what is going on in stellar evolution, as the environment around the star influences how we observe and interpret explosive events. For this section we concentrate on ages ranging from stellar birth to the time it takes the star to reach the zero-age main sequence (ZAMS), at which point the star is in stable hydrostatic equilibrium, and there is negligible contribution to the stellar luminosity from any accretion-related processes. This time scale is a function of stellar mass, being approximately 50 Myr for a solar-mass star, and longer than 160 Myr for a star of 0.5 $M_\odot$ or less. For the purposes of discussion, and because stellar ages can be uncertain by factors of two or more, we include stars of ages up to $\sim$100 Myr. The main point is to differentiate factors of stellar youth affecting explosive events from those associated with accretion and the remnants of the planet-forming disk.

A star is considered born when the fragmented molecular cloud has contracted to the point where it forms a hydrostatic core. Phases of star and planet formation are still continuing, and the star can be referred to generally as a pre-main-sequence star, because its location in the Hertzsprung–Russell diagram is above that of the ZAMS. A solar-mass pre-main-sequence star, for instance, has a larger stellar radius and lower gravity than an equal-mass star on the main sequence. At these early times the star is still enshrouded by a massive disk out of which it formed. Once the star + disk system has proceeded to the point that both components contribute to the integrated light, the object can be referred to as a classical T Tauri star, or class-II object (see Evans *et al.*, 2009, for a "Diskionary" explaining different terms relevant to young stars). Accretion processes are still occurring, and the star is rotating rapidly due to its youth. As the disk disperses, the total light becomes dominated by the star itself, and these objects are referred to as weak-lined T Tauri stars or class-III objects. These stars might not show any evidence for a massive, gas-rich disk but it is still possible for them to host a weaker disk that is gas poor. The lifetime of the gas-rich disk has a systematic dependence on the mass of the star, with solar-mass stars losing their disk in $\sim$5 Myr. However, there is a range of disk lifetimes within a given mass range: other factors such as multiplicity, proximity to the intense radiation field of O stars, and residing in a dense stellar cluster can each shorten the disk lifetime. The gas-poor or debris disks, are composed of rock, dust, and ice, which represent the end phase of the process of star formation. This phase can last for a long period of time, however; while the small particles (typically 1–100 μm in size) are subject to removal by radiation pressure from the star, or spiral in towards the star through Poynting–Robertson drag, these time scales are relatively short (of order 10 Myr), and other processes such as a collisional cascade of larger particles can replenish the disk. Increasingly sensitive astronomical detectors can also see evidence for faint increases in mid-infrared light above that expected from the stellar photosphere, signaling the presence of a debris disk at much older ages. Our solar system's asteroid and Kuiper belts would be considered late-time examples of debris disks.

Magnetic activity in general is at a high level in these young stars because of their rapid rotation, but the interpretation can be confused by other processes occurring in the system that have similar observational characteristics to magnetic reconnection processes. While explosive events produce temporal changes in luminosity, so too can accretion produce variable signatures of similar diagnostics. Feigelson and Montmerle (1999) discuss high-energy processes in these young stars. Observations of flares at X-ray and radio wavelengths are dominated by reconnection processes. The thermal coronal radiation showing up in X-rays does appear to follow patterns seen in older more active stars (temporal trends of intensity, temperature, VEM), and non-thermal radio gyrosynchrotron emission has a different spectral and temporal signature from radio thermal bremsstrahlung emission from an ionized wind of outflowing material from the star + disk. In contrast, observations of optical variability are more problematic to interpret, because accretion-related variability can also produce changes in continuum and Balmer-line emission similar to what is seen during flares on older more active stars. Because young stars are born in clusters, they tend to be spatially concentrated, which can make crowding in the field of view an issue, but the advantage of multiplexing hundreds of stars in the field of view of an astronomical telescope to study stochastically occurring and relatively infrequent explosive events outweighs this concern for the most part.

Flares some 100–1000 times more energetic than the biggest solar flares occur roughly once a week on these young, rapidly spinning stars. Most studies have concentrated on singling out particular stars for further study, but there have been a few systematic studies of flaring in clusters. Wolk *et al.* (2005) studied X-ray flares on young suns in the Orion Nebula Cluster, with masses $0.9 \leq M \leq 1.2\ M_{\odot}$; these are pre-main-sequence stars with an age near 2 Myr, so for the most part they still have their accretion disk. The radiated energies span 10^{34}–10^{36} erg (to be contrasted with the largest solar flares approaching 10^{33}; Schrijver *et al.*, 2012), with a power-law distribution characterized by an index α of $\sim$1.7. The flare frequency in this sample is 1 flare per star every 650 ks after correcting for data gaps. The flare durations span 1 h to 3 days, with amplitudes generally less than about 10 times the underlying level. Stelzer *et al.* (2007) looked at X-ray flaring in young stars in the Taurus Molecular Cloud (with ages spanning 1–3 Myr). They also found energetic flares, with flares spanning the range of a few times 10^{33}–5×10^{35} erg. Flares with energies in excess of 10^{35} erg happen once per 800 ks per star, with a duration of 10 ks. They also found a power-law distribution of flare energies, with an index $\alpha = 2.4 \pm 0.5$. The level of flares to which such studies are sensitive is limited to those events which are far more energetic than the largest event observed on the Sun, due to the intrinsic limitation of astronomical detector sensitivity.

For these very energetic flares from stars in the range of a few Myr, stars with and without a gas-rich disk show no strong difference in flare rate, and there also does not appear to be a change in flare rate from solar mass to low mass stars. Stelzer *et al.* (2000) found no difference between X-ray flares on stars with and without an accretion disk in clusters with ages 1–3 Myr. Caramazza *et al.* (2007) compared the coronal flare frequency distributions of flares on solar-mass stars in Orion (0.9–1.2 $M_{\odot}$) with those of low mass stars (in the mass range 0.1–0.3 $M_{\odot}$) and found no difference in flaring behavior for stars as a function of mass. From a compilation of flares seen in Balmer emission lines and optical continuum, with flare energies in excess of 10^{33} erg, Guenther and Ball (1999) found a factor of two difference in the flare rate of stars with and without an accretion disk, with the flare rate for classical T Tauri stars (i.e., those with disks) about a factor of two lower than weak-lined T Tauri stars.

Over slightly longer evolutionary time scales there is a decrease in flare rate. The optical flare study of Guenther and Ball (1999) also included one pre-main-sequence star, and the flare rate of the ZAMS star was significantly reduced, by about a factor of 10 compared to the weak-lined T Tauri stars. Stelzer *et al.* (2000)

did a systematic study of the X-ray flare rates of T Tauri stars at ages 1–10 Myr, and studies of stars in the Pleiades (age ~100 Myr) and the Hyades (age ~600 Myr). They noted that the G stars in their group had the lowest flare rates by spectral type. Corrected for sensitivity biases, their flare rate for the Pleiades was 0.67±0.13%, for the Hyades 0.32±0.17%, expressed as a fraction of time in the flaring state compared to total observation time.

For the youngest stars, the loop lengths inferred from X-ray flare analysis may be much larger than seen on the Sun, and can provide a connection from the star to the disk. This is potentially important, as the flares can provide a source of turbulence to the disk and may affect planet formation. However, the measurement of flare loop lengths on stars, lacking spatial resolution, is model-dependent and relies of the applicability of the model. Favata *et al.* (2005) analyzed some of the largest flares from young stars in the Orion Nebula Cluster, without selecting mass ranges. From an analysis of the flare decays, and using a one-dimensional hydrodynamic loop model, which returns constraints on loop length, they found that a substantial number of large flares appeared to originate from flaring loops with lengths sufficient to connect the star to the planet-forming disk (and thus potentially increase the turbulence in the disk and affect planet formation). The loop lengths implied are several tens of stellar radii, compared with a fraction of solar radius for the largest flaring loops observed on the Sun. This idea has been investigated further with different samples of flares from pre-main-sequence stars. Just as other flare parameters show a distribution, there is a distribution of flare loop lengths. McCleary and Wolk (2011) found that the longest loop lengths were of order several stellar radii and tended to occur on stars with an infrared excess, indicating that they still possessed their planet-forming disk. Whether this is a cause and effect, or a result affected by over-extrapolation of a model beyond its bounds of applicability, is still being debated.

Studies of radio flaring on young stars occur more sporadically, but indicate a population of much more energetic electrons than usually seen in solar flares. One particular well-studied pre-main-sequence star, V773 Tau, shows energetic *synchrotron* flares, and this has also been seen on a few other young stars (Kóspál *et al.*, 2011; Getman *et al.*, 2011). In contrast, solar flares show less energetic gyrosynchrotron emission from lower harmonics of the gyrofrequency, indicating less energetic electrons. The periodic nature of the flares studied in detail has been explained as arising in an eccentric binary system where periastron passage of the secondary, and two extended magnetospheres, causes reconnection to high energies with a recurrence of approximately the orbital period. Whether this behavior is commonplace among young stellar objects or limited to binary systems satisfying these conditions is a current topic of investigation. It does indicate a role for

MeV-level electrons in the explosive event, which may be important in explaining meteoritic abundance anomalies.

Observations of isolated young stars confirm the general pattern of elevated activity in young stars seen in clusters, but show some marked differences from solar behaviors: steeper flare frequency distributions, flares limited in temperature/wavelength region, and multi-wavelength flare signatures exhibiting the opposite trend to the standard solar flare scenario. The nearby G star EK Dra is a solar analog with an estimated age of 50–70 Myr, placing it in the age range under consideration. The distribution of its high-energy flares have been characterized as a power-law with α of 2.08 ± 0.34 above an energy of $10^{30.2}$ erg (Audard *et al.*, 2000). Guedel *et al.* (1995) found radio variability in this object that was interpreted as flares, but the observations were not sensitive enough to perform a detailed study of their characteristics. Ayres and France (2010) found a transient UV event on this star seen only in Si IV but not in a coronal line found in the far-ultraviolet bandpass. Whether this last example can even be counted as a component of an explosive event is not clear, because its formation in a narrow temperature range in the upper atmosphere (and without any apparent contribution from high-energy emission) suggests a limited physical range for energy deposition. Feigelson *et al.* (1994) studied V773 Tau, a classical T Tauri star, in which a radio flare was observed without any apparent X-ray variations. Similar behavior was noted in a cool star serendipitously observed in outburst at mm wavelengths in the Orion Nebula Cluster, in which the X-ray flare preceded an accompanying radio outburst from the star (see Fig. 2.7).

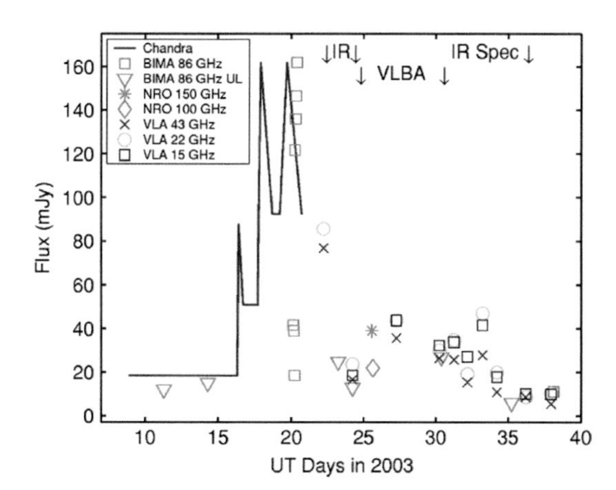

Fig. 2.7 Unusual radio/X-ray flare observed on a pre-main-sequence K-type star in the Orion Nebula. The strong outburst was observed first at X-ray wavelengths, then at radio wavelengths, in contradiction to the pattern usually seen for solar explosive events. Figure from Bower *et al.* (2003).

2.2.2 Stellar teenage years: ZAMS–1 Gyr

At this phase in a star's evolution, rapid rotation is still an important factor, although it has declined since the star's youth. According to the Skumanich $t^{-1/2}$ relation between age and rotation, a solar-mass star would have a rotation rate that is only a factor of 2–7 above the Sun's present-day rotation; activity that accompanies the faster rotation should be enhanced, but below the extremes represented by the youngest stars. Earlier-type cool stars spin-down faster than lower-mass stars (Stauffer, 1991), so by these ages M dwarfs dominate the samples of active stars. The general decrease in activity levels compared to the extremes seen at young ages means that capturing flaring activity on stars of this age range (with the exception of M dwarfs) is more difficult to do systematically, and consequently there is a heavy bias towards the lower-mass end in observations of flares on stars of this age range. The fact that M dwarfs are the most common type of star based on mass functions also contributes to this bias. There are open clusters (notably the Hyades at an age of ~800 Myr) which are nearby enough for sensitive studies of explosive events, although they are spatially dispersed compared to star-forming regions and this makes it difficult to capture more than one or two objects in the field of view of typical astronomical telescopes.

The possible dependence of stellar flare rate on evolutionary age can be explored by combining scaling relations between flare frequency and underlying coronal emission with those relating coronal and chromospheric emission, and others describing the decline of chromospheric emission with time. Audard *et al.* (2000) found a scaling between coronal flare rate and underlying stellar X-ray luminosity for a sample of stars including several with ages in this age range. They quantified this as

$$N(> E_c = 10^{32}) = 1.9 \times 10^{-27} L_X^{0.95} \text{ flares per day,} \quad (2.9)$$

where $N(> E_c = 10^{32})$ is the coronal flare rate above a threshold flare energy of 10^{32} erg, and L_X is the star's coronal X-ray luminosity. Ayres *et al.* (1995) and Piters *et al.* (1997) established scalings between coronal emission and different chromospheric emission indicators for cool main-sequence dwarfs, $L_X \propto L_{\text{chrom}}^y$, where y ~1.5 for C IV emission (Ayres *et al.*, 1995), $y \sim 2$ for Ca II HK emission (Piters *et al.*, 1997), and y ~3 for Mg II h emission (Ayres *et al.*, 1995). Skumanich (1972) showed that chromospheric emission declines roughly with stellar age as $L_{\text{chrom}} \propto t^{-1/2}$. Simplifying these relations to

$$N(> E_c) \propto L_X, \quad (2.10)$$

$$L_X \propto L_{\text{chrom}}^y, \quad (2.11)$$

where y takes on different values depending on the chromospheric emission being considered, and

$$L_{\mathrm{chrom}} \propto t^{-1/2}, \tag{2.12}$$

suggests that the flare rate may decline with age anywhere from $N(> E_c) \propto t^{-0.75}$ to $N(> E_c) \propto t^{-1.5}$. However, note that Schrijver *et al.* (2012) found that the above scaling between flare rate and coronal luminosity cannot be used to "correct" the flare rate of active stars like those considered in Audard *et al.* (2000) to the solar flare rate via their coronal luminosity. This suggests a breakdown in the validity of a scaling-relation approach.

Single G stars in this age range exhibit flares at least as powerful as the largest solar flares, but occurring several times per day. Audard *et al.* (2000) presented an analysis of EUV flare activity for a sample of late-type stars, including κ Cet, a G5V star with an age of 300–400 Myr (Mamajek and Hillenbrand, 2008). They determine an index to κ Cet's flare frequency distribution, with α greater than 2 (2.2) but with large error bars ($\geq$0.5); using the scaling above between flare rate and X-ray luminosity for κ Cet's X-ray luminosity of 10^{29} erg s^{-1} translates to 6.7 flares per day, above a threshold lower flare energy of 10^{32} erg. κ Cet was also listed as a "superflaring" star in the compilation by Schaefer *et al.* (2000), with an estimated flare energy in the He I D3 line of about 2×10^{34} erg. The dynamic range in the measurements that constitute the flare frequency distribution is not large (only about a factor of 20): some criticisms of the large α index returned by these analyses suggest that other factors (such as a downturn in the intrinsic flare-frequency distribution due to a rollover or cutoff) may be affecting the results.

Solar-neighborhood M dwarfs have historically been the most studied flaring stars: the propensity for dramatic blue–optical flare increases initially brought them to the attention of stellar astronomers but now their flares have been studied across the electromagnetic spectrum. Ages of individual M dwarfs are difficult to ascertain, but most objects are considered to have ages of at least a few hundred Myr to 1 Gyr, so they fall squarely within the age range being considered. There is a range of flare energies observed on dMe (i.e., dwarf M-type stars with emission line features) flare stars: the lowest level detectable X-ray flares from the closest star to the Sun (a flare star – Proxima Centauri), are comparable to M-class solar flares (Güdel *et al.*, 2004; flare energy of $10^{28}-10^{29}$ erg). At the other extreme, the most energetic flares from dMe flare stars are in the $10^{34}-10^{35}$ erg range (Kowalski *et al.*, 2010; Osten *et al.*, 2010). Flares are also very frequent, with flare frequencies of the least energetic flares as often as a few per hour. Figure 2.8 compares the flare frequency distributions for flares of different M spectral type, as well as broken up by activity level (Hilton, 2011). Inactive stars (classed by their low or undetectable quiescent Hα emission levels) have a lower flare rate compared to active stars of

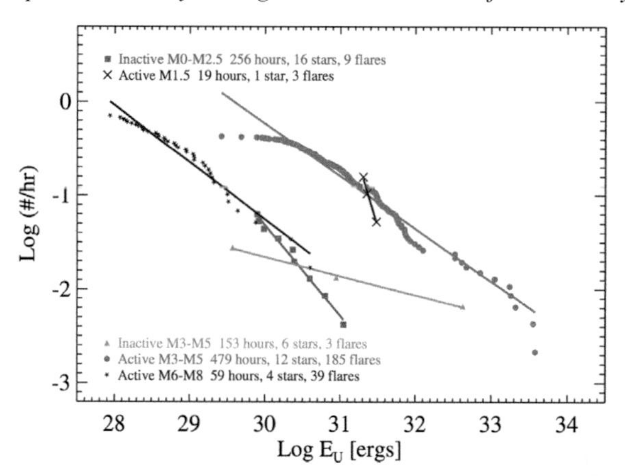

Fig. 2.8 Cumulative flare frequency distributions for different categories of M dwarfs, from the PhD thesis of Eric Hilton. Through long stares at different classes of M dwarfs, the difference in the occurrence rate of flares of different U band energies can be determined.

similar spectral type, with maximum flare energies about an order of magnitude less than those observed on the active M dwarfs.

Flares are even observed to extend into the realm of the ultracool dwarfs: late M dwarfs with photospheric temperatures less than 3000 K (Schmidt *et al.*, 2014). X-ray studies of flares on M dwarfs have considered the flare frequency distribution of individual coronal flares; several studies (Audard *et al.*, 2000; Kashyap *et al.*, 2002) have determined that the shape of the distribution is a power-law with index α greater than 2, which would suggest that flares may be the source of energy required to maintain a corona.

The Hyades cluster is a relatively nearby open cluster of stars with an age of ~800 Myr and has been the subject of multiple flare studies. The flare rate should be lower than for the younger stars considered in the previous section, and the sparseness of flare detections and inability to connect flares in different wavelength regions confounds interpretation. Stelzer *et al.* (2000) studied X-ray flares from a sample of objects in the Hyades as well as nearby stars from younger associations. Figure 2.9 depicts the combined flare rates of objects as a function of age, up to about 1 Gyr, separated by evolutionary status at the youngest ages. The flare rate by 800 Myr is lower than for younger stars. Total integrated energies could not be determined due to the sparse temporal sampling of the data, but the flares represented factors of a few to 20 times increase in the underlying stellar X-ray luminosity. These observations span a range of cool spectral types, and small numbers of detected events prevents exploration of other relevant parts of parameter space like stellar mass and rotation. There is a clear decrease in the flare rate, but whether this is biased by the types of objects detected is not clear. White *et al.*

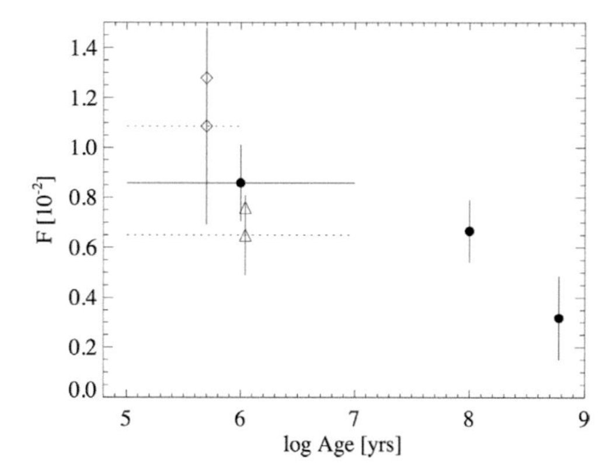

Fig. 2.9 X-ray flare rate expressed as a percent of observing time for stars of different ages, from the study of Stelzer *et al.* (2000). The sensitivity to flares is set to a uniform value to account for the differing distances. Diamonds indicate classical T Tauri stars, triangles indicate weak-lined T Tauri stars. The horizontal line gives the age spread of the young stellar objects. There is a clear drop in flare rate by the age of the Hyades (800 Myr).

(1993) performed a radio study of stars in the Hyades cluster, finding no detections from M dwarfs which have a similar age and rotation rate to the M dwarfs in the solar neighborhood, and with sensitivities sufficient to detect the largest events observed from nearby M dwarfs. They conclude that by the age of the Hyades the activity of most dwarf stars has decreased to a low level where flare rates and energies have fallen below the levels of the nearby population of M dwarf stars. Their results are consistent with expected declines in activity from the age–rotation–activity connection. Browne *et al.* (2009) probed for ultraviolet flares on stars in the Hyades and Pleiades (age $\sim$100 Myr) clusters, using the 1.2° field of view of the GALEX satellite. Most of the variable sources inferred to be flaring are M dwarfs, and only one was a bona fide Hyades cluster member (and an M dwarf). It had a flare with an NUV (1800–2800 Å) energy release of 4.5×10^{29} erg. The event rate is one flare in $\approx$15 000 s, or a flare duty cycle of 3%, which is clearly elevated compared to the X-ray flare rates in Fig. 2.9. Given the singular nature of the detection it is not possible to extrapolate further on the comparison of X-ray versus ultraviolet flare rates. Concentrated optical observations have returned a range of flare rates for one of the more active Hyades M dwarfs: flare frequencies ranging from 0.09 flares h^{-1} up to 1.25 flares h^{-1} have been measured (Haro and Parsamian, 1969; Rodono, 1974).

2.2.3 Stellar adulthood: 1–5 Gyr

The solar system, and thus the Sun's, age measurement of 4568 Myr (Bouvier and Wadhwa, 2010) fits squarely within the "stellar adulthood" phase of its life.

Detections of flares on stars in this age range are much fewer. The decline of flaring with age is generally assumed to follow the trends of other activity indicators, but whether this is in fact the case is an open question. Evidence that magnetic activity may not decline monotonically at Gyr ages comes from a few sources: Silvestri *et al.* (2005) concluded that chromospheric activity in M dwarfs did not decline in the 1–10 Gyr range as fast as predicted based on extrapolating from objects with ages < 1 Gyr. Studies of chromospheric activity and its dependence on age and rotation led Pace and Pasquini (2004) to conclude that they found no evidence of decay in quiescent chromospheric activity for stars older than a few Gyr; instead, the major decline in activity was in objects at ages of the Hyades and earlier (0.6 Gyr), for clusters of 1.7 Gyr and older (up to 4.5 Gyr) the same activity level was seen. This result was updated by Pace (2013) who found no evidence for a decay of chromospheric activity after about 2 Gyr.

Binarity influences activity and thus flaring rates may remain high for old stars in close binary systems (Rocha-Pinto *et al.*, 2002) which can overestimate the flare rate for stars in this age range, if binarity has not been ruled out. The influence of rotation on activity means that tidally locked binary close systems have significantly enhanced levels of activity. They are frequently observed to undergo enormous outbursts at radio (Mutel *et al.*, 1998) and X-ray wavelengths (Osten *et al.*, 2007), even being picked up by satellites tuned to finding the transient hard X-ray radiation from gamma-ray bursts. The radiated energies approach 10^{38} erg, with flare frequency distributions with a power-law index of near 1.8 (Osten and Brown, 1999). Time scales for these flares can be staggering, often lasting longer than the rotational/orbital period of the system. Owing to their frequent and extreme flares, close binary systems have been popular subjects of study; in contrast with flares on M dwarfs, these flares are most readily studied at radio and X-ray wavelengths due to the small contrast at optical bands. Osten *et al.* (2004) provided a comprehensive discussion of multi-wavelength flare campaigns on one active binary system, HR 1099. Although several trends in these flares suggest an extension of the solar flare analogy, the large difference in observed properties of these flares may give pause to the extent to which the solar/stellar analogy can be stretched (and whether it may break). These stars already produce steady levels of high-energy radiation near the limit of what stars seem capable of maintaining, but flares increase the luminosity above that, often approaching the bolometric luminosity of the system. Correlations between flare frequency and underlying stellar luminosity as found for single active dwarfs in Audard *et al.* (2000) have been found among flares on active binary stars (Osten and Brown, 1999).

Because the flare rate is expected to be low on older stars, a systematic search for flares in an older stellar population needs a large number of stars, and involves

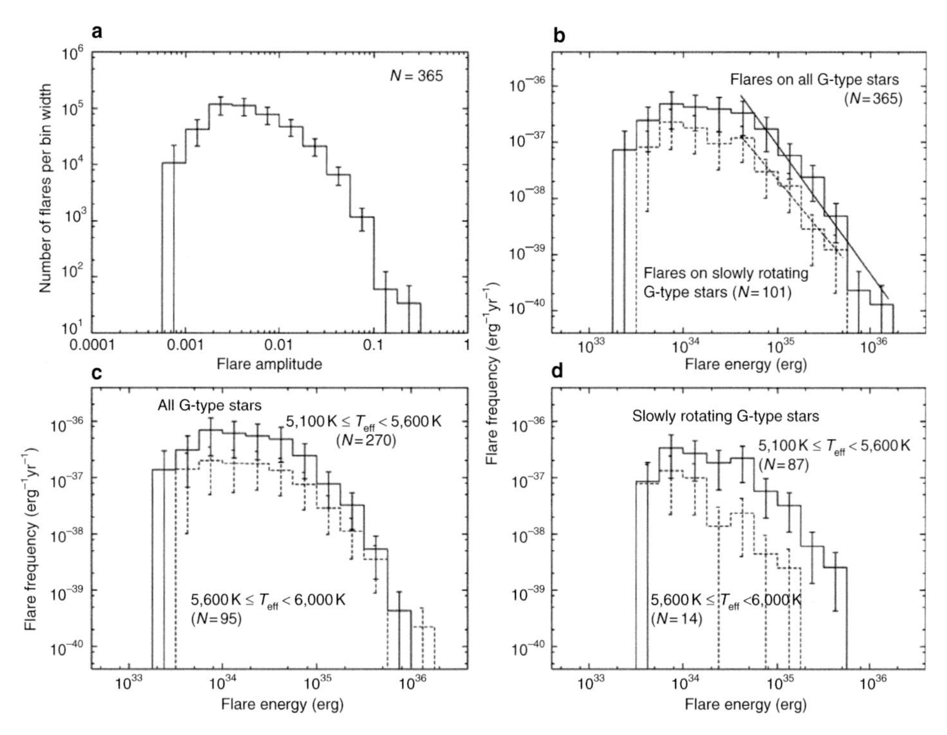

Fig. 2.10 Summary of flare characteristics seen on G-type stars from Maehara *et al.* (2012). (a) The peak flare amplitudes are 10% or less of the star's underlying brightness. (b) The distribution of flares with energy on slowly rotating stars appears similar to that of all G-type stars surveyed. (c) The distribution of flares with energy appears similar when breaking the stars up in hotter vs. cooler stars. (d) The slowly rotating hot stars show a decrease in flare frequency.

a relatively long stare coupled with fast cadence to detect and resolve the flaring emission from any other variability. The Kepler spacecraft's exquisite photometry can be re-purposed from finding evidence of transiting extrasolar planets around stars to looking for rare short-time-scale flaring events on the stars themselves. Maehara *et al.* (2012) reported on energetic flares found from a sample of G-type stars in the Kepler field of view (see Fig. 2.10). While the ages of the individual stars are not known, the spread of rotation periods, particularly the continuing trend of flares on stars with rotation periods longer than 10 days, suggests an intermediate to old age. Further follow-up by Nogami *et al.* (2014) finds flares on two G-type stars with rotation periods of 21.8 and 25.3 days, near the solar value, and thus approximately solar age. The energetics of these flares is large, with minimum flare energies in the range 10^{33} erg, and extending up to 10^{36} erg. The distribution has a power-law index of -2.3 ± 0.3 for superflares found on all G stars, and -2.0 ± 0.2 for slowly rotating G stars, above 4×10^{34} erg.

2.2.4 Stellar old age: beyond 4.5 Gyr

The age of the Sun is pegged to the age of the solar system, and so consideration of stars older than the Sun technically is beyond the scope of this chapter if we look at the history of the Sun. For completeness we include these old stellar ages because there are observations demonstrating the continuation of flaring activity into this regime, and it is of interest to know what the Sun may have in store for the solar system over the coming billions of years.

That magnetic activity can survive to such old ages is evidenced by the flare observed serendipitously on the old solar neighborhood M dwarf Barnard's star, at an estimated age of 11–12 Gyr (Paulson *et al.*, 2006). Because M dwarfs are presumed to carry magnetic activity for longer time scales this is perhaps not surprising. The event had incomplete coverage due to its serendipity, and the flare was identified through the increase in Balmer lines and continuum enhancements. Although noting that such events are likely rare in such old stars, Paulson *et al.* (2006) point out two other references in which observations of flares on Barnard's star are noted, so they may not actually be that uncommon.

Osten *et al.* (2012) re-purposed an observation designed to find eclipsing extra-solar planets in a 10-Gyr stellar population in the Galactic bulge, and used it to study the incidence of optical flares in an old stellar population (see Fig. 2.11). The majority of the flaring stars could be associated with stars at the distance and age of the Galactic bulge, indicating that flaring variability can be maintained. A large fraction of the stars exhibited large-scale photometric modulations (in addition to

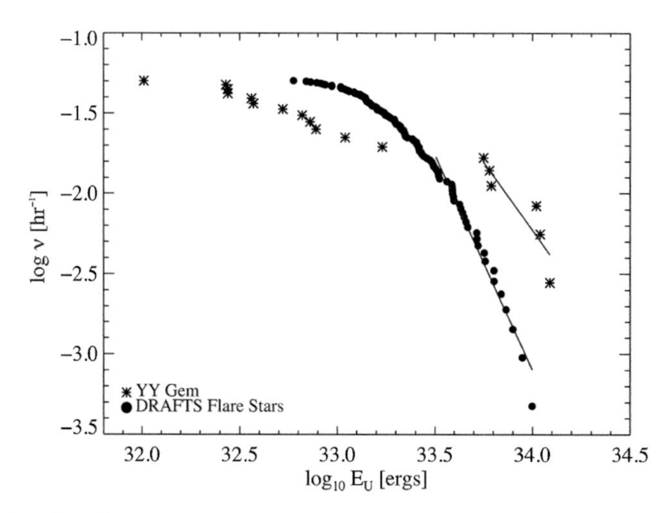

Fig. 2.11 Flare frequency distribution of flares observed on 10-Gyr stars in the Galactic bulge. (From Osten *et al.*, 2012.)

the flaring variations), which indicate rotation periods of a few days. The supposition is that both types of stellar variability can be explained due to the survival of magnetic activity on stars found in tidally locked binary systems, in which the space motions or kinematics indicate an old age but the activity indicators suggest youth.

2.3 Take-away points

The central thesis of this chapter is that studies of flares on stars other than the Sun can be used to give insight into flares on the Sun, and thus extend the time baseline for examining the influence of the Sun on its environs, as well as studying the influence that other host stars may have on their planets as a function of evolutionary time. This approach suffers from some biases in the observations of stars: the types of stars which are the best studied for their flares tend not to be exact replicas of the Sun. M dwarfs, tidally locked active binaries, or young stars with disks have the best characterization of their flares. The wavelength ranges used for studies of stellar flares are often not the same as those used for solar-flare studies, and so an intercomparison requires a conversion to the same wavelength range, after accounting for energy partition. Because of the astronomical difference between our closest star, the Sun, and the distances of even the most nearby stars, there will always be sensitivity differences between solar- and stellar-flare studies; at best, stellar-flare studies can hope to probe the region of the largest solar-flare events.

Nevertheless, despite the possible pitfalls in such an approach, detailed studies of the flares on these stars show agreement with solar flares, indicating that the same physical processes are occurring during solar and stellar flares. Flares are tracers of time-dependent magnetic activity, and exhibit trends with some diagnostics of steady-state magnetic activity, such as the correlation between coronal flaring and underlying coronal emission noted earlier. Assuming that such scaling laws are valid enables an opening of the parameter space. Observations of ensembles of stars can gain insight into the gross flaring properties of these stars, where such conclusions may not be evident from observations of a single star due to the low flaring rate per star.

While we still do not have detailed characteristics (including all relevant flare processes) of a true solar analog at various stages in stellar evolution, the vast astronomical discovery space does enable a reasonable picture to be painted of how the Sun's flaring activity changed over the course of its time, and even predicts what lies in its future. Outstanding issues include incomplete sampling of flares in both wavelength space, and as a function of stellar type, for stars of different ages and rotation rates.

3

Astrospheres, stellar winds, and the interstellar medium

BRIAN E. WOOD AND JEFFREY L. LINSKY

3.1 The spatial extent of the solar wind

While orbiting the Sun, the Earth and the other planets in our solar system are exposed to an outflow of plasma from the Sun. This is the solar wind, which largely determines the particle environment of the planets, as well as their magnetospheric properties. Long-term exposure to the solar wind may have had significant effects on planetary atmospheres in some cases. Mars is a particularly interesting example, as the Martian atmosphere was once much thicker than it is today, and the erosive effects of solar wind exposure are commonly believed to be responsible for its depletion (see Ch. 8; also Jakosky *et al.*, 1994; Atreya *et al.*, 2013).

The solar wind does not expand indefinitely. Eventually it runs into the interstellar medium (ISM), the extremely low-particle-density environment that exists in between the stars. Our Sun is moving relative to the ISM that surrounds the planetary system, so we see a flow of interstellar matter in the heliocentric rest frame, coming from the direction of the constellation Ophiuchus. The interaction between the solar wind and the ISM flow determines the large-scale structure of our heliosphere, which basically defines the extent of the solar wind's reach into our Galactic environment. Other stars are naturally surrounded by their own "astrospheres" (alternatively "asterospheres") defined by the strength of their own stellar winds, the nature of the ISM in their Galactic neighborhoods, and their relative motion.

Much as the Earth's magnetosphere plays a role in protecting the Earth's atmosphere from direct and potentially damaging exposure to the solar wind, the global heliosphere has its own role to play in shielding the planets that exist inside it from high-energy particles, called Galactic Cosmic Rays (GCRs), that permeate the Galaxy (see Vol. II, Ch. 9; and Vol. III, Ch. 9). Some of these GCRs do penetrate into the inner heliosphere and encounter Earth. The extent to which they have influenced our planet and the evolution of life is debatable, but cosmic rays cause radiation damage and mutations in DNA (Dartnell, 2011). There are also claims that cosmic rays may be involved in atmospheric processes such as lightning formation and cloud formation (Carslaw *et al.*, 2002; Shaviv, 2005a; Svensmark *et al.*, 2009). In any case, the global heliosphere does reduce the fluxes of GCRs that reach Earth and other planets, providing some level of protection from GCR influence within the solar system.

Given their effects on planets in our solar system, including Earth, it is of interest to understand how the solar wind and heliosphere have changed during the course of the Sun's 4.6 billion year lifetime. Likewise, it is also important to understand the evolution of stellar winds and their effect on exoplanets. Not only does the solar wind evolve as the Sun ages, but the Sun encounters different ISM environments as it orbits the Galactic Center, leading to changes in heliospheric structure. The focus of this chapter is on describing the evolution of the solar wind and heliosphere, or more generally stellar winds and astrospheres, on these very long timescales. We begin by describing what is known about the current structure and extent of our heliosphere.

3.2 Observed properties of the local interstellar medium

The structure of our heliosphere depends on the properties of both the solar wind and the surrounding ISM. The basic properties of the hot and fully ionized solar wind are described in other chapters in the Heliophysics book series (cf., Table 1.2), such as in Ch. 9 in Vol. I. In this section we describe what we know about the ISM surrounding the Sun.

Our knowledge of the local ISM is based on observations of high-resolution spectra of nearby stars that contain absorption lines produced by interstellar gas, and on measurements of interstellar gas entering the heliosphere. An example of absorption line data is shown in Fig. 3.1, where these particular data are UV spectra from the *Hubble Space Telescope* (HST). Absorption towards the nearby star 36 Oph is shown from singly ionized Mg and Fe atoms (the spectra of which are referred to as Mg II and Fe II), and from neutral deuterium (D I). Crutcher (1982) first noticed that the radial velocities of interstellar absorption lines in the spectra of nearby stars are consistent with interstellar gas flowing toward the Sun from the direction of the Scorpio–Centaurus Association. Subsequent investigations (e.g., Lallement and Bertin, 1992; Frisch *et al.*, 2002) found that the interstellar gas flow has a number of components with slightly different flow directions and speeds. Redfield and Linsky (2008) identified 15 different velocity components, which they called clouds, by analyzing interstellar absorption lines in HST spectra of 157 stars located within 15 parsecs (pc) of the Sun.

The sky map in Fig. 3.2 shows the locations of four of the closest clouds. The Local Interstellar Cloud (LIC) is so-named because it occupies nearly half of the

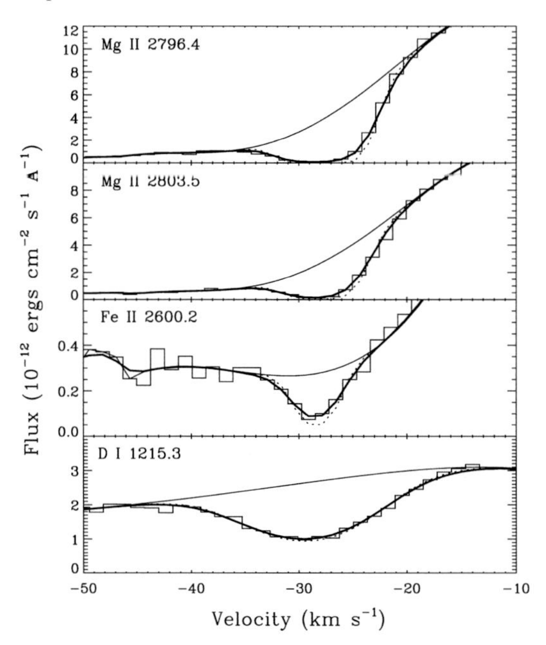

Fig. 3.1 High-resolution spectra of the star 36 Oph showing interstellar absorption in Mg II, Fe II, and D I lines, with wavelengths provided in ångström (Å) units. The spectra are plotted in a heliocentric rest frame. Note the different absorption widths, with D I being broadest due to deuterium atoms having the lowest mass and therefore the highest thermal speeds. (From Wood *et al.*, 2000.)

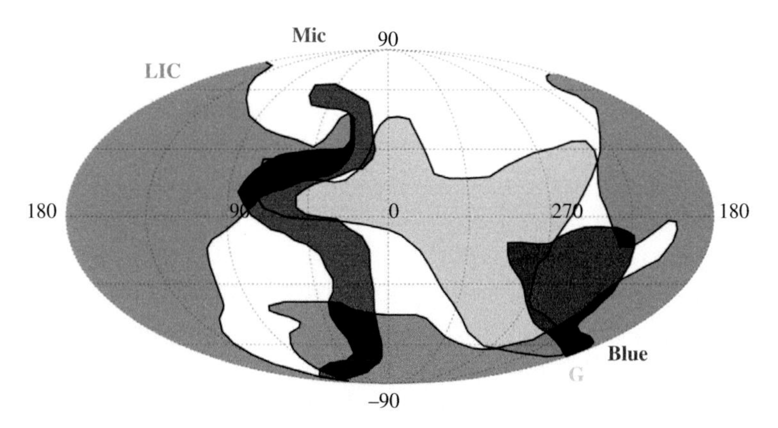

Fig. 3.2 Map in Galactic coordinates of the four partially ionized warm interstellar clouds that are closest to the Sun, mapped using ISM absorption lines. The Sun is likely located inside of the Local Interstellar Cloud (LIC), but very near the edge, explaining why LIC absorption is not detectable in all directions and therefore the ISM absorption lines by the LIC do not fill the sky. (From Redfield and Linsky, 2008.)

sky, implying that the Sun is located just inside of the LIC or immediately outside. The decision as to which option is more likely true requires the second type of data: measurements of interstellar gas flowing into the heliosphere as described below. The second largest cloud by angular area is the G cloud, so-named because it is centered in the Galactic Center direction. The line of sight to α Centauri, the nearest star to the Sun at a distance of 1.3 pc, shows interstellar absorption only at the projected velocity of the G-cloud vector with no absorption at the LIC cloud velocity. The upper limit of the neutral hydrogen (H I) column density (the total number of atoms or ions between the source and the telescope per unit cross section) at the LIC velocity in this direction indicates that the Sun will leave the LIC in less than 3000 years if it has not already done so. The Blue Cloud is also very near because absorption at the projected velocity of its flow vector is seen toward the star Sirius (at 2.6 pc). Finally, the nearby Mic Cloud shows a filamentary structure that may be typical of many clouds. A further question is whether the picture of discrete interstellar clouds described by Redfield and Linsky (2008) is the best representation of the interstellar kinematics or whether a single cloud with a continuous gradient in flow velocity is a more realistic model for the interstellar gas surrounding the Sun. Gry and Jenkins (2014) provide arguments for a continuous flow pattern rather than many discrete clouds, but Redfield and Linsky (2015) argue that the discrete cloud model best fits the data.

Absorption line data like those in Fig. 3.1 provide measurements of ISM column densities, projected flow velocities onto the line of sight, and Doppler width parameters (b). The column densities provide one crucial indication of ISM density outside the heliosphere for various atomic species. Electron (and hence proton) density can be estimated by certain density sensitive line ratios (Frisch, 1994; Wood and Linsky, 1997; Redfield and Falcon, 2008). The Doppler parameters can be related to temperature (T) and non-thermal speed (ξ) through the equation

$$b^2 = \frac{2kT}{m} + \xi^2, \tag{3.1}$$

where k is the Boltzmann constant and m is the mass of the atomic species in question. Note the broader width of the D I line in Fig. 3.1 compared to the heavier species Mg II and Fe II. Measurements of b from different species with different masses (and hence different thermal widths) can provide measurements of T and ξ, as shown explicitly in Fig. 3.3.

Besides absorption data, the other principle sources of direct information about the ISM properties in the solar neighborhood are observations of neutral ISM species streaming through the solar system with hardly any collisional interactions. Most important are observations of neutral He from particle detectors on the *Ulysses* spacecraft, which operated from 1990 until 2007 (e.g., Witte, 2004), and on the later *Interstellar Boundary Explorer* (*IBEX*) mission (e.g., McComas

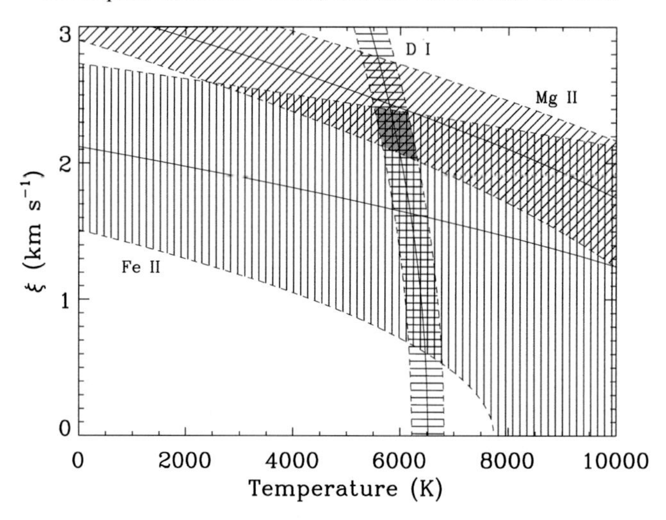

Fig. 3.3 Non-thermal ISM speeds (ξ; see Eq. (3.1)) plotted versus temperature, based on the Doppler width parameters measured from the lines in Fig. 3.1. The shaded area where the three curves overlap indicates the best temperature and ξ value. (From Wood *et al.*, 2000.)

et al., 2012); and observations of neutral H in the inner solar system from Lyman-α backscatter measurements (e.g., Lallement *et al.*, 2005).

Based largely on the observational constraints described above, the ISM flow speed in the solar rest frame is known to be in the range $v_{\mathrm{ISM}} = 23\text{–}27\,\mathrm{km\ s^{-1}}$, and the temperature is $T = 6000 - 9000$ K. The dominant constituents of the ISM by mass and number are protons and neutral hydrogen atoms, and typical estimates of proton and neutral hydrogen densities are $n_p \approx 0.06\ \mathrm{cm^{-3}}$ and $n_H \approx 0.18\ \mathrm{cm^{-3}}$ (e.g., Müller *et al.*, 2008; Slavin and Frisch, 2008; Frisch *et al.*, 2011).

3.3 Introduction to heliospheric structure

Speculation about the nature of the heliosphere's structure began shortly after the solar wind was first discovered (Parker, 1961). Figure 3.4 shows the prevailing picture of the global heliosphere that has existed for several decades (Holzer, 1989; Zank, 1999). The structure is characterized by three flow discontinuities: the termination shock (TS), the heliopause (HP), and the bow shock (BS). The solar wind density and therefore its ram pressure fall off as r^{-2}, where r is the distance from the Sun. The wind speed is supersonic and super-Alfvénic, so when the ram pressure falls to the pressure of the ambient ISM, the result is a shock, specifically the ellipsoidal TS in Fig. 3.4, where the flow is decelerated.

If the ISM flow is super-Alfvénic, it also encounters a shock as it approaches the Sun, specifically the roughly hyperboloid shaped BS in Fig. 3.4, where the

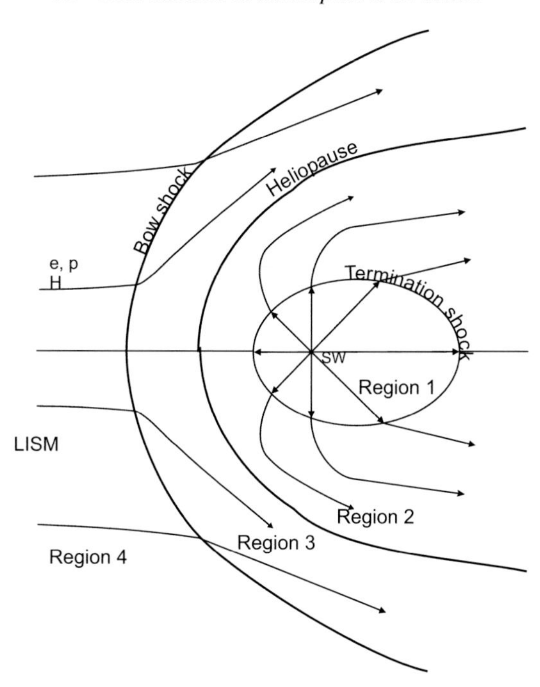

Fig. 3.4 Basic structure of the global heliosphere, with three discontinuities: the termination shock (TS), heliopause (HP), and bow shock (BS). These discontinuities separate the heliosphere into four regions, with significantly different plasma properties: (1) supersonic solar wind; (2) decelerated, heated, and deflected solar wind; (3) decelerated, heated, and deflected ISM; and (4) undisturbed local ISM (LISM) flow, here assumed supersonic, resulting in a bow shock. (From Izmodenov *et al.*, 2002.)

ISM flow is decelerated to subsonic speeds. However, the $v_{\rm ISM} = 23{-}27\,{\rm km\ s^{-1}}$ interstellar flow happens to yield an Alfvénic Mach number of $M_A \approx 1$, making the existence or nonexistence of a BS very much an open question. Much depends on the strength and orientation of the ISM magnetic field, $B_{\rm ISM}$. The higher $B_{\rm ISM}$ is (and the more perpendicular to the ISM flow), the lower M_A should be, and less likely that there is a bow shock.

Even the seemingly small uncertainty in $v_{\rm ISM}$ is enough to make a difference. For many years the best assessments were believed to be the $v_{\rm ISM} = 26.3 \pm 0.4\,{\rm km\ s^{-1}}$ measurement of the ISM neutral He flowing through the solar system by *Ulysses* (Witte, 2004) and the $v_{\rm ISM} = 25.7 \pm 0.5$ km s^{-1} measurement of Lallement and Bertin (1992) from ISM absorption lines. With these relatively high values, heliospheric modelers favored $M_A > 1$, implying the existence of a BS. However, later He flow measurements and a new analysis of ISM absorption line data have yielded lower velocities. Specifically, measurements of neutral He flow from *IBEX* suggest

$\upsilon_{\rm ISM} = 23.2 \pm 0.3$ km s^{-1} (McComas *et al.*, 2012), and Redfield and Linsky (2008) find $\upsilon_{\rm ISM} = 23.84 \pm 0.90$ km s^{-1} from ISM absorption lines.

This has been enough for many to argue that $M_A < 1$ should be preferred (McComas *et al.*, 2012; Zieger *et al.*, 2013), though Scherer and Fichtner (2014) argue that including He^+ density in the calculation of sound and Alfvén speeds instead of just assuming a pure proton plasma would still suggest $M_A > 1$ even if $\upsilon_{\rm ISM} \approx 23$ km s^{-1}. With M_A so close to 1, it is possible that the issue will not be fully resolved until an interstellar probe mission of some sort is sent out to this region (McNutt *et al.*, 2004). However, with M_A so close to 1 it is also possible that secondary physical processes (e.g., charge-exchange interactions with neutral particles) make it fundamentally ambiguous whether any boundary that may exist out there should be called a true BS, or whether we should instead refer to it as a "bow wave" (Zank *et al.*, 2013).

Regardless of whether or not a bow shock exists, strong plasma interactions prevent the ISM plasma from mixing with the solar-wind plasma. The roughly paraboloid heliopause in Fig. 3.4, lying between the termination shock and the bow shock (or wave) is the contact discontinuity separating the two plasma flows. Representing the boundary between solar wind and ISM plasma, the heliopause is generally considered the true boundary of the heliosphere.

3.4 Observational constraints on the global heliosphere

Until relatively recent times, the heliospheric structure shown in Fig. 3.4 was a purely theoretical one, with few observational constraints to test models of this structure. This changed dramatically on 2004 December 16 when *Voyager 1* crossed the TS at a distance of $r = 94$ Sun–Earth distances (astronomical units, or AU; Stone *et al.*, 2005). The two *Voyager* spacecraft, *Voyager 1* (V1) and *Voyager 2* (V2), were launched in 1977. After exploring the outer planets, both ended up in advantageous trajectories roughly in the upwind direction of the heliosphere (relative to the ISM flow), allowing them to cross the TS while still operational, and probe regions of the heliosphere never before explored. However, V1's identification of the TS crossing was not trivial. The instrument on the spacecraft designed to study the ambient bulk plasma failed in 1980, limiting V1's diagnostics of its plasma environment to those provided by its magnetometer, and instruments designed to study energetic particles and plasma waves. In 2002, V1 began seeing enhancements in energetic particles that created uncertainty as to whether the TS had been crossed (Krimigis *et al.*, 2003) but, with the benefit of hindsight, 2004 December 16 is now considered the true crossing, which came with a clear jump in magnetic field. Thanks in part to the fact that V2 has an operating plasma instrument, V2 observed a less-ambiguous TS crossing on 2007 August 30 at

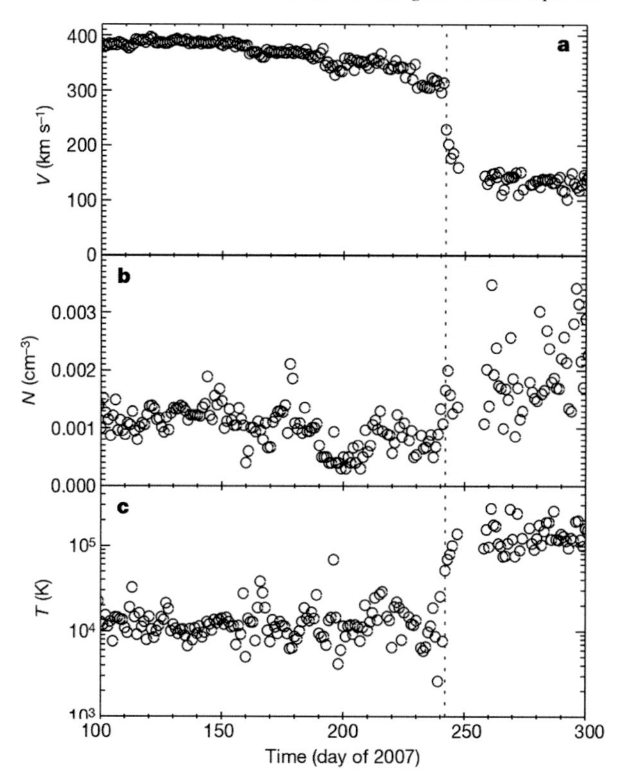

Fig. 3.5 *Voyager 2* observations of solar wind (a) speed, (b) proton number density, and (c) temperature. The dotted line indicates the termination shock (TS) crossing on 2007 August 30 at a distance of 84 AU from the Sun. (From Richardson *et al.*, 2008.)

$r = 84$ AU, which is shown explicitly in Fig. 3.5 (Richardson *et al.*, 2008; Stone *et al.*, 2008).

Much as identifying V1's TS crossing proved problematic, identifying the HP crossing is also proving difficult. On August 25, 2012, at $r = 122$ AU, V1 observed an increase in magnetic field, a dramatic increase in GCR intensities, and a disappearance of lower energy ions that had been observed since the TS crossing (Stone *et al.*, 2013; see Fig 4.6). Plasma-wave data from V1 have since suggested a substantial increase in plasma density, consistent with this being the HP crossing (Gurnett *et al.*, 2013). This would make V1 the first spacecraft to leave the heliosphere and enter interstellar plasma. However, the lack of any change in magnetic-field direction and the unexpectedly close HP distance have led to alternative, more complex interpretations (Schwadron and McComas, 2013; Borovikov and Pogorelov, 2014). Presumably time and a supporting HP crossing by V2 will ultimately resolve the issue.

The basic structure in Fig. 3.4 is mostly defined by plasma interactions. The local ISM is partly neutral, but collisional mean free paths for neutrals are large compared to the size of the heliosphere, so their effects on heliospheric structure were long ignored. In essence, the assumption was that neutrals pass through the heliosphere unimpeded, feeling only the Sun's gravity and photoionizing flux. However, in reality neutrals do participate in heliospheric interactions through charge exchange (CX). The CX interactions end up providing ways to remotely explore the heliosphere that would be impossible if the local ISM were fully ionized.

A CX interaction is a rather simple process by which an electron hops from a neutral atom to a neighboring ion (e.g., $H^0 + H^+ \rightarrow H^+ + H^0$). Mean free paths for CX for most neutral ISM atoms are short enough that they do experience significant CX losses on their way through the heliosphere. The exceptions are the noble gases, which have low CX cross sections, explaining why neutral He flowing through the solar system is considered the best local probe of the undisturbed ISM flow (Witte, 2004).

Modeling neutrals in the heliosphere is very difficult because CX sends the neutrals wildly out of thermal and ionization equilibrium with the ambient plasma. Including neutrals in hydrodynamic models of the global heliosphere therefore requires either a fully kinetic treatment of the neutrals, or at least a sophisticated multi-fluid approach. The earliest models that could treat neutrals properly were from the 1990s (Baranov and Malama, 1993, 1995; Zank *et al.*, 1996). These models demonstrated that through CX, neutrals could have significant effects on heliospheric structure. Figure 3.6 shows one of these models (Müller and Zank, 2004). The ISM protons are heated, compressed, deflected, and decelerated as they approach the HP, and thanks to CX the proton properties are at least partially imprinted on the neutral hydrogen (H I) as well, creating what has been called a "hydrogen wall" of higher density H I (see Fig. 3.6) around the heliosphere, in between the HP and BS. The importance of this hydrogen wall is that it is actually detectable in UV spectra from HST, not only around the Sun but around other stars as well.

The effect of heliospheric and astrospheric absorption on stellar H I Lyman-α spectra is described by Fig. 3.7 (Wood, 2004), showing the journey of a Lyman-α photon from the star to the observer. Most of the absorption is by interstellar gas in the line of sight from the star to the Sun, but the astrosphere and heliosphere provide additional absorption on the left and right sides of the interstellar absorption, respectively. The effect of the hydrogen wall around the Sun is to provide additional red-shifted absorption on the right side of the interstellar absorption feature because the neutral hydrogen gas in the solar hydrogen wall is slowed down and deflected relative to the inflowing interstellar gas. Conversely, the absorption by the

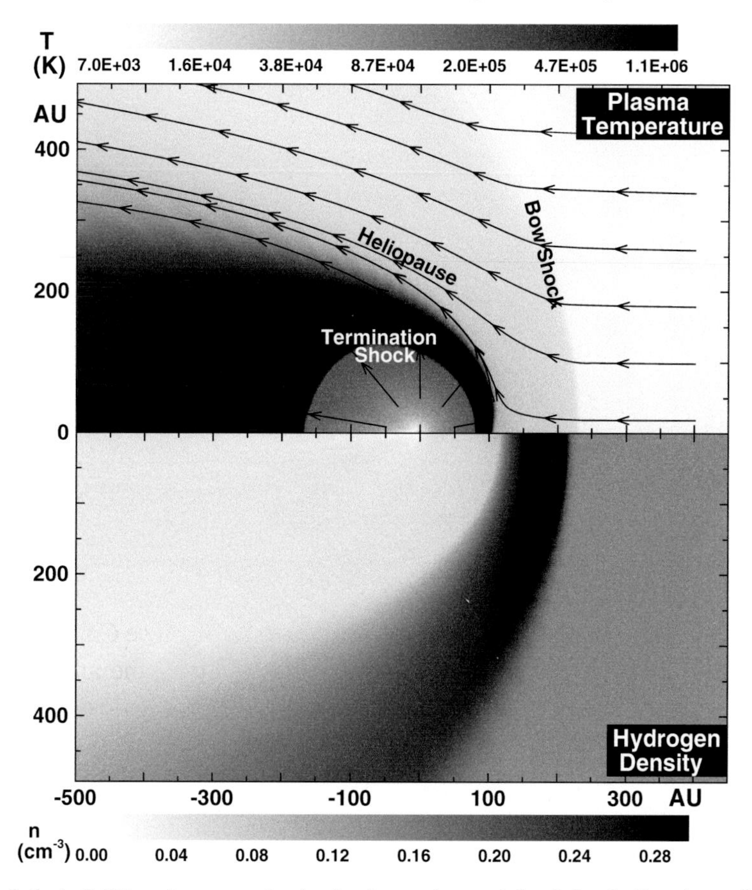

Fig. 3.6 A 2.5D axisymmetric, hydrodynamic model of the heliosphere from Müller and Zank (2004). The upper panel shows plasma temperature, and the bottom panel shows neutral hydrogen density.

hydrogen wall gas around the star is seen as blue-shifted relative to the interstellar flow from our perspective outside the astrosphere, and is therefore seen on the left side of the absorption line. The first detection of hydrogen wall absorption was in HST observations of the two stars in the very nearby binary α Cen (with spectral types G2 V+K1 V; i.e., with one component very comparable to our Sun also of spectral type G2 V, see, for example, Fig. 2.8 in Vol. III for a Hertzsprung–Russell diagram for spectral type characterization). The bottom panel of Fig. 3.7 shows the H I (and D I) Lyman-α spectrum of the lower-brightness component of this binary, known as α Cen B (Linsky and Wood, 1996). Most of the intervening H I and D I between us and the star is interstellar, but the ISM cannot account for all of the H I absorption. As mentioned above, the red-shifted excess on the right side is heliospheric and the blue-shifted excess on the left is astrospheric (Gayley *et al.*, 1997).

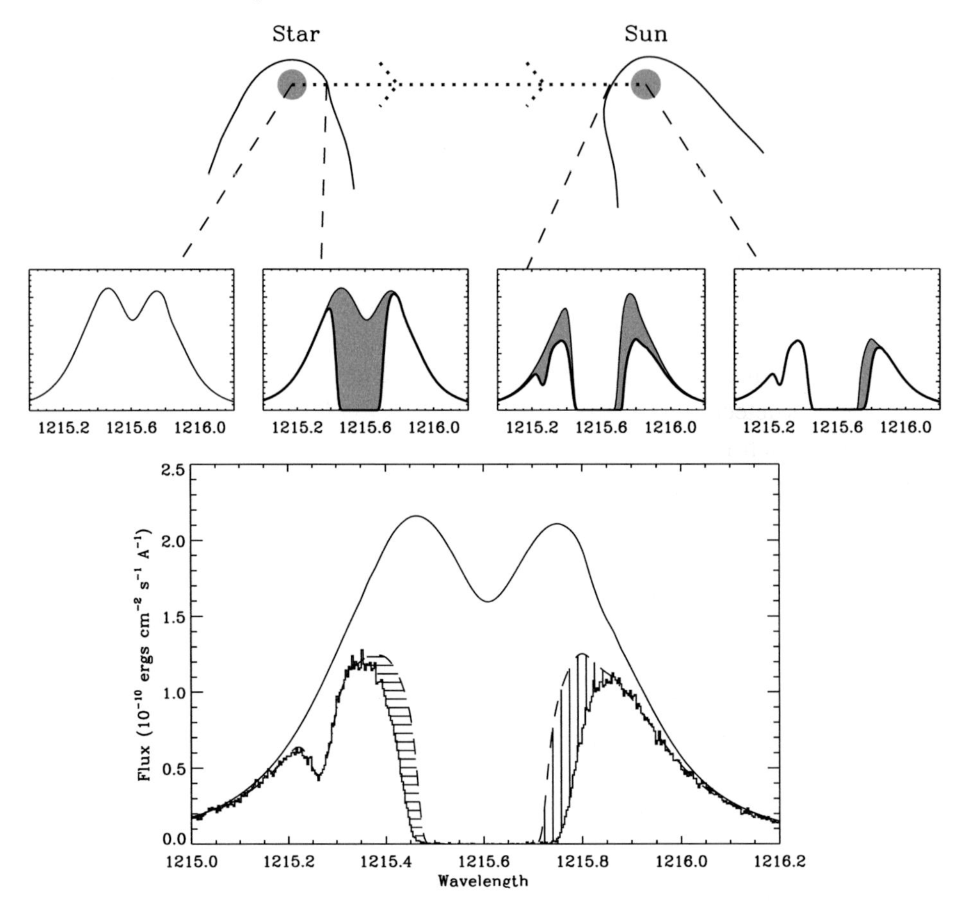

Fig. 3.7 Top panel: the journey of a Lyman-α photon from a star through its astrosphere, the interstellar medium, and the heliosphere. Middle panel from left to right: the Lyman-α emission line emitted by the star, absorption due to the stellar astrosphere, additional absorption due to the interstellar medium, and additional absorption due to the heliosphere. Bottom panel: HST Lyman-α spectrum of α Cen B, showing broad H I absorption at 1215.6 Å and D I absorption at 1215.25 Å. The upper solid line is the assumed stellar emission profile and the dashed line is the ISM absorption alone. The excess absorption is due to heliospheric H I (vertical lines) and astrospheric H I (horizontal lines). (From Wood, 2004.)

Besides the hydrogen wall, CX in the heliosphere has other observable consequences. Within the TS, outflowing solar wind plasma can CX with neutral H that penetrates inside the TS, yielding solar-wind neutral H atoms, and inflowing ISM protons that are then picked up by the solar wind. These "pickup ions" (PUIs) create a suprathermal particle population within the solar wind that is energetically very important beyond the orbit of Saturn ($\sim$ 10 AU from the Sun). The PUI population is energized at the TS. It was long believed that this is the source of anomalous cosmic rays (ACRs) (see Vol. II, Ch. 9), but V1 and V2 found that the ACRs that they observed near the TS were not coming from nearby (Stone *et al.*, 2005). This has led to other suggestions for the source of ACRs somewhere beyond the TS (Fisk and Gloeckler, 2008), but the issue is far from resolved.

The *IBEX* mission, launched in 2008, was designed primarily to study energetic neutral atoms (ENAs) streaming into the solar system from beyond the TS. The globally distributed ENAs observed by *IBEX* are presumed to be created by CX with the post-TS PUI population (Schwadron *et al.*, 2011). Charge exchange with the bulk solar-wind plasma beyond the TS creates a less-energetic population of neutrals that is detectable in HST Lyman-α spectra. In contrast to the hydrogen wall population, which is detectable primarily in upwind directions, it is only in very downwind directions, with very long path lengths through the post-TS region 2 (see Fig. 3.4), that very broad but very weak absorption from the region 2 neutrals can be discerned against the Lyman-α emission from background stars. Detections of this absorption represent the first detection of the heliotail (Wood *et al.*, 2007b, 2014a).

The most unexpected result from *IBEX* does not relate to the neutralized post-TS PUI ENAs but instead to a ribbon of ENA flux that stretches across the sky (McComas *et al.*, 2009). The ribbon's origin is uncertain, but most believe it must be due to heliospheric asymmetries induced by the ISM magnetic field (Heerikhuisen *et al.*, 2010), which is poorly constrained by any direct ISM observations. Modeling heliospheric asymmetries induced by the ISM field requires sophisticated, fully three-dimensional (3D) MHD state-of-the-art models (e.g., Pogorelov *et al.*, 2008; Opher *et al.*, 2009).

3.5 Effects of a variable ISM on past heliospheric structure

The Sun is now traveling through the ISM at a rate of 16–20 pc per million years (Myr) compared to the average motion of nearby stars about the Galactic Center. The ISM has densities ranging from 10^4 cm^{-3} or higher in dense molecular clouds down to about 0.005 cm^{-3} in very-low-density hot-gas regions. Because the heliosphere will contract or expand by large factors when the Sun enters such high- or low-density regions, it is important to investigate when such environmentally driven changes could have occurred and will possibly occur by considering the Sun's historic and future path through the ISM.

At present, the heliosphere resides inside of the partially ionized LIC (see Section 3.2 of this chapter), with properties likely similar to other warm partially ionized clouds within 15 pc of the Sun. The Sun likely entered this cluster of local warm clouds about 1 Myr ago. However, on a larger scale, the Sun actually lies in a region called the Local Cavity, or Local Bubble, which is $\sim$ 200 pc across and is filled mostly with fully ionized, low density ISM (Welsh *et al.*, 2010; Frisch *et al.*, 2011). Wyman and Redfield (2013) measured the Na I and Ca II column densities towards stars along the path that the Sun has traveled for the past 30 Myr (500 pc). They found no evidence for the Sun having encountered an environment significantly denser than the LIC until about 7.5 Myr ago (120 pc) when the Sun was at the edge of the Local Cavity. If the discrete cloud model of Redfield and Linsky (2008) is accurate, the Sun will soon leave the LIC in less than 3000 yr. What will be the properties of this new environment?

The classic papers of McKee and Ostriker (1977) and Wolfire *et al.* (1995) predict that the ISM should consist of three components: the cold neutral medium ($T \leq 50$ K), the warm neutral or ionized medium with $T \sim 8000$ K (e.g., like the LIC), and the million-degree low-density ionized medium. These models assume that the three components coexist in pressure and thermal equilibrium, which is actually unlikely. Numerical simulations by Berghöfer and Breitschwerdt (2002), which include supernova explosions and realistic thermal processes, predict a very wide range of densities and temperatures in the ISM but no pressure equilibrium and no identifiable thermal phases. Observations are needed to determine which of these two very different pictures is more realistic.

The nearby warm partially ionized gas clouds have properties roughly consistent with the warm component (neutral or ionized) predicted by the classic models. Cold dense molecular clouds are observed typically by CO and H I 21-cm emission in many parts of the Galaxy. The nearest cold gas with a temperature of 15–30 K is the Leo cloud located at a distance between 11.3 and 24.3 pc from the Sun (Peek *et al.*, 2011).

Our ideas concerning the properties of the gas located between the warm local ISM clouds have undergone a radical change in the past 20 years. The gas between the clouds, extending out to roughly 100 pc from the Sun in what is now called the Local Cavity was originally assumed to be hot (roughly 10^6 K), fully ionized, and low density (roughly 0.005 cm^{-3}). This conclusion was based upon the predictions of the classical models and observations of diffuse soft X-ray emission consistent with the properties of the hot gas. This picture has since been complicated by the realization that X-ray emission from CX reactions between the solar wind ions and inflowing interstellar neutral hydrogen can explain much of the observed diffuse X-ray emission, except for the Galactic pole regions (Cravens, 2000; Koutroumpa *et al.*, 2009). Welsh and Shelton (2009) have instead proposed that the Local Cavity is an old supernova remnant with photoionized gas at a temperature of about 20 000 K. The likely photoionizing sources are the hot stars ϵ CMa and β CMa (Vallerga and Welsh, 1995) and nearby hot white dwarfs like Sirius B (Linsky and Redfield, 2014).

How will the heliosphere change as the Sun passes through very different regions of the interstellar medium? To answer this question, Müller *et al.* (2006) computed a set of heliosphere models under the assumption that the solar wind retains its present properties. Figure 3.8 compares today's heliosphere properties with the Sun located inside of the partially ionized warm LIC to a model computed for the Sun surrounded by 10^6 K fully ionized interstellar plasma. The main difference between these models is that the hydrogen wall does not exist when the inflowing interstellar gas contains no neutral hydrogen atoms. The locations of the termination shock (TS), heliopause (HS), and bow shock (BS) are determined by pressure balance between the solar-wind ram pressure and the thermal and ram pressure of the surrounding interstellar gas. In this comparison, the locations of the TS, HP, and BS are about the same in the two models because the high temperature and

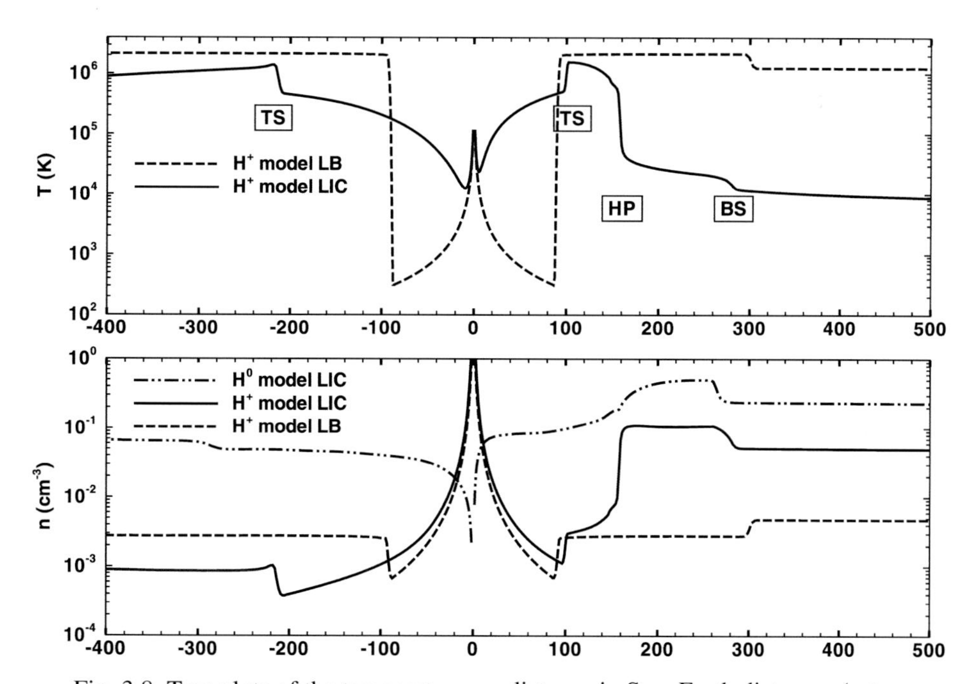

Fig. 3.8 Top: plots of the temperature vs. distance in Sun–Earth distances (astronomical units, or AU) relative to the Sun (interstellar flow upwind direction to the right and downwind to the left) for a heliosphere model with the Sun located inside of the Local Interstellar Cloud (LIC; solid line) or inside a 10^6 K hot interstellar medium (dashed line). The heliosphere in the LIC model has a termination shock (TS), heliopause (HP) and bow shock (BS) structure. Bottom: density structures for the LIC neutral hydrogen (solid line), LIC protons (dot-dash line), and hot interstellar model protons (dashed line). Note that the hydrogen wall at 150–280 AU exists when the heliosphere is located inside partially neutral interstellar gas but not when it is inside fully ionized interstellar gas. (Figure from Müller *et al.*, 2009.)

low density of the interstellar gas produce a pressure that is about the same as in the LIC.

When the Sun enters a region of much higher density or speed, and therefore higher ram pressure, the effect is to compress the heliosphere. For example, a model for $n_{HI} = 15$ cm^{-3}, roughly 100 times that of the LIC (Müller *et al.*, 2009), has a TS at 9.8 AU such that Uranus would move in and out of the TS and Neptune would be surrounded by hot, shocked plasma beyond the HP (upwind) or heliotail (downwind). Models of the heliosphere inside a high-speed interstellar wind with corresponding high ram pressure would compress the heliosphere in a similar way. Smith and Scalo (2009) define a "de-screening event" to be a cloud encounter that results in a stellar astrosphere being compressed to less than the size of the star's habitable zone (defined as the distance from the star where liquid water can be present on a planet's surface), which they find should happen when a star encounters an interstellar cloud with a number density of $600(M/M_{\odot})^{-2}$ cm^{-3}, where M is the mass of the star. Only the densest ISM clouds are capable of this de-screening, with such clouds being relatively rare. The densest clouds are cold ($T \sim 100$ K) molecular clouds, with many of the refractory elements depleted onto dust grains. In addition to increased GCR exposure (see Ch. 4), a de-screening event caused by a molecular cloud encounter would also expose planetary atmospheres to high fluxes of interstellar dust, with potentially dramatic consequences (Pavlov *et al.*, 2005). In any case, the mass dependence in the above equation leads Smith and Scalo (2009) to conclude that habitable zone planets orbiting stars significantly less massive than the Sun (with spectral types of late K to M) are virtually never exposed to de-screening events, but de-screening may happen occasionally for stars with the Sun's mass or larger. However, these calculations assumed that the relative velocity of these encounters is only 10 km s^{-1}. Assuming a faster encounter speed would increase the estimated frequency of de-screening events.

Fields *et al.* (2008) calculated the first gas dynamic models for the effects of the blast wave from a nearby supernova on the heliosphere assuming the present day properties of the solar wind. Pressure balance between the supernova shock wave and the solar wind produces extreme heliosphere models that have the same physical structures as the Müller *et al.* (2009) models with the heliopause at 1.4 times the distance of the termination shock in the upwind direction but with both located very close to the Sun. Fields *et al.* (2008) find that a supernova located at ≈ 9 pc from the Sun would create a heliopause that penetrates to within 1 AU, subjecting the Earth to an infusion of supernova debris including iron and other heavy atoms. The discovery of the radioisotope ^{60}Fe with a half-life of 1.5 Myr in a deep-sea ferromanganese crust and dated to 2.8 ± 0.4 Myr ago by Knie *et al.* (2004) indicates that a nearby supernova explosion likely occurred only 2.8 Myr ago.

The effect on the Earth of a nearby supernova and the effect on more distant planets from supernovae at distances up to 30 pc will include an increase in the amount of neutral hydrogen atoms, dust, supernova metals, and Galactic Cosmic Rays reaching the planet's atmosphere. The latter would influence the planet's magnetosphere and change the planet's atmospheric chemistry, including the important molecule ozone (see Smith and Scalo, 2009 and references therein).

3.6 Detecting astrospheres

The heliosphere is obviously just one example of stellar wind/ISM interaction available for study. All stars with winds will have analogous astrospheres around them. Unfortunately, they are not easy to detect. At least for stars with coronal winds like that of the Sun, the only successful astrospheric detection method so far is the Lyman-α detection method illustrated in Fig. 3.7. Other proposed approaches for detecting winds include the search for CX-induced X-ray emission (Wargelin and Drake, 2001) and free–free emission at radio wavelengths (Gaidos *et al.*, 2000), but these methods have not been successful. Figure 3.7 illustrates that the same HST Lyman-α spectrum can be used to study both heliospheric and astrospheric Lyman-α absorption. Table 3.1 lists all the stars with detected astrospheric absorption (Wood *et al.*, 2005, 2014b). All are very nearby. Interstellar H I column densities increase with distance, and at sufficiently large distances the resulting ISM H I absorption becomes too broad to detect astrospheric (or heliospheric) absorption.

The amount of astrospheric Lyman-α absorption will depend on the thickness of the hydrogen wall, and more generally the size of the astrosphere, which is largely determined by the stellar-mass-loss rate. The absorption can therefore be used as a diagnostic of the stellar wind, which is crucial because coronal winds have so far proved impossible to detect directly around other solar-like stars. This topic is discussed in detail in the next section.

Although we are here mostly interested in the astrospheres of cool main-sequence stars like the Sun, with coronal winds like the solar wind, it is worth at least briefly mentioning other types of astrospheres created by stars very different from the Sun with much more massive winds. Red giant stars have cool, slow ($v_w \sim 30$ km s^{-1}) winds with $\sim 10^3$ times the mass-loss rate of the Sun. The winds are easily detected spectroscopically by observing wind absorption features within emission lines formed close to the star, but there is so far only one reported detection of a red giant's wind/ISM interaction. For α Tau (K5 III), a Mg II absorption feature has been identified as being due to stellar-wind material compressed and heated behind a radiative TS (Robinson *et al.*, 1998; Wood *et al.*, 2007a).

Table 3.1 *Astrospheric detections from Lyman-α absorption*

Star	Spectral type	d (pc)	v_{ISM} (km s^{-1})	θ (deg)	$\dot{M}$ ($\dot{M}_\odot$)	Log L_X	Surf. area ($A_\odot$)
			Main-sequence stars				
Sun	G2 V	...	26	...	1	27.30	1
Proxima Cen	M5.5 V	1.30	25	79	< 0.2	27.22	0.023
α Cen	G2 V+K0 V	1.35	25	79	2	27.70	2.22
ϵ Eri	K2 V	3.22	27	76	30	28.32	0.61
61 Cyg A	K5 V	3.48	86	46	0.5	27.45	0.46
ϵ Ind	K5 V	3.63	68	64	0.5	27.39	0.56
EV Lac	M3.5 V	5.05	45	84	1	28.99	0.123
70 Oph	K0 V+K5 V	5.09	37	120	100	28.49	1.32
36 Oph	K1 V+K1 V	5.99	40	134	15	28.34	0.88
ξ Boo	G8 V+K4 V	6.70	32	131	5	28.90	1.00
61 Vir	G5 V	8.53	51	98	0.3	26.87	1.00
π^1 UMa	G1.5 V	14.4	43	34	0.5	28.96	0.97
			Evolved stars				
δ Eri	K0 IV	9.04	37	41	4	27.05	6.66
λ And	G8 IV-III+M V	25.8	53	89	5	30.82	54.8
DK UMa	G4 III-IV	32.4	43	32	0.15	30.36	19.4

Spectroscopic diagnostics of astrospheres are useful, but directly perceiving the size and global structure of an astrosphere requires an image. Imaging astrospheres is generally possible only for the largest astrospheres surrounding stars with the most massive or energetic of winds. Examples include pulsar bow shocks, which are a result of relativistic winds emanating from the rapidly rotating neutron stars (e.g., Brownsberger and Romani, 2014). Bow shocks are also observed around some red supergiants, with slow ($v_w \sim 10$ km s^{-1}) but very massive winds having $\sim 10^7$ times the mass-loss rate of the Sun (e.g., Martin *et al.*, 2007; van Marle *et al.*, 2014). Hot stars, with O and B spectral types, have similarly massive winds driven by radiation pressure, which can also lead to visible bow shocks (e.g., Kobulnicky *et al.*, 2010). Finally, a high-density ISM environment can allow an astrosphere to be detectable. An example is shown on the covers of the *Heliophysics* series (lower right-hand side), in which the wind of a very young star is interacting with the high-density nebula from which it has just been born (see, e.g., Bally *et al.*, 2006, for more information).

3.7 Long-term evolution of stellar winds

3.7.1 Inferences from short-term solar wind variability

A complete understanding of how winds affect planets and their atmospheres must start with assessments of how stellar winds evolve with time. It is well established that stellar activity decreases with age (see Vol. III, Ch. 2, and also Ch. 2 in this volume; Ribas *et al.*, 2005). Young stars rotate rapidly and have very active coronae that emit copious X-ray and EUV radiation. This coronal emission declines as stars age and their rotation rates slow down. Therefore, it is natural to expect that the winds emanating from these coronae should evolve significantly as well. One might expect winds to evolve in concert with the coronal emission, because winds originate in stellar coronae. Thus, young stars might be expected to have more massive winds. There is undoubtedly more material heated to coronal temperatures for young, active stars, and therefore more material available to be accelerated into a wind. However, coronal X-ray emission arises from regions of closed magnetic field where the densities are large, while wind emanates along open field lines. Furthermore, coronal X-ray emission from the Sun varies by a factor of 5–10 over the course of the solar activity cycle (Judge *et al.*, 2003), but the solar mass-loss rate of $\dot{M}_{\odot} = 2 \times 10^{-14}\ M_{\odot}\ \mathrm{yr}^{-1}$ does not vary much on these time scales (Cohen, 2011). Thus, a more active corona clearly does *not* automatically lead to a stronger coronal wind.

On the other hand, active stars might also have substantial mass loss associated with coronal-mass ejections (CMEs). Strong flares on the Sun are usually accompanied by fast CMEs. These CMEs do not account for most of the solar mass loss (cf., Ch. 4), but flare rates and energies are known to increase dramatically with stellar activity. Young, active stars have stronger and more frequent flares (see Ch. 2), meaning that the mass loss associated with flare-associated CMEs could be much higher (although we do not have the means to empirically establish this; cf., Ch. 2).

Drake *et al.* (2013) computed the CME mass-loss rate expected for young, active stars by extrapolating a known correlation between solar-flare energy and CME mass to more-active stars with more-frequent and energetic flares. The resulting prediction is shown in Fig. 3.9, implying that the most active stars with X-ray luminosities $L_X \sim 3 \times 10^{30}$ erg s^{-1} (compared with $L_X \sim 2 \times 10^{27}$ erg s^{-1} for the Sun) could have mass-loss rates of $2 \times 10^4 \dot{M}_{\odot}$ due to CMEs alone, although limits on the percentage of a star's bolometric luminosity L_{bol} that can reasonably be expected to be converted to flares and CMEs ($\sim 10^{-3} L_{bol}$) lead Drake *et al.* (2013) to suggest a lower ceiling of $2500\dot{M}_{\odot}$. However, this prediction of a strong wind for young, active stars is only valid if the solar flare-energy/CME-mass connection can actually be extrapolated to more active stars.

3.7.2 *Inferences from stellar observations*

Observations are required to truly establish how winds evolve with time. Currently, the only way coronal winds can be detected around other stars is through astrospheric Lyman-α absorption, but the number of astrospheric Lyman-α detections is

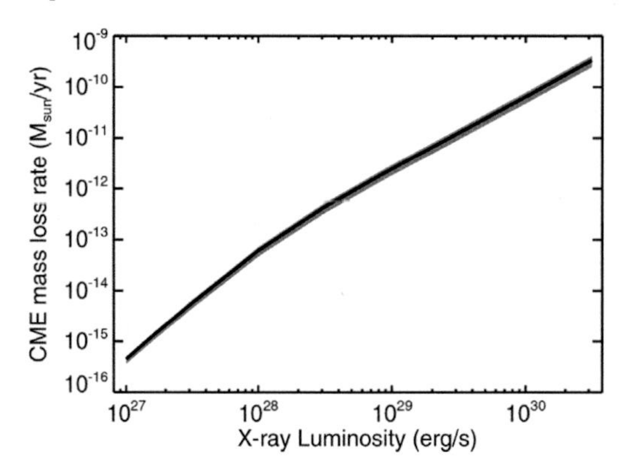

Fig. 3.9 Plausible stellar mass-loss rate due to CMEs as a function of coronal X-ray luminosity, computed by assuming that the solar-flare/CME–mass correlation can be extrapolated to younger, more active stars. Note that the solar X-ray luminosity is $L_X \sim 2 \times 10^{27}$ erg s^{-1}. (From Drake *et al.*, 2013.)

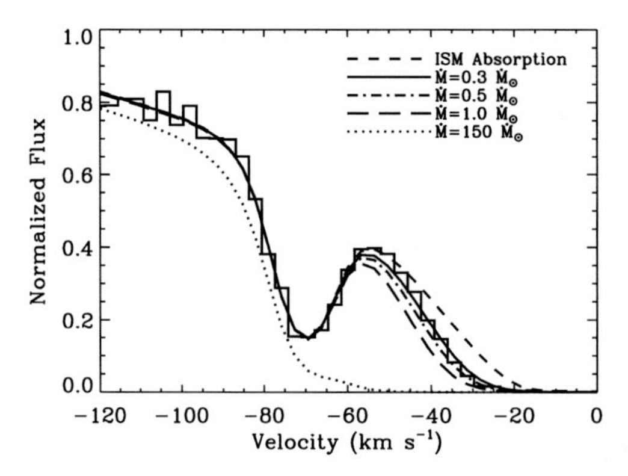

Fig. 3.10 The blue side of the Lyman-α absorption line of π^1 UMa, plotted on a heliocentric velocity scale. The absorption at -70 km s^{-1} is from interstellar D I. Because the ISM absorption cannot explain all of the H I absorption, the excess is assumed to be from the stellar astrosphere. The astrospheric absorption signature is compared with absorption predictions from four hydrodynamic models of the astrosphere, assuming four different mass-loss rates for π^1 UMa (after the astrospheric absorption is added to that of the ISM). (From Wood *et al.*, 2014b.)

still very limited. Table 3.1 provides the complete list. Among these stars, the best analog for a very young Sun is π^1 UMa, a 500 Myr old G1.5 V star. Figure 3.10 shows the Lyman-α spectrum for π^1 UMa, zooming in on the blue side of the line where the astrospheric absorption is detected. The amount of absorption correlates with the strength of the stellar wind, but extracting a stellar mass-loss rate from the Lyman-α data requires the assistance of hydrodynamic models of the astrosphere, such as the one in Fig. 3.11.

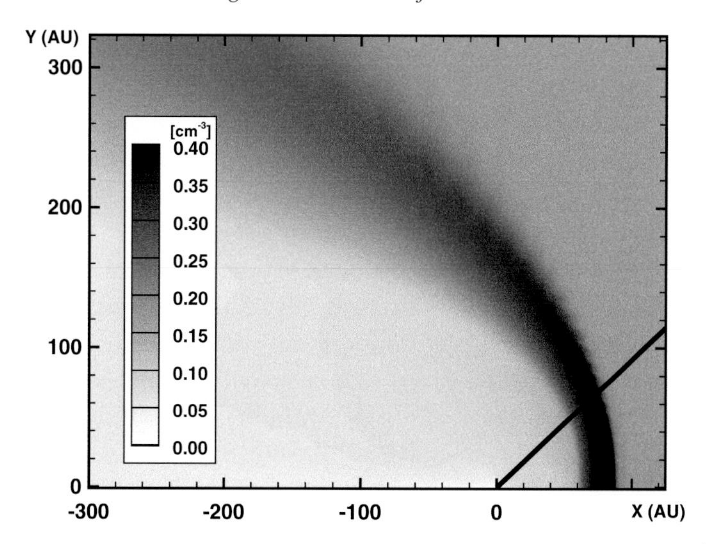

Fig. 3.11 The H I density distribution of a hydrodynamic model of the π^1 UMa astrosphere, assuming $\dot{M} = 0.5\dot{M}_\odot$, which leads to the best fit to the data in Fig. 3.10. The star is at the origin, and the ISM is flowing from the right in this figure. The "hydrogen wall" is the parabolic shaped high-density region stretching around the star. The black line indicates our line of sight to the star. (From Wood *et al.*, 2014b.)

Astrospheric models like that in Fig. 3.11 are extrapolated from a heliospheric model that successfully reproduces heliospheric absorption, specifically a multi-fluid model described by Wood *et al.* (2000). These models assume the same ISM characteristics as the heliospheric model, with the exception of the ISM flow speed in the stellar rest frame, $v_{\rm ISM}$, which can be computed using our knowledge of the local ISM flow vector and each star's unique space motion vector. Values of $v_{\rm ISM}$ are listed in Table 3.1, and θ indicates the angle of our line of sight to the star relative to the upwind direction, shown explicitly in Fig. 3.11 for the π^1 UMa case.

The astrospheric models are computed assuming different stellar wind densities, corresponding to different mass-loss rates, and the Lyman-α absorption predicted by these models is compared with the data to see which best matches the observed astrospheric absorption. Figure 3.10 shows the astrospheric absorption predicted by four models of the π^1 UMa astrosphere, assuming four different stellar mass-loss rates. The model with half the solar mass-loss rate shown in Fig. 3.11 is deemed the best fit to the data (Wood *et al.*, 2014b). Thus, this 500-Myr-old young-Sun analog appears to have a wind comparable in strength to that of the current Sun. Given that π^1 UMa is much more active than the Sun, with $L_X = 9 \times 10^{28}$ erg s^{-1}, this implies that the prediction of Fig. 3.9 must be incorrect, meaning that the relation between flare energy and CME mass seen for the Sun cannot be extrapolated to younger, more active stars.

Mass-loss-rate estimates have been made in this way for all of the astrospheric detections, and Table 3.1 lists these $\dot{M}$ measurements (Wood *et al.*, 2005, 2014b). Table 3.1 also lists logarithmic X-ray luminosities ($\log L_X$) and surface areas for the observed stars, which can be used to compute X-ray surface flux densities, F_X (the ratio of X-ray luminosity to surface area). In order to look for some correlation between coronal activity and wind strength, Fig. 3.12 shows mass-loss rates (per unit surface area) plotted versus F_X, focusing only on the main-sequence stars. For the low-activity stars, mass loss increases with activity in a manner consistent with the $\dot{M} \propto F_X^{1.34\pm0.18}$ power law relation shown in the figure. For the ξ Boo binary, in which (like α Cen) the two members of the binary share the same astrosphere, Fig. 3.12 indicates how the binary's combined wind strength of $\dot{M} = 5\dot{M}_\odot$ is most consistent with the other measurements if 90% of the wind is ascribed to ξ Boo B, and only 10% to ξ Boo A.

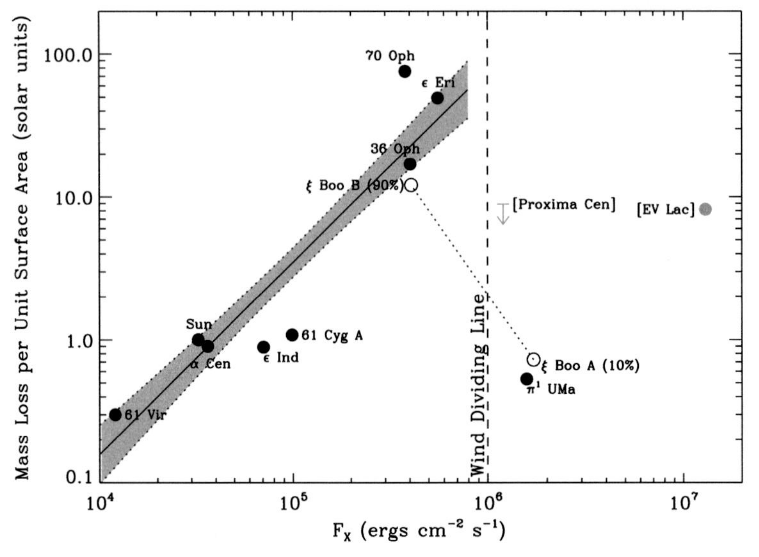

Fig. 3.12 A plot of mass-loss rate (per unit surface area) versus X-ray surface flux density for all main-sequence stars with measured winds. Most of these have spectral types of G (like the Sun) or (cooler) K, but the two with square-bracketed labels are (much cooler) tiny M dwarf stars. Separate points are plotted for the two members of the ξ Boo binary, assuming ξ Boo B accounts for 90% of the binary's wind, and ξ Boo A only accounts for 10%. A power law, $\dot{M} \propto F_X^{1.34\pm0.18}$, is fitted to the less active stars where a wind/corona relation seems to exist, but this relation seems to fail for stars to the right of the "Wind Dividing Line" in the figure. (From Wood *et al.*, 2014b.)

For $F_X < 10^6$ erg cm^{-2} s^{-1}, mass loss appears to increase with activity. However, above $F_X = 10^6$ erg cm^{-2} s^{-1} (i.e., for more active, and thus generally younger stars) this relation seems to fail, a boundary identified as the "Wind Dividing Line" in Fig. 3.12. Highly active stars above this limit appear to have surprisingly weak winds. This is suggested not only by the two solar-like G stars above the limit, ξ Boo A and π^1 UMa, but also by the two active M dwarfs above the limit, which have very modest mass-loss rates. (For Proxima Cen we only have an upper limit of $\dot{M} < 0.2\dot{M}_\odot$, while for EV Lac $\dot{M} = 1\dot{M}_\odot$.) The apparent failure of the wind/corona correlation to the right of the "Wind Dividing Line" may indicate a fundamental change in magnetic field topology at that stellar activity level.

It is not easy to study magnetic fields on stars that are unresolvable point sources. Nevertheless, estimates can be made on the basis of empirical stellar age/rotation relations and theoretical magnetic braking models (Schrijver *et al.*, 2003). Sophisticated spectroscopic and polarimetric techniques are also available for studying stellar magnetic fields (Donati and Landstreet, 2009). One interesting discovery is that very active stars usually have stable, long-lived polar starspots (Schrijver and Title, 2001; Strassmeier, 2002), in contrast to the solar example where sunspots are only observed at low latitudes. Perhaps the polar spots are indicative of a particularly strong dipolar magnetic field that envelopes the entire star and inhibits stellar wind flow, thereby explaining why very active stars have surprisingly weak winds. Strong toroidal fields are also often observed for active stars (Petit *et al.*, 2008).

Given that young stars are more active than old stars (e.g., Ribas *et al.*, 2005), the correlation between mass loss and activity indicated in Fig. 3.9 implies an anticorrelation of mass loss with age. Ayres (1997) finds the following relation between stellar X-ray flux and age for solar-like stars: $F_X \propto t^{-1.74\pm0.34}$. Combining this with the power law relation from Fig. 3.12 yields the following relation between mass loss rate and age:

$$\dot{M} \propto t^{-2.33\pm0.55} \tag{3.2}$$

(Wood *et al.*, 2005). Figure 3.13 shows what this relation suggests for the history of the solar wind, and for the history of winds from any solar-like star for that matter. The truncation of the power law relation in Fig. 3.13 near $F_X = 10^6$ erg cm^{-2} s^{-1} leads to the mass-loss/age relation in Fig. 3.13 being truncated as well at about $t = 0.7$ Gyr. The plotted location of π^1 UMa in Fig. 3.13 indicates what the solar wind may have been like at times earlier than $t = 0.7$ Gyr.

Although winds may have been surprisingly weak for $t < 0.7$ Gyr, at later times Fig. 3.13 does indicate that solar-like coronal winds can be up to two orders of magnitude stronger than the current solar wind at $t \approx 1$ Gyr. This makes it more

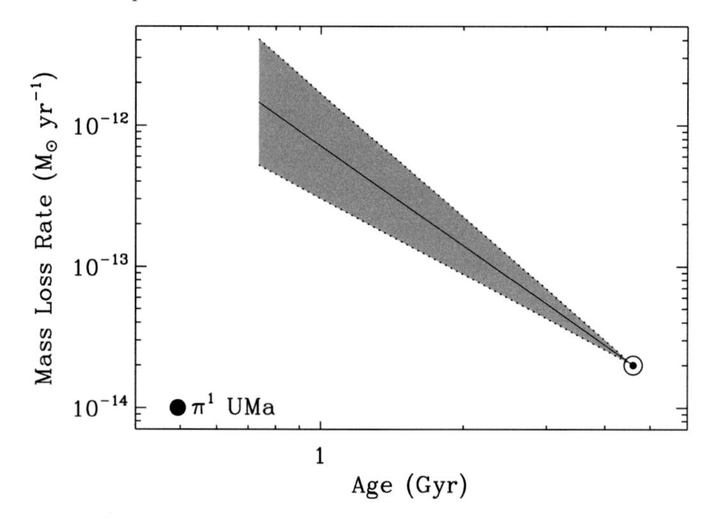

Fig. 3.13 The mass-loss history of the Sun inferred from the power law relation in Fig. 3.12. The truncation of the relation in Fig. 3.12 means that the mass-loss/age relation is truncated as well. The low mass-loss measurement for π^1 UMa suggests that the wind weakens at $t \approx 0.7$ Gyr as one goes back in time. (From Wood *et al.*, 2005.)

likely that the erosive effects of stellar winds play an important role in planetary atmosphere evolution at these later ages (e.g., Luhmann *et al.*, 1992; Lammer *et al.*, 2003; Brain *et al.*, 2010b; Khodachenko *et al.*, 2012).

The wind history suggested by Fig. 3.13 also has relevance for the Faint Young Sun problem (see Vol. III, Ch. 2), which relates to our understanding that at $t \approx 0.7$ Gyr the Sun would have been $\sim$ 25% fainter than today. This would imply temperatures for Earth too cool for liquid water to exist on the surface, inconsistent with our understanding of Earth's geologic history (Sagan and Mullen, 1972). One proposed solution has to do with the possibility that the Sun was at least a few percent more massive in the past, and therefore more luminous (Sackmann and Boothroyd, 2003). This requires a stronger solar wind in the past to allow it to lose that excess mass. Figure 3.13 does indicate a generally stronger solar wind in the past, but not strong enough, unfortunately, as the cumulative solar mass loss suggested by Fig. 3.13 is only $\sim 10^{-3} M_\odot$, at least an order of magnitude too low. Alternative explanations involving a stronger greenhouse effect may instead be required to resolve the Faint Young Sun problem (Kasting and Ackerman, 1986; see also Chs. 2 and 11 in Vol. II).

In closing, we note that our current understanding of stellar wind and astrosphere evolution for Sun-like stars is relying heavily on a relatively small number of detections of astrospheric Lyman-α absorption. More detections of such absorption will hopefully become available in the future. But there is also a need to explore completely different ways of studying winds and astrospheres. The Atacama Large Millimeter/submillimeter Array (ALMA) may provide detections of free–free emission from winds, or at least provide much lower upper limits for stellar mass-loss rates. An even more sensitive radio telescope will ultimately be required to directly detect emission from winds as weak as that of the Sun.

4

Effects of stellar eruptions throughout astrospheres

OFER COHEN

Stars like the Sun evolve from young, fast-rotating and very active stars to older, slowly rotating main-sequence stars like our own Sun. The changes in stellar magnetic fields and stellar activity of such solar analogs with stellar evolution and the change in their rotation period are described in Ch. 2 of Vol. III. In this chapter, we review how the changes in stellar activity of Sun-like stars over stellar evolution translate to changes in their stellar winds, the structure of their interplanetary space and of their astrospheres, the transport of particles, and the propagation and evolution of coronal mass ejections (CMEs). We also review the consequences of CMEs in stellar systems other than our own and their role in planet habitability.

Since the dawn of the space exploration era, great knowledge has been acquired about the solar system's interplanetary space and the heliosphere. The growing amount of spacecraft in-situ measurements of the interplanetary medium (direct measurements of solar-wind particles; see reviews by McComas *et al.*, 2007, and Owens and Forsyth, 2013), as well as increasing amount of global remote-sensing observations monitoring the Sun's photospheric magnetic field, EUV and X-ray coronal radiation, radio emissions, and energetic particles (reviewed by, e.g., Lang, 2009) have revealed a clear dependence of the state of the heliosphere and the interplanetary space on the solar activity level and on the solar magnetic-field structure. These long-term observations also revealed how the frequency of solar eruptions change over the solar cycle (see the review by Webb and Howard, 2012).

4.1 Astrospheres in time

4.1.1 Astrospheric structure and evolution with time

The extent and structure of astrospheres are determined by the radially expanding super-Alfvénic stellar wind that drags the stellar magnetic field from the stellar corona through the interplanetary medium, until the wind is stopped by the interstellar medium (ISM; see Ch. 3). It is also determined by the rotation of the star. As a result, each astrospheric magnetic field (AMF) line has one end (or "footpoint") attached to the stellar surface, while its location at each point in the astrosphere, $\mathbf{r}(r, \theta, \phi)$ (for co-latitude θ), is given by the following formula. It describes a spiral shape and is known as the "Parker Spiral" (Parker, 1958):

$$\mathbf{B}(r) = B_0 \left(\frac{r_0}{r}\right)^2 \left[\hat{r} + \frac{(r - r_0)\Omega \sin\theta}{v}\hat{\phi}\right]. \tag{4.1}$$

Here Ω is the stellar rotation rate (angular velocity), v is stellar-wind speed (which is here assumed to be radial and fixed in time); r_0 is the actual base point of the AMF, and is at a reference distance from the stellar surface at which we assume the stellar wind is fully developed and has achieved its asymptotic speed and radial direction; B_0 is the magnetic field magnitude at that point. We can see that the radial component of the AMF has an r^{-2} dependence, while the azimuthal component has only a r^{-1} dependence. As a result, through most of the astrospheres, the AMF is dominated by the azimuthal field, which is a function of Ω, except for high latitudes (small θ) where the AMF lines are nearly radial.

Over time, stellar-rotation periods vary from less than one day for very active, young stars to about 20–100 days for older, main-sequence stars like the Sun. For very fast rotating stars, the AMF spiral is completely dominated by the azimuthal component: the field is highly compressed, and its azimuthal component dominates even at relatively small distances from the star and inside the stellar corona, which typically extends to 10–20 stellar radii (Cohen *et al.*, 2010a). In this case, even extended closed magnetic loops can be bent as a result of the fast rotation. This effect can have implications for the triggering of very strong stellar flares, and for the mass-loss rate of the star to the stellar wind (see e.g., Maggio *et al.*, 2000; Matranga *et al.*, 2005; Cohen *et al.*, 2010a,b). The left-hand panel in Fig. 4.1 shows how the compression of the AMF spiral changes for different stellar rotation periods. The other two panels show the AMF lines close to the star (up to 24 stellar radii). It can be clearly seen that the field lines are nearly radial for the slow, solar-like rotation period of 25 days, while the field lines are strongly bent in the azimuthal direction for fast rotation period of half a day.

Equation (4.1) describes how a given magnetic field line changes with distance for a given value of B_0 at its base (r_0), and a given asymptotic stellar wind speed v. However, the AMF is formed by a collection of field lines that are defined by some spherical distribution of B_0 at the base of the stellar corona. This distribution depends on the topology of the stellar magnetic field at a given time. In addition, the value of v also varies as it empirically depends on the expansion of the magnetic flux tubes and on the non-uniform distribution of B_0 (Wang and Sheeley, 1990). For the Sun, the distribution of B_0 changes dramatically over the solar cycle. This

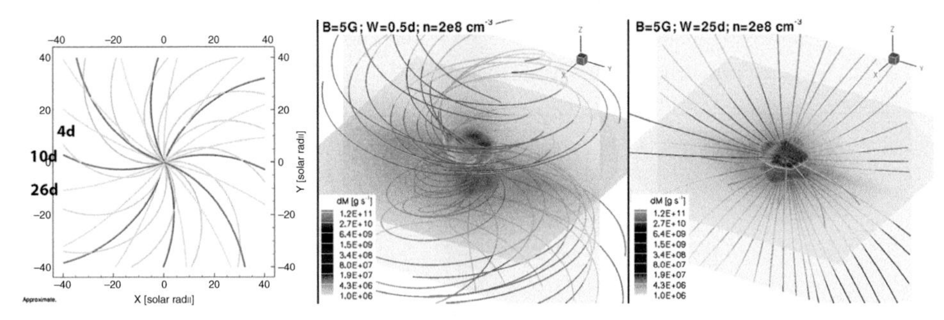

Fig. 4.1 Left: conceptual display of different stellar-wind magnetic field spirals for a Sun with a 4.6-day rotation period, a 10-day period, and a 26-day period, as a function of distance in solar radii (from Cohen *et al.*, 2012). Center/right: results from numerical simulations for the stellar coronae of solar analogs with rotation period of 0.5 day (A) and 25 days (B). The astrospheric field lines are shown in gray. Also shown is the surface at which the Alfvénic Mach number equals unity. In the original figure (Cohen and Drake, 2014) the meridional and equatorial planes are colored with contours of the mass-loss rate, but these are not adequately reproduced in this gray-scale rendering.

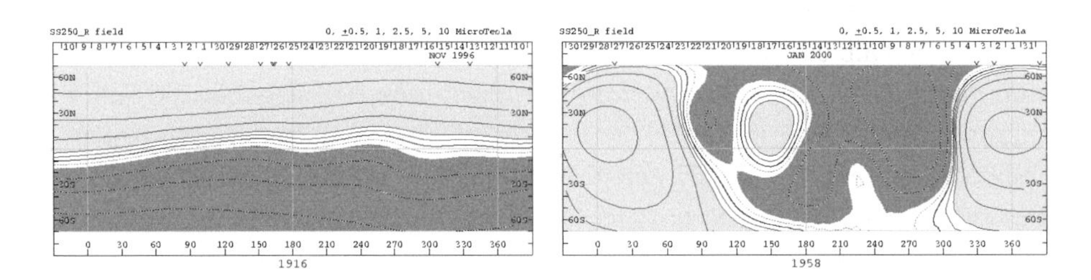

Fig. 4.2 The distribution of the solar magnetic field at $r = 2.5R_{\odot}$ (B_0) for solar minimum period (left, November 1996), and solar maximum period (right, January 2000) obtained by the Wilcox Solar Observatory (WSO, wso.stanford.edu).

can be seen in Fig. 4.2, which shows the distribution of the solar magnetic field at a distance of $r_0 = 2.5R_{\odot}$ for solar minimum period (November 1996) and for solar maximum period (January 2000).

Over time, stellar activity appears at different latitudes, while changing in magnitude as the behavior of surface magnetic activity is highly tied to the rotation rate. Young active stars seem to have very strong large-scale magnetic fields with magnitude of several kilo-gauss. For reference, the Sun's dipole field strength is of the order of 5–10 G, and while the magnetic flux density within active regions can be high (ranging up to well over a kilo-gauss in sunspots), solar active regions are rather small in size. In addition, magnetic activity in active stars tends to appear at high-latitude, polar regions (Strassmeier, 1996, 2001, Donati and Collier Cameron, 1997). This behavior is most likely related to the role of the fast stellar rotation

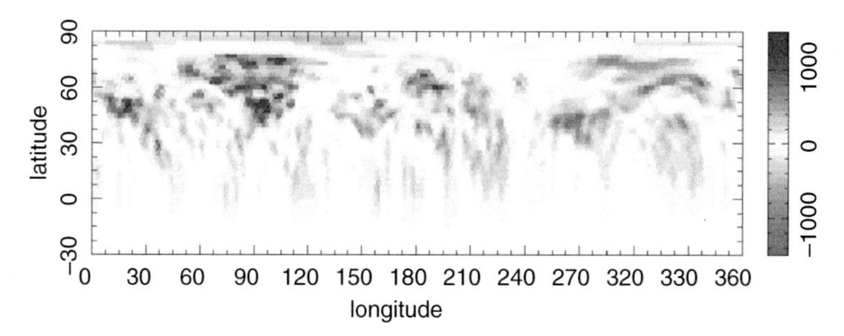

Fig. 4.3 Longitude–latitude map of the photospheric radial magnetic field of AB Doradus. Note that field in the deep southern hemisphere cannot be observed owing to the tilt of the spin axis. (From Hussain *et al.*, 2007.)

in the stellar dynamo and meridional magnetic flux circulation (Schuessler and Solanki, 1992; Solanki *et al.*, 1997; Schrijver and Title, 2001), see also Ch. 2.6, Vol. III. An example of such a young active stars is AB Doradus, which is a 50-Myr-old K0 dwarf star rotating with a half a day period. Figure 4.3 shows the photospheric distribution of the radial stellar magnetic field taken from Hussain *et al.* (2007). One can see that there is a great coverage of magnetic field of over a kilo-gauss in magnitude, and that these strong field regions appear at latitudes higher than 45 degrees and up to 75–80 degrees from the equator, in contrast to solar active regions, which do not appear above 30 degrees from the equator (cf., Fig. 2 in Vol. III). See also Fig. 4.4.

The appearance of stellar activity described above reflects a change in the distribution of B_0. Therefore, it affects the shape of the AMF and the astrosphere volume. It is not clear how v changes for young stars as we cannot directly measure stellar winds of “cool stars”, i.e. stars with a convective envelope beneath their surfaces such as in the case of the Sun. Some techniques to estimate mass-loss rates from cool stars are described in Ch. 3. However, these estimates do not separate the stellar-wind speed from the density, so it cannot be obtained independently. Another cause for the lack of estimates for stellar wind speeds of cool stars is the incomplete theory about the solar wind acceleration (see Vol. I, Ch. 9). In order to demonstrate how the change in the photospheric field affects the three-dimensional structure, Fig. 4.5 shows the distribution of the photospheric magnetic field and the shape of the three-dimensional magnetic field close to the Sun. The left panel is obtained using actual data of the photospheric field during high solar activity period. In the other two panels, the original data were manipulated, so that the active regions have been shifted by 30 and 60 degrees, respectively, towards higher latitudes in order to mimic the activity distribution of young active stars. It can be seen that the field topology changes dramatically even if only the positions of the active regions are changed.

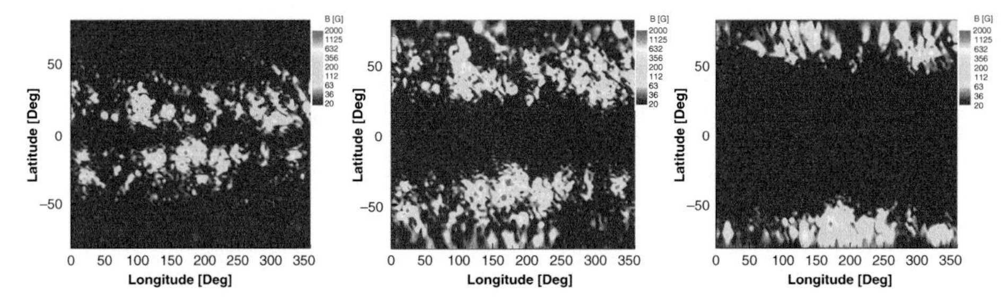

Fig. 4.4 A map of the solar photospheric radial magnetic field (magnetogram) during Carrington Rotation 1958 (January 2000, solar maximum period) shown on the left. The middle and right panels show manipulation of the original map, where the active regions have been shifted by 30 and 60 degrees towards the pole, respectively. (From Cohen *et al.*, 2012.)

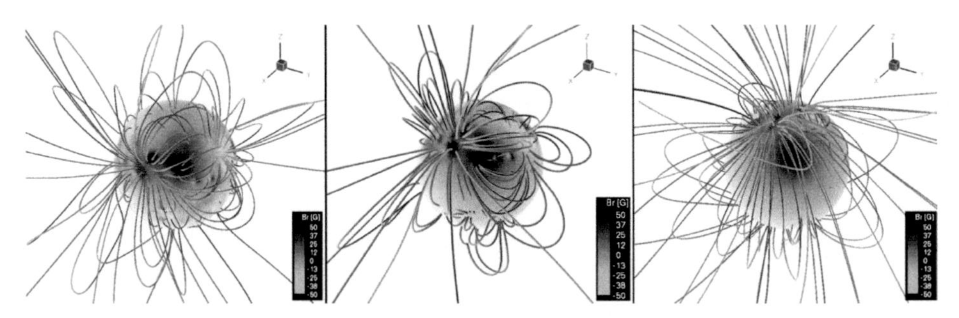

Fig. 4.5 The three-dimensional magnetic field corresponding to the surface distribution of the photospheric radial magnetic field (shown on a sphere of $r = R_{\odot}$) during solar maximum (left), and for manipulated photospheric filed with the active regions shifted by 30 degrees (middle) and 60 degrees (right) towards the poles, as shown in Fig. 4.4. (From Cohen *et al.*, 2012.)

4.1.2 Astrospheric evolution and particle transport

Ch. 9, Vol. II and Ch. 9, Vol. III describe transport processes of energetic particles in the solar system and in particular, Galactic Cosmic Rays (GCRs), which carry energies of up to 10^{21} eV. It has been known for many years that the lower part of the GCR energy spectrum (about 1 GeV), is modulated by the state of the AMF and the solar wind. This happens because GCR transport depends on the state of the AMF via two terms in the transport equation described in Ch. 9, Vol. II. The drift term, which depends on the magnetic field magnitude and direction, dictates whether particles travel inwards or outwards near the heliospheric ecliptic plane, and whether GCRs approach the ecliptic plane (and the vicinity of the Earth) from equatorial regions or from polar regions, depending on the polarity and magnitude of the AMF. The diffusion term (and the diffusion coefficient) depend on the field magnitude, and has components parallel and perpendicular to the mean local magnetic field. The drift and diffusion terms are responsible for the clear GCR modulation over the solar cycle, where the GCR intensity anti-correlates with the sunspot number (see Fig. 9.4 in Vol. III). The reduction in GCR intensity during high solar activity (solar maximum) is a result of the increase in the AMF magnitude as the Sun sheds more magnetic flux into the heliosphere, the increase in the number of interplanetary shocks (i.e., increase in CME rate), and the overall increase in the level of turbulence in the solar wind. All of the above make it harder for GCRs to penetrate deep into the heliosphere all the way to the Earth. During solar minimum, the AMF reduces to its floor value, and the CME rate decreases as well. This improves the ability of GCRs to penetrate into the heliosphere so their intensity increases during low solar activity periods. Voyager 1 has provided us with a direct observation of how the GCR intensity changes with distance from the Sun inside the heliosphere. Figure 4.6 shows the dramatic increase in GCR intensity accompanied by a similar sharp drop in the intensity of solar wind particles. This is one of the indications that Voyager 1 has indeed, left the solar system and is currently in the ISM.

The Earth is shielded from most energetic GCRs by its own magnetic field, and by the AMF. Nevertheless, some cosmic rays with very high energies can reach the top of the Earth's atmosphere and generate a cascade reaction with atmospheric particles (air shower, see Ch. 11 in Vol. III). There are a number of ways that GCRs have played a role in the evolution of Earth. They can be an ionization source for the production and creation of complex organic molecules and nucleotides (e.g., Court *et al.*, 2006; Simakov *et al.*, 2002), can cause cellular mutation through direct and indirect processes (e.g., Nelson, 2002; Dartnell, 2011), and they can play a role in triggering lightning (e.g., Gurevich *et al.*, 1999; Dwyer *et al.*, 2012). It has been suggested that GCRs can contribute to global climate change periods in the Earth's history as they may change the Earth's albedo by affecting cloud condensation (Svensmark and Friis-Christensen, 1997; Shaviv, 2003, 2005b; Wallmann, 2004; Medvedev and Melott, 2007; Kirkby *et al.*, 2011). This subject is still under debate, where the argument focuses on the magnitude of this effect and whether it is significant or not.

In the previous section, we have shown how the AMF changes with time as a result of the increase in stellar rotation and stellar magnetic topology. Let us focus on a particular period of time in Earth's history called the Archean eon, which spanned from about 3.8 to 2.5 billion years ago. This period of time occurred right after the Late Heavy Bombardment (the time when the Earth was continuously hit by solar system small bodies), and when it is believed that simple life forms began to emerge (see Ch. 4, Vol. III). With the above effects of GCRs on the Earth atmosphere, it is useful to estimate the GCR intensity near Earth during the Archean eon.

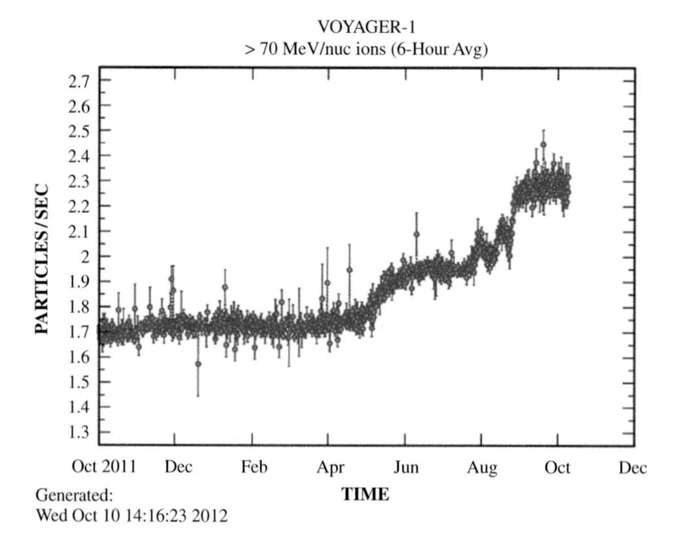

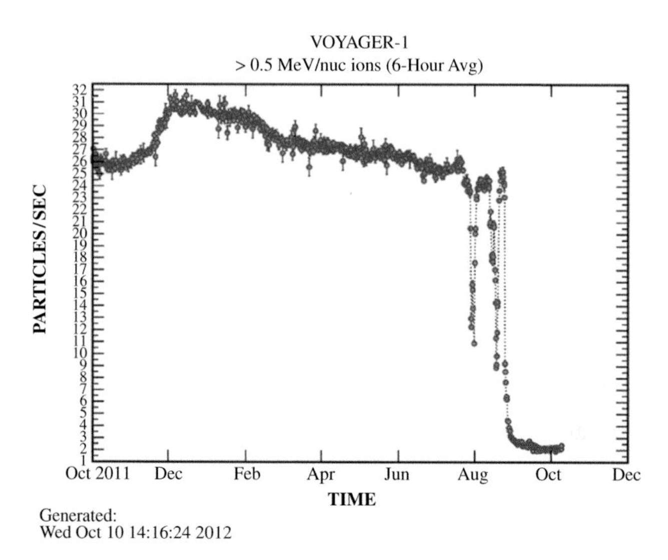

Fig. 4.6 The dramatic increase in cosmic-ray flux (top) and decrease in solar-wind particle flux (bottom) observed by Voyager 1 around September 2012. This observation strongly suggests that *Voyager 1* reached the ISM at the time of the change. (Figures from science.nasa.gov.)

During that time, the solar rotation was 2–4 times faster, with rotation period of about 6–15 days in contrast to the current 25 day period. In addition, based on astronomical observations of solar analogs of that age, the magnetic activity seems to appear at higher latitudes. Based on this information, Cohen *et al.* (2012) used the solar magnetohydrodynamic (MHD) model and GCR transport model to calculate the GCR intensity near the Archean Earth as a function of a solar rotation period, and the topology of the Sun's magnetic field. They used solar magnetic field data and split them into a weak, dipole component, and a strong, "spots" component which represents active regions. They then shifted the spots component in latitude (as seen in Fig. 4.4), and also modified the magnitude of each component to study the effect of the spot location and magnitude of the weak and strong solar field on the GCR transport and intensity.

Figures 4.7 and 4.8 show the results for the GCR intensity near the Archean Earth. They show that the dominant effect is the change in solar rotation rate, which dramatically reduces the intensity peak around 1 GeV by two orders of magnitude. The results also show that in the case of fast rotation (2 days), the Earth is completely shielded from GCRs with energies of less than about 20 MeV. The intensity reduction is enhanced even further if the magnitude of the solar active regions is increased, where almost all the GCRs are prevented to reach Earth, except for those with energies above 1 GeV or so.

The results shown here quantify and clearly demonstrate how the increase in solar rotation and solar magnetic activity lead to a significant reduction of GCR penetration to the inner heliosphere during the early solar system. Therefore, the role of GCRs in the evolution of the Earth's atmosphere and the evolution of life on Earth (via the processes mentioned above) has increased with time.

4.1.3 Stellar activity and disk evolution

In Section 4.1.2, we discussed how fast rotation and strong magnetic fields, in particular in the polar regions, of young stars significantly increase their AMF. Pre-main-sequence stars hosting accreting disks (also known as Classical T-Tauri Stars – CTTS; cf., Ch. 3 in Vol. III) have relatively fast rotation, with rotation periods ranging from 7 to 10 days, strong surface flux intensity observed to be more than 0.5 kG and, in many cases, frequent large flaring that may be due to the interaction between the disk and the stellar magnetosphere (see the review by Hussain, 2012).

In protoplanetary disks of such young stars, angular momentum transport controls the transfer of material to and from different regions of the disk (Bodenheimer, 1995). Therefore, it is crucial to understand the evolution of disk angular momentum transfer in order to understand the evolution and formation of planetary systems and the origin of planets. Some of the more popular mechanisms to explain disk angular momentum transfer involve an interaction between magnetic fields in the disk and the disk's gas. Among them are turbulence as a result of the so-called magneto-rotational instability (MRI, Balbus and Hawley, 1991), large-scale magnetic field driving outflow that causes stress (Blandford and Payne, 1982), or disk shearing (Turner and Sano, 2008) (see also Ch. 3, Vol. III). However, these processes require strong coupling between the disk's gas and the magnetic field, and

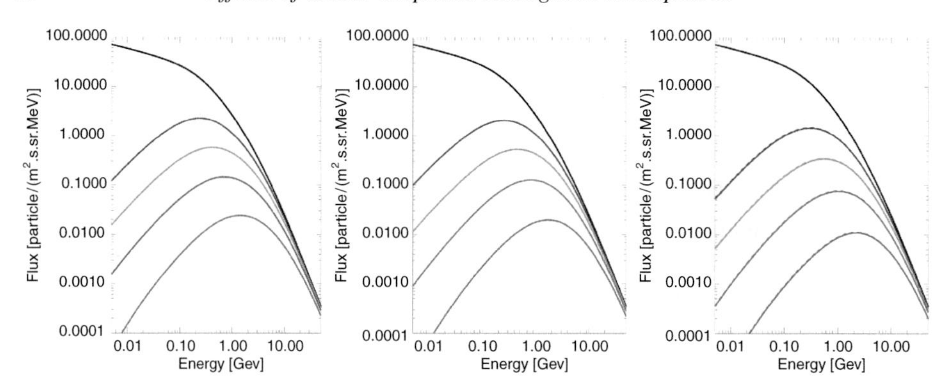

Fig. 4.7 Cosmic-ray energy spectrum for modeled solar rotation periods of 26 days (current rotation), 10 days, 4.6 days, and 2 days, along with the local ISM spectrum. Plots are for the current Sun (left), spots shifted towards the pole by 30 degrees (middle), and spots shifted towards the pole by 60 degrees (right). This plot is similar to Fig. 11.11 in Vol. III, except that the x-axis unit is in GeV instead of MeV, and the flux is normalized by the additional steradian (sr) geometrical factor.

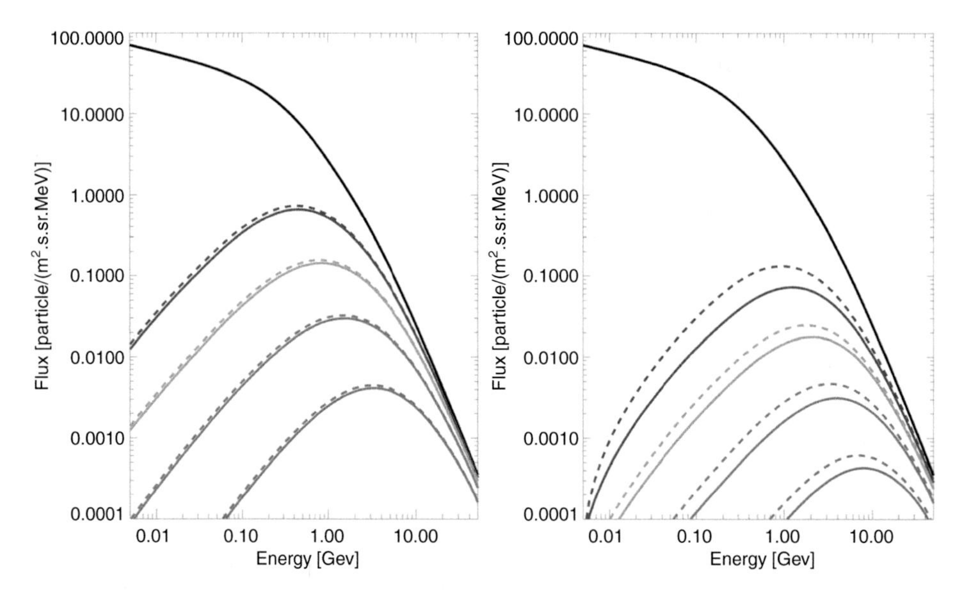

Fig. 4.8 Same as the right-hand panel of Fig. 4.7, but with the dipole component enhanced by a factor of 10 (left), and the spot component enhanced by a factor of 10 (right). Solid lines represent the spectrum with the Termination Shock (TS) scaled with the solar-wind dynamic pressure, while dashed lines represent the spectrum for the TS fixed at 90 AU.

such a coupling requires the gas in the disk to be sufficiently ionized: if the disk's gas is neutral it cannot interact with the electromagnetic force.

One potential source of disk ionization is GCRs, which penetrate from the edge of the stellar system to the vicinity of the disk near the equator. Other possible sources for disk ionization are heat, X-ray radiation originating from the corona of the central star, and the decay of radionuclides within the gas (Turner and Drake, 2009; Cleeves *et al.*, 2013). GCRs and X-ray radiation are expected to penetrate up to certain depth from the top of the disk, creating a disk "skin", which is sufficiently ionized to couple the magnetic field and the gas, and an inner "dead zone", where the gas is neutral and the magnetic field is essentially irrelevant (see e.g., Gammie, 1996; Sano *et al.*, 2000; Ilgner and Nelson, 2006; also compare Fig. 5.9). Because the magnetic fields of CTTS are strong, and their rotation periods are rather short, the intensity of GCRs may not be sufficient to ionize the disk. Owing to this, the impact of the stellar wind and the AMF on disk ionization by GCRs is usually neglected, where ISM GCR intensities are used to estimate ionization rates.

Turner and Drake (2009) have tested all of the above disk-ionization mechanisms, as well as ionization by energetic protons originating from the corona of the central star. They found that a dead zone is created in all scenarios, as well as an undead zone at which resistivity is high enough to allow shear-generated large-scale magnetic field, but it hampers MRI turbulence. However, they found that while the necessary conditions to couple the magnetic field and the gas are feasible, they are in some conflict with the conditions necessary for planet formation in the disk as the solution without a dead zone requires surface density below the minimum mass defined by a protosolar model (Fromang *et al.*, 2002). In other words, the density necessary for sufficient gas ionization should be low, but the density necessary for planet formation in the disk should be high. Cleeves *et al.* (2013) calculated the GCR intensity and disk-ionization rates while taking into account the GCR flux reduction and modulation by the stellar wind and AMF. They found that the ionization rate by GCR is one order of magnitude less than the standard value used for disk ionization and chemistry. Therefore, it is not clear whether the extensive study of disks using the MHD formalism is valid everywhere in the disk.

In summary, while some sources of disk ionization, such as heat close to the central star, X-ray radiation, and energetic protons originating from shocks in the stellar corona, can ionize the inner part of the disk and are enhanced in CTTS, global ionization is more likely to occur due to GCRs. However, the high magnetic activity level of CTTS actually reduces the amount of ionizing GCRs that reach the disk in its inner parts and up to about 100 AU (the GCR intensities can be higher at the outer edge of the disk).

4.2 Coronal mass ejections in time

In general, stellar activity is quantified by the stellar total X-ray luminosity, L_x, which is an indication for the amount, strength, and temperature of hot coronal loops, and by the X-ray/EUV flaring rate and magnitude, which provide an insight for coronal dynamic activity and its time scales. These "activity indicators" are known to be correlated with the stellar rotation rate and age (see Ch. 2, and Ch. 2, Vol. III). CMEs and large X-ray flares on the Sun are known to be correlated to each other (see Ch. 6, Vol. II). The traditional view on the generation of solar flares is that particles are accelerated down from the top of the CME flux-rope (as it propagates out) and hit the chromosphere, leading to "evaporation" of heated plasma, generating strong X-ray emissions. Because we cannot observe CMEs on other stars, stellar flares serve as proxies for CME activity on other stars.

In this section, we review the possible role of CMEs in stellar evolution, based on the known flaring activity (see Ch. 2 of this volume). We also review how the change in the interplanetary medium may affect the propagation and evolution of CMEs. As very little work has been done on studying stellar CMEs, this section is in part necessarily qualitative.

4.2.1 Initiation, propagation, and evolution of CMEs through different astrospheres

We can divide our overview on CMEs over time into two different categories. First, we can estimate how the change in stellar activity over time may impact the initiation, rate, and angular distribution of CMEs. Second, we can estimate how the change in the interplanetary medium over time affects the propagation and evolution of CMEs.

Let us first discuss how CMEs may be initiated differently in young, active stars. On the Sun, active regions appear within latitudes of about ±30 degrees or less from the equator. These active region are the source location of many CMEs and indeed, most CMEs appear to originate from within these latitudes (Gopalswamy *et al.*, 2008). As mentioned in Section 4.1.1, in young active stars, the magnetic activity appears at much higher latitude. Therefore, if we assume a similar relation between the location of the active regions and the source point of the CMEs, it is possible that most of the CMEs in such stars are launched from, and propagate into, the polar regions of the astrosphere (Fig. 4.9 shows a conceptual schematic of this difference). In any case, it is not possible to determine the latitudinal distribution of stellar flares due to the fact that observations represent the source integrated photon flux of what is essentially a point source.

Another aspect that is important to discuss is whether stellar flares, in particular large ones, are triggered by traditional, solar-like CMEs. The general CME initiation mechanism, while not completely understood, can be associated with a slow storage of magnetic energy via twisting of the CME flux-rope (by some kind of motions at the base of the magnetic loops), followed by a sudden release of

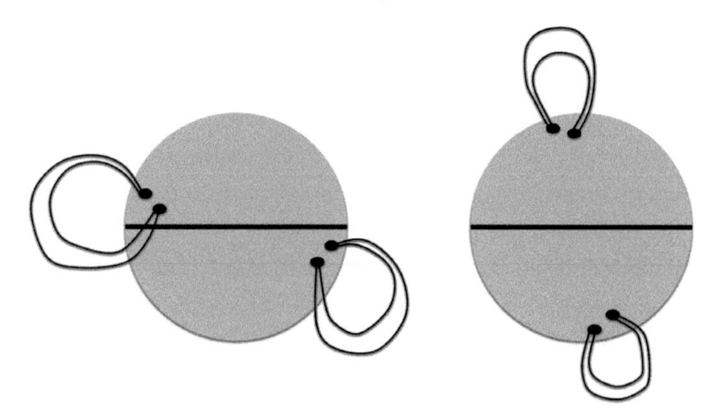

Fig. 4.9 On the Sun, CMEs are launched from active regions that emerge at low latitudes near the equator (left). In active young stars, magnetic activity appears at high, polar latitudes. Therefore, CMEs may be launched mostly towards the polar astrospheric regions (right).

the energy, most likely due to magnetic reconnection (see Ch. 6, Vol. II). We can imagine other, similar scenarios to trigger flaring activity. For example, the coronal loops of fast rotating stars are highly tangled in the azimuthal direction (as shown in Fig. 4.1). Such a tangling could build magnetic energy in the loops due to the increased magnetic tension. Similar to the way it triggers CMEs, magnetic reconnection could trigger a sudden release of the magnetic tension, triggering a very large flare as a result of the large size of the loop (these stretched loops can be of the order of the star size due to the strong stellar magnetic field). Cohen *et al.* (2010b) used an MHD model to simulate the corona of FK Comae, a rapidly rotating (2.4 days) late-type giant G star. These simulations showed that the azimuthal tangling of the large coronal loops indeed builds up high magnetic tension. While the steady-state simulation could not provide any dynamic triggering for reconnection, it is more than possible that such a triggering could occur due to footpoint motions on the photosphere. In CTTS, magnetic energy could slowly build up due to twisting or stretching of the field by the interaction between the stellar magnetosphere and the accretion disk (Hussain, 2012). As mentioned in Section 4.1.3, for such an interaction to occur, the disk gas should be sufficiently ionized. However, the interaction between the magnetosphere and the disk in the context of the flaring activity occurs at the inner part of the disk, where ionization levels are most likely sufficient.

Now we discuss how the state of the astrosphere itself affects CMEs. We keep in mind that CMEs carry with them new magnetic flux that is injected into the pre-eruption Astrospheric Magnetic Field (AMF). Therefore, we need to consider how changes in the state of the ambient AMF over time may impact the propagation and evolution of CMEs. The AMF changes as a result of the change in stellar rotation rate, where the AMF spiral is more compressed for faster rotating, young stars and it is less compressed for older stars. The AMF also changes with the reduction in stellar activity over time (indicated by the reduction in flare rate and total X-ray/EUV flux, which is a consequence of the stellar spin down). In both cases, the end result is a reduction in the strength of the AMF.

To date, we do not have much information about CMEs on other stars (with the exceptions discussed in the next section). CMEs are the result of a buildup of non-potential energy that is released as the system relaxes to a lower energy state. The erupted magnetic flux is then carried by the CMEs and is added to the AMF flux via interchange reconnection (see Section 8.8, Vol. III). Therefore, with the addition of CME magnetic flux, the AMF strength is higher during solar maximum than its "floor" value during solar minimum when very little CME magnetic flux is added (Owens and Forsyth, 2013). It has been suggested that the reason for the record-low AMF strength during the extended solar minima between solar cycle 23 (1996–2007) and 24, i.e., from 2008 to about 2012, is due to the record-low number of CMEs during that period of time (Owens *et al.*, 2008).

We assume that CMEs, due to their role in regulating the system's energy, scale with the overall available magnetic energy. In other words, if the overall stellar field is much stronger for young active stars, then we expect the magnetic flux in CMEs in these stars to be high accordingly. In this case, the role of CMEs in the evolution of the AMF (as discussed in Section 8.8, Vol. III) is probably similar to the case of the Sun, unless the CME rate in active stars is much greater so that the AMF never falls to its floor level. Alternatively, it is also possible that CMEs in active stars carry magnetic flux of similar magnitude to that of solar CMEs. This is possible if the large-scale strong fields observed on active stars are actually composed of many small-scale active region that are smeared out by the lack of high resolution. In this case, the role of CMEs in the evolution of the AMF is weaker for young active stars, the AMF is dominated by the floor value of the stellar ambient field, and the role of CMEs increases over time.

Active young stars have their AMF spiral more compressed with the azimuthal component of the AMF being dominant even at relatively close distances from the stars. Therefore, it is more likely that the radially and fast propagating CME will shock a slower stream of ambient stellar wind moving within the azimuthally tangled magnetic field. This process is similar to shocks driven by interacting solar wind streams known as Co-rotating Interaction Regions (CIRs, see Section 8.5.2, Vol. III). Based on this scenario, many more shocks are expected in the astrospheres of young, active, and fast rotating stars, with consequences for particle transport as discussed in Section 4.1.2. However, as mentioned above, it is possible that CMEs on such stars are launched at very high latitudes, where the AMF is nearly radial.

In that case, the interaction between CMEs and the AMF should be similar to that of the Sun.

4.2.2 The role of CMEs in stellar mass loss and stellar spin down

Over their lifetime, cool stars lose angular momentum and spin down (from rotation periods of less than a day to 20–100 days). The conventional mechanism for stellar spin down is that stars lose angular momentum to the magnetized stellar wind in the concept called "magnetic breaking" (Weber and Davis, 1967). In this process, the mass flux carried by the accelerating stellar wind drains angular momentum as long as the wind speed is below the Alfvén speed, $v_A = B/\sqrt{4\pi\rho}$ (in cgs units of cm s^{-1}), where B is the local magnetic field strength, and ρ is the local mass density. Once the wind speed equals the Alfvén speed at a point called the "Alfvén point", coronal magnetic field lines that are carried and stretched by the wind open up, and all the mass at this point is considered lost from the star. Another way to look at this process is to think of the magnetic field lines as rods that are attached to the spinning star at one end, where the other ends of the open field lines are radially stretched beyond the Alfvén point. As a result, each field line applies torque on the star and spins it down. This torque is proportional to the momentum of the wind at the Alfvén point, to the stellar rotation rate, and to the distance of the Alfvén point (the lever arm that applies the torque). The imaginary surface that represents all the Alfvén points is called the "Alfvén surface" and the integral of the mass flux through this surface is the mass-loss rate, $\dot{M}$, of the star to the stellar wind. For a spherically symmetric wind, and a dipole stellar magnetic field, we can calculate the total torque on the star and the total angular momentum loss rate, $\dot{J}$:

$$\dot{J} = \frac{2}{3}\Omega \dot{M} r_A^2, \tag{4.2}$$

where Ω is the stellar rotation rate, r_A is the average distance to the Alfvén surface, and we assume constant moment of inertia. From Eq. (4.2) we see that the mass-loss rate is necessary to estimate the spin-down rate of a star. However, stellar winds of cool, Sun-like stars are very weak and cannot be directly observed (see previous chapter of this volume), which makes it challenging to estimate $\dot{J}$ as a necessary input for stellar evolution models. It is also important to determine the mass-loss rates of young active stars in the context of the *Faint Young Sun paradox* (see Section 2.3.1 in Vol. III). The paradox arises from the stellar evolution models prediction that the young Sun was about 30% less luminous than current day, so the Earth's surface temperature should have been below the freezing temperature of water. Nevertheless, we find geological evidences for the existence of liquid water on the surface. There are a number of solutions to the paradox, such as the existence of atmospheric greenhouse gases that can increase the surface temperature. In the context of stellar mass-loss rates, a solution for the paradox is possible if we can demonstrate that the young Sun was about 10% more massive in the past, and it had a high mass-loss rate that led to its current mass (Graedel *et al.*, 1991; Wood *et al.*, 2002).

Based on indirect measurements mentioned in the previous chapter, theoretical models (Cranmer and Saar, 2011), and numerical models (see, e.g., Matt *et al.*, 2012; Cohen and Drake, 2014) have shown that mass-loss rates in Sun-like stars seem to fall in the range between 10^{-15}–10^{-11} $M_\odot$ yr^{-1} (the present-day solar mass-loss rate is (2–3) $\times$ 10^{-14} $M_\odot$ yr^{-1}; Cohen, 2011). However, stars can also lose mass via CMEs. In the case of the Sun, each CME carries some 10^{13}–10^{17} g into space (Yashiro and Gopalswamy, 2009), with an annual integrated mass loss via CMEs of several percents of the ambient mass loss (Vourlidas *et al.*, 2010). Therefore, CMEs on the Sun play very little role in the solar mass loss. This role could become significant if the CME rate were higher by a factor of 10 or more. In this case, CMEs could even dominate the stellar mass loss.

In the section above, we discussed the possibility that not every stellar flare is a result of a CME. Nevertheless, stellar flares are still our only indication for CMEs on other stars. Keeping this in mind, let us assume that stellar X-ray flare rate also represents CME rate. In this case, we can investigate the relation between solar CMEs and solar flares, and extrapolate this information to other stars. Both Aarnio *et al.* (2012) and Drake *et al.* (2013) performed quite similar calculations to estimate stellar mass-loss rates due to CMEs; here we follow the formalism by Drake *et al.* (2013).

Based on observations from the Large Angle and Spectrometric Coronagraph Experiment (LASCO) on board the Solar and Heliospheric Observatory (SOHO) mission, and 1–8 Å X-ray data obtained by the GOES satellite, Yashiro and Gopalswamy (2009) and Aarnio *et al.* (2011) obtained a power-law relation for CME mass, m_c, as a function of the CME flare energy, E, in the GOES X-ray bandpass:

$$m_c(E) = \mu E^\beta, \tag{4.3}$$

with $\mu \approx 0.002 - 0.02$ and $\beta \approx 0.6$. Yashiro and Gopalswamy (2009) also found a power-law for the CME kinetic energy, E_k, as a function of the flare energy:

$$E_k(E) = \eta E^\gamma, \tag{4.4}$$

with $\eta \approx 10 - 30$ and $\gamma \approx 1$. Figure 4.10 shows the scatter and fit for the above power laws from Drake *et al.* (2013). The dashed line in the bottom panel represents constant ratio of CME kinetic energy to GOES X-ray energy loss. There is a constant factor of about 200 between the X-ray energy fit (dashed line) and the

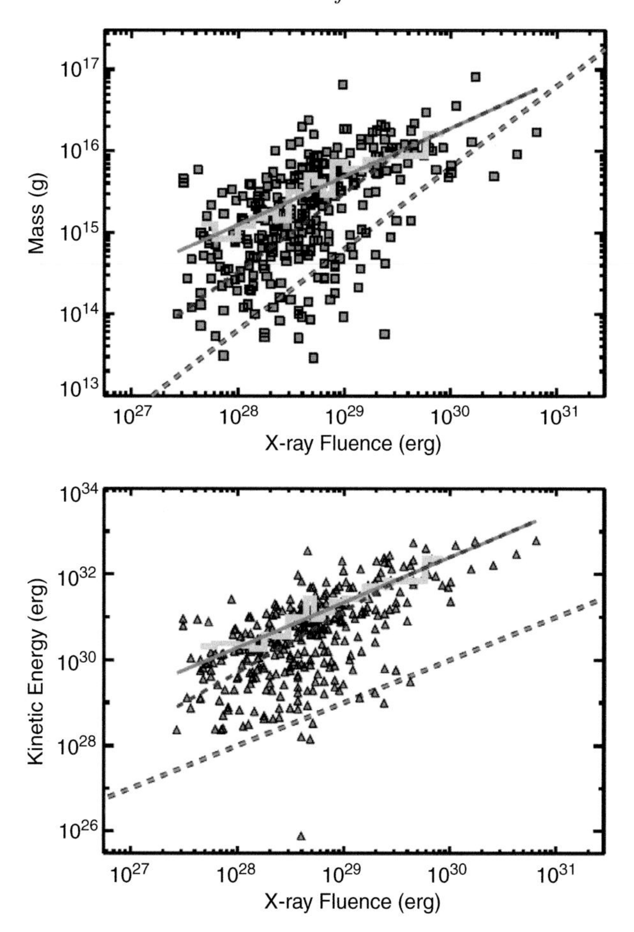

Fig. 4.10 Distribution of solar CME mass (top) and kinetic energy (bottom) as a function of flare energy. The light-gray histograms are the means over 20 data points and the solid lines are linear fits to these means. (From Drake *et al.*, 2013.)

kinetic energy fit (gray line), which suggests that the CME energy release is dominated by the mass ejection itself, and the flare energy as measured in the GOES X-ray pass band represents only a small fraction (about 1%) of the total CME energy (Schrijver *et al.*, 2012).

Solar-flare observations also reveal a power-law relation between the occurrence rate of CMEs and the associated flare energy (Drake *et al.*, 2013):

$$\frac{dn}{dE} = kE^{-\alpha}. \tag{4.5}$$

The index α is found to be between 1.5–2.5 for all stellar types and k is a normalization factor. The total flare power, P, can be obtained from the following integral:

$$P = \int E dN = \int E \frac{dN}{dE} dE = \int_{E_{min}}^{E_{max}} E k E^{-\alpha} dE = \frac{k}{2-\alpha}\left[E_{max}^{2-\alpha} - E_{min}^{2-\alpha}\right]. \tag{4.6}$$

Because for very active stars the total power in flares is assumed to dominate coronal emission, the total power P should equal the total available X-ray flux (e.g., the total X-ray luminosity, L_x), we find that the normalization constant is:

$$k = \frac{L_x(2-\alpha)}{E_{max}^{2-\alpha} - E_{min}^{2-\alpha}}. \tag{4.7}$$

Another bit of information we obtain from Yashiro and Gopalswamy (2009) is the association fraction as a function of X-ray energy, $f(E)$. This function tells us what the probability is that a CME actually erupts for a given flare energy (not every solar flare is associated with a CME: the more energetic the flare, the more likely it is there is a CME associated with it). In general, $f(E)$ can be expressed as a power-law:

$$f(E) = \zeta E^{\delta}, \tag{4.8}$$

where $\zeta = 7.9 \times 10^{-12}$, and $\delta = 0.37$ for $E \leq 3.5 \times 10^{29}$ erg, and $f(E) = 1$ for X-ray total energies higher than 3.5×10^{29} (every flare above this energy is associated with a CME). The total mass-loss rate can then be estimated by the following integral:

$$\dot{M}_{\mathrm{CME}} = \int_{E_{min}}^{E_{max}} m_c(E) f(E) \frac{dn}{dE} dE. \tag{4.9}$$

Combining Eqs. (4.2), (4.5), (4.7), and (4.8), we obtain an expression for the stellar mass-loss rate to CMEs:

$$\dot{M}_{\mathrm{CME}} = \mu \zeta L_x \left(\frac{2-\alpha}{1+\beta+\delta-\alpha}\right)\left[\frac{E_{max}^{1+\beta+\delta-\alpha} - E_{min}^{1+\beta+\delta-\alpha}}{E_{max}^{2-\alpha} - E_{min}^{2-\alpha}}\right]. \tag{4.10}$$

Figure 4.11 shows the range of mass-loss rates due to CMEs from Drake *et al.* (2013). It can be seen that this range is for very high mass-loss rates of $(2-4) \times 10^{-10}$ $M_\odot$ yr^{-1}. Aarnio *et al.* (2012) found similar high mass-loss rates in the range of 10^{-11}–10^{-9} $M_\odot$ yr^{-1}. These mass-loss rates are higher than the upper limit estimated for the ambient stellar wind, while the associated CME energy can reach a tenth of the total bolometric energy. Therefore, while it is possible that mass-loss processes in very active stars are dominated by CMEs, it is more likely that the solar CME-flare relation used here breaks down for higher flare energies. Drake *et al.* (2013) concluded that a more reasonable value for the associated CME energy is 1% of the total bolometric energy, which corresponds to mass-loss rate value of 5×10^{-11} $M_\odot$ yr^{-1}.

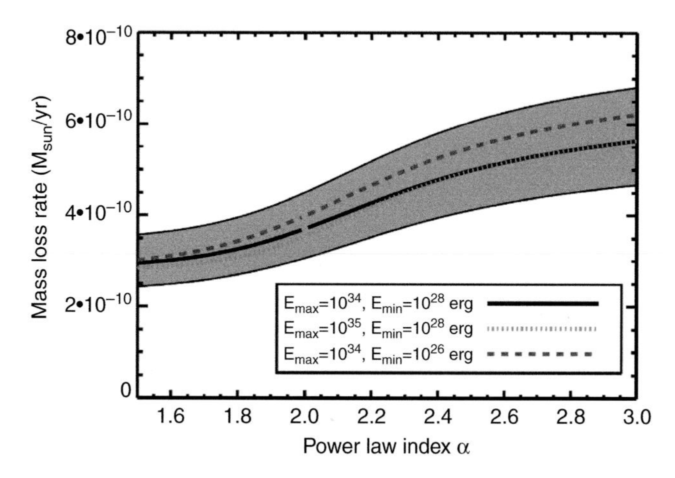

Fig. 4.11 The expected mass-loss rate due to CMEs as a function of the power index α for $L_x = 10^{30}$ erg s^{-1} and for different maximum and minimum event energies. The gray area represents the uncertainty in the other power-law indices. (From Drake *et al.*, 2013.)

Aarnio *et al.* (2012) also estimated the torque on T-Tauri Stars (TTS) as a result of stellar mass loss to CMEs. They used the same method used to calculate the stellar spin-down rate due to the mass-loss to the ambient wind (Matt and Pudritz, 2008; Matt *et al.*, 2012), but replaced the mass-loss rate with that of CMEs. They found that the torque on the star due to CMEs, τ is:

$$\tau = k^2 \left(\frac{M_\star}{\dot{M}_{\mathrm{CME}}} \right) \left(\frac{R_\star}{r_A} \right), \tag{4.11}$$

where $M_\star$ is the stellar mass, $R_\star$ is the stellar radius, r_A is the average distance to the Alfvén point, and k is a constant. They have estimated that the Alfvén radius for TTS could range between 3 $R_\star$ and 70 $R_\star$, and that the torque could be efficient in spinning down stars when considering an upper limit mass-loss rate due to CMEs of $\dot{M}_{\mathrm{CME}} > 10^{-10}\ M_\odot\ \mathrm{yr}^{-1}$.

In the context of the Faint Young Sun paradox, it is interesting to mention a scenario at which the CME rate is very high for young, fast-rotating, active stars. If the stellar magnetic activity in such stars is concentrated at high latitudes, and the CMEs are launched at these latitudes, then we can have an efficient way to remove mass from the star without spinning the star down quickly (the torque applied on the star has a latitudinal dependence that goes to zero above and below the stellar poles). Therefore, the star can maintain its high level of activity and high CME rate, while losing a large amount of mass. This way, we may be able to demonstrate that the young Sun has been more massive in the past and solve the paradox. While the continuous mass loss cannot be high enough, transient mass-loss rate scenarios, such as the one presented here may be sufficiently high. This is an active topic that is currently studied both theoretically and observationally using the Chandra and XMM-Newton X-ray observatories, as well as helioseismic and (*Kepler* mission) asteroseismic data.

4.3 Coronal mass ejections and close-in exoplanets

Since the mid 1990s, and through the era of the *Kepler* mission, hundreds of exoplanets have been discovered. Many of these planets are so-called "hot jupiters" – gas giant planets that orbit their parent star within a distance of less than one tenth of the Sun–Earth distance (0.1 AU). Additionally, the current search for habitable, Earth-like, rocky exoplanets is focused on planets orbiting M-dwarf stars, which are the most common in the Universe, and are very faint so their habitable zone (see Ch. 4, Vol. III) is located very close to the star, close enough that these planets can be detected by the current techniques.

Many interesting processes can arise from the close proximity of a planet to its parent star. Particularly, if the planet is magnetized, and it resides within the Alfvén point of the stellar corona. In this case, interaction between the planet and the star/stellar corona may be possible (known as star–planet interaction, SPI). Shkolnik *et al.* (2003, 2005a,b, 2008) observed an increase in coronal activity in the Ca II K line attributed to SPI in several planetary systems. There is also growing evidence that stars harboring close-in planets have excess angular momentum (Pont, 2009; Lanza, 2010), X-ray activity (Kashyap *et al.*, 2008), and EUV activity (Shkolnik, 2013). In other words, stars with close-in planets rotate faster than they should for their age, and they are also more active for their age. The excess in angular momentum can be due to tidal interaction between the star and the planet, which spins up the star, or due to a reduction in stellar magnetic breaking, because the planet and its magnetosphere serve as an obstacle in the stellar corona, so the stellar spin down decreases. These findings are very important for stellar activity evolution, because they shuffle the common rotation–age–activity relation which has been used for many years. In contrast, the observational evidence for SPI is still debated. For example, Miller *et al.* (2015) performed a statistical study to test the hypothesis that planets can boost their host's activity and found that this behavior is biased to planets that are both very massive and are extremely close to the host star, where most other cases did not show consistent increase in activity. This means that the interaction may be dominated by tidal effects and not magnetic SPI.

While all the aspects above involve very interesting plasma physics processes, in this section we focus on the unique features of the interaction between CMEs and close-in planets, and the resulting effects on both the planets and the CMEs.

4.3.1 The impact of CMEs on close-in exoplanets

Because of their close proximity to the stars, close-in planets can be eroded by CMEs and lose a significant fraction of their atmospheres (it is possible that CMEs have played a role in the loss of the Martian atmosphere; see Ch. 7 in this volume). In order to sustain an atmosphere, a planet should have a strong internal force to resist the stripping force of the CME, i.e., a strong internal magnetic field. Alternatively, it needs a thick atmosphere so that it can survive longer. While the interaction between CMEs and planets in the solar system has been studied extensively and in detail, it is not clear how CMEs impact close-in planets due to two factors. First, it is not clear how CME properties and frequencies scale with stellar properties. Second, it is hard to predict whether close-in planets would have a strong or weak internal magnetic field; there are conflicting arguments for and against each of the options (see some discussion on planetary dynamos in Ch. 6 in this volume).

Khodachenko *et al.* (2007) and Lammer *et al.* (2007) have estimated the planetary magnetic field necessary to protect the atmospheres of planets located at distances of less than 0.2 AU from erosion by CMEs. The CME density was scaled to close-in orbits based on statistical characterization of solar CMEs with possible minimum and maximum extremes as follows:

$$n^{min}_{eject} = n^{min}_0 \left(\frac{d}{d_0}\right)^{-2.3}, \tag{4.12}$$

$$n^{max}_{eject} = n^{max}_0 \left(\frac{d}{d_0}\right)^{-3.0}, \tag{4.13}$$

with $n^{min}_0 = 4.88$ cm^{-3}, and $n^{max}_0 = 7.0$ cm^{-3}. Here d is the orbital distance and d_0 and n_0 are the distance and density of the CME at the point of eruption. They then calculated the magnetosphere standoff distance, $R_{\rm mp}$, described by the balance between the planetary magnetic pressure and the ram pressure of the CME (cf., Ch. 10 in Vol. I), $P_{\rm CME} = n_{\rm CME}\, m v^2_{\rm CME}$. Here $n_{\rm CME}$ ranges between n^{min}_{eject} and n^{max}_{eject}, m is the average mass of the CME particles (taken here to be the proton mass), and $v_{\rm CME} = 500$ km s^{-1} is the CME speed. The expression for $R_{\rm mp}$ is then (Vol. I, Eq. (10.1)):

$$R_{\rm mp} = \left(\frac{\mu_0 f_0^2 M^2}{8\pi^2 P_{\rm CME}}\right)^{1/6}, \tag{4.14}$$

with μ_0 being the magnetic permeability, $f_0 = 1.16$ is a numerical factor that accounts for the non-spherical shape of the magnetosphere, and M is the planetary magnetic moment. Based on these scaling laws, and estimations of the strength of the planetary magnetic moment for a given stellar mass and orbital separation, Khodachenko *et al.* (2007) have estimated the range at which the planetary magnetopause is far enough to protect the planetary atmosphere from erosion by

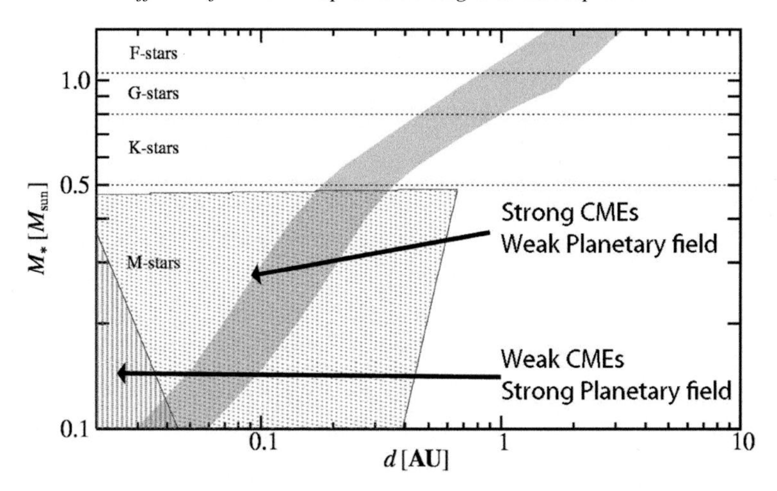

Fig. 4.12 Comparison between the habitable zone (HZ; shaded area) and the areas where strong magnetospheric compression is possible by CMEs (lightly and heavily dotted areas). The lightly dotted area indicates Earth-like exoplanets with a minimum value of the magnetic moment exposed to strong (dense) CMEs. This area denotes the region where CMEs compress the magnetosphere down to 1.15 Earth radii or less (i.e., 1000 km above the planetary surface). The heavily dotted area indicates Earth-like exoplanets with a maximum value of the magnetic moment exposed to weak (sparse) CMEs. In this region, CMEs compress the magnetosphere to less than 2 Earth radii. (From Khodachenko *et al.*, 2007.)

CMEs. Figure 4.12 shows a plot for stellar masses as a function of the orbital distance. It shows the habitable zone as shaded area, and it also shows that CMEs can erode the planetary atmosphere across a significant fraction of this area. Overall, both Khodachenko *et al.* (2007) and Lammer *et al.* (2007) concluded that planets located at an orbital distance of 0.2 AU or less would lose a significant fraction of their atmosphere unless they have a significant internal magnetic field.

Cohen *et al.* (2011) performed a numerical simulation of a CME event hitting a close-in exoplanet in order to study atmospheric protection by the planetary magnetic field for a range of field strengths. They used an MHD model that is used to simulate solar space weather events and simulated the extra-solar CME in the same manner that CMEs on the Sun are obtained. The parameters for the CME were selected based on the parameters of the May 2005 real solar CME event (a typical solar CME), where the planet was embedded in the simulation domain. Figure 4.13 shows the calculated penetration of the CME estimated by the mass flux through three spheres around the planet at distances of 0.5, 1, and 2 planetary radii above the surface. A negative value of the mass flux means that the CME has penetrated the sphere while positive flux means that the CME has not reached that height (this is a planetary outflow flux). It can be seen that for a field strength of 0.5 G (slightly larger than the Earth's magnetic field), the CME strongly penetrates at 2 planetary radii, but at 1 planetary radii or below, there is no penetration. For

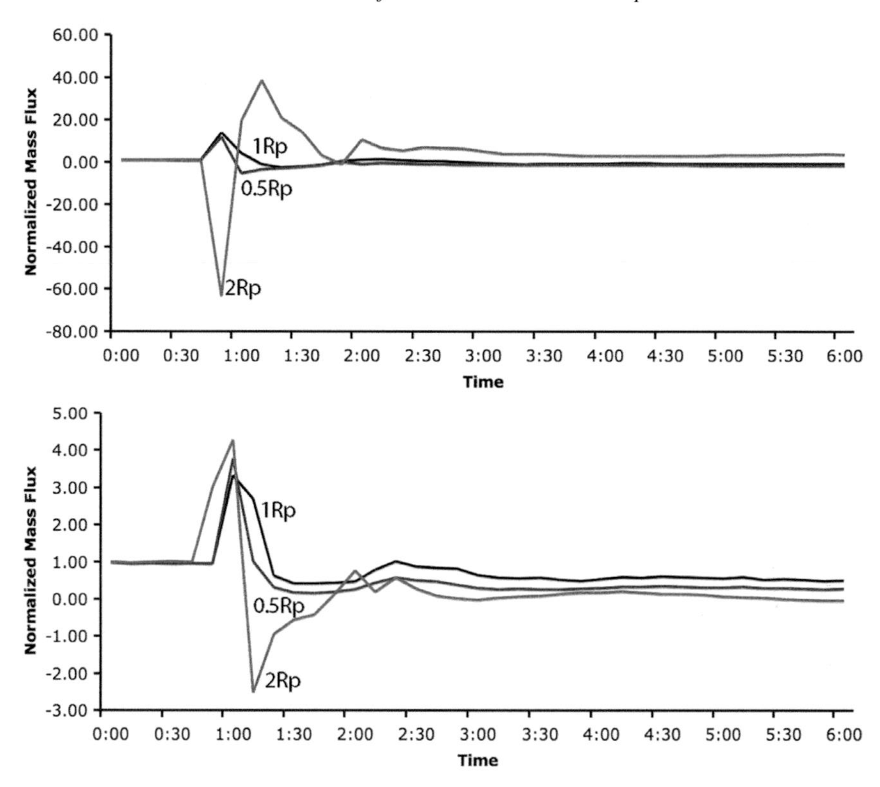

Fig. 4.13 Mass flux for a CME simulation integrated over three spheres around the impacted planet at heights of 0.5, 1, and 2 planetary radii above the surface, shown as a function of time. Fluxes are normalized to the value of the initial state at $t = 0$ (which is positive). The top panel shows results for planetary field strength of 0.5 G, and the bottom panel shows results for planetary field strength of 1 G. (From Cohen *et al.*, 2011.)

a stronger planetary field strength of 1 G, the CME barely penetrates even to 2 planetary radii above the surface.

Another interesting aspect that was noted by Cohen *et al.* (2011) is the change of magnetospheric orientation during the CME event, as shown in Fig. 4.14. Owing to their fast orbital motion, close-in magnetized planets may have their magnetotails stretched in the azimuthal direction (similar to a cometary tail). Therefore, the general orientation of the planetary magnetosphere is tilted by 45–90 degrees with respect to the radially flowing stellar wind. However, once the CME hits the planet, the magnetosphere is rotating to be radially aligned with the direction of the CME propagation trajectory. This rotation of the whole magnetosphere within a time period of less than an hour can have implications of induced currents and deposition of energy to the planetary upper atmosphere. Based on the numerical simulation, Cohen *et al.* (2011) estimated that the energy deposited onto the planet in such an event is about a thousand times higher than the energy deposited to the Earth in a typical CME event. Therefore, the interaction of CMEs with close-in planets

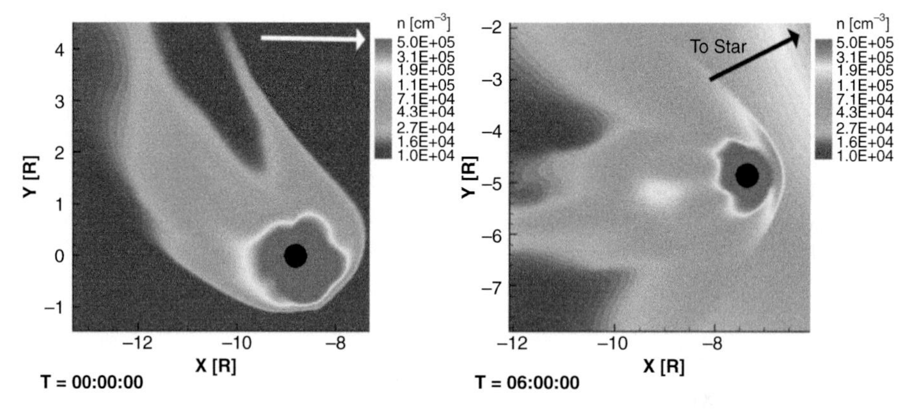

Fig. 4.14 Renderings of the number density around a close-in exoplanet shown on the equatorial plane for the initial, pre-eruption state (left), and during the CME event, 6 h after the eruption. The plot is in the Astrocentric coordinate system at which the star is located at the origin of the coordinate system. (From Cohen *et al.*, 2011.)

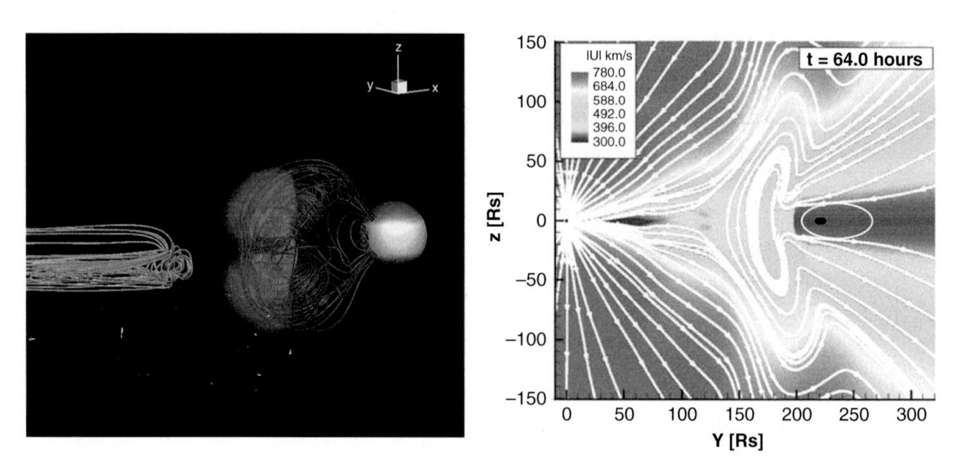

Fig. 4.15 Left: a CME approaching a planet. The star is shown on the right with selected CME field lines. The shaded volume represents an iso-surface for speed of 1500 km s^{-1}. The planet is shown as a small sphere with magnetospheric field lines shown as well. Right: meridional cut shows contours of speed between the Sun and the Earth with a small black ellipse representing the Earth's magnetosphere. (Left-hand image from Cohen *et al.*, 2011; right-hand image from Manchester *et al.*, 2004.)

could be very violent, where a significant amount of the planetary magnetosphere is stripped by the CME.

4.3.2 The effect of close-in planets on CME evolution

Another unique feature of CME–planet interaction in close-in planets is the impact on the CMEs themselves. Let us look at Fig. 4.15. On the left, we see the CME

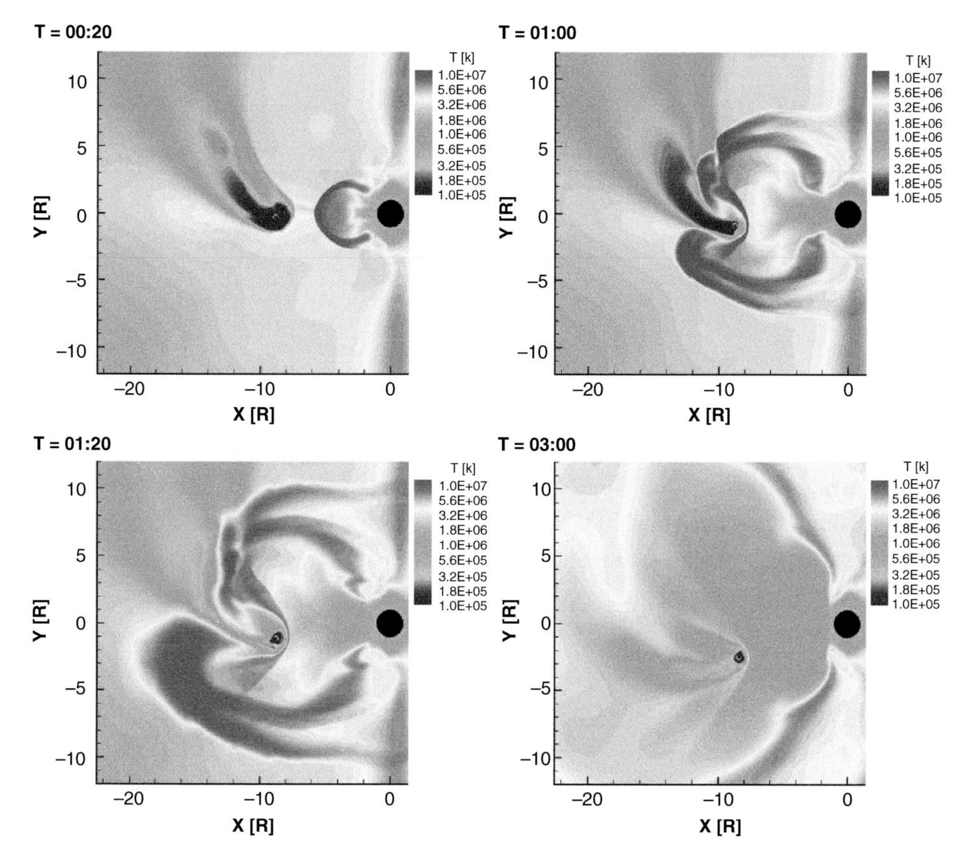

Fig. 4.16 Contours of the temperature displayed on the equatorial plane during a CME event at a close-in planet for four phases of the interaction. Coordinates are shown in units of stellar radius. (From Cohen *et al.*, 2011.)

approaching the magnetosphere of the close-in planet based on the simulation by Cohen *et al.* (2011). On the right, we see a meridional cut in the space between the Sun and the Earth taken from a simulation by Manchester *et al.* (2004). The plot shows a CME that was launched from the Sun approaching the Earth, where the small black ellipse represents the Earth's magnetosphere. Figure 4.15 demonstrates the difference in scales between the CME and the magnetosphere. By the time a CME reaches 1 AU, it is so much bigger than the Earth's magnetosphere that the interaction between them affects only the Earth. However, on close-in planets, the CME and the magnetosphere are comparable in size. As a result, the CME itself is affected by the interaction and it breaks in the middle as shown in Fig. 4.16. Because close-in planets orbit their host stars in periods of only a few days, the chance of a CME to hit a planet is rather high. If this is the case, it is possible that the interaction of CMEs and the close-in planets accelerates the dissipation of the CME as it moves into the outer astrosphere.

5

Characteristics of planetary systems

DEBRA FISCHER AND JI WANG

Philosophical musings that other worlds might exist date back more than 2000 years to the ancient Greeks. We live in a fortunate time, when the discovery of exoplanets has the potential to address questions about how planetary systems form and evolve. In what ways do exoplanetary systems mirror our solar system? How are they different? Does the presence of a binary star affect planet formation? Are Earth analogs common? Does the energy from other stars give rise to life?

Confirmed and candidate exoplanets number in the thousands and search techniques include Doppler measurements, transit photometry, microlensing, direct imaging, and astrometry. Each detection technique has some type of observational incompleteness that imposes a biased view of the underlying population of exoplanets. In some cases, statistical corrections can be applied. For example, transiting planets can only be observed if the orbital inclination is smaller than a few degrees from an edge-on configuration. However, with the reasonable assumption of randomly oriented orbits, a geometrical correction can be applied to determine the occurrence rate for all orbital inclinations. In other cases, there is simply no information about the underlying population and it is not possible to apply a meaningful correction. For example, the number of planets with a similar mass (or radius) and a similar intensity of intercepted stellar flux as our Earth is not secure at this time because the number of confirmed detections for this type of planet is vanishingly small.

As a result of the sample biases and observational incompleteness for each discovery technique, our view of exoplanet architectures is fuzzy at best. There are no cases beyond the solar system where the entire parameter space for orbiting planets has been observed. Instead, we piece together an understanding of exoplanet architectures by counting planets in the regimes where techniques are robust and then we estimate correction factors when possible. When drawing conclusions about the statistics of exoplanets, it is helpful to understand completeness in this underlying patchwork of orbital parameter space.

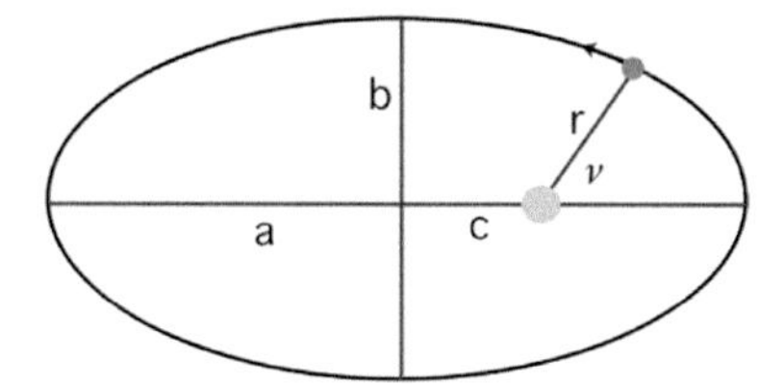

Fig. 5.1 Geometry of an elliptical orbit with semi-major axis a, semi-minor axis b, and eccentricity $e = c/a$.

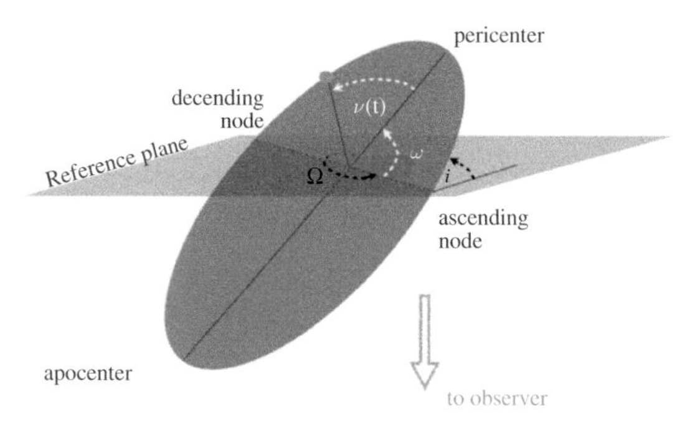

Fig. 5.2 Orbital angles i, ω, and Ω define the orientation of the orbit with respect to the plane of the sky.

We begin by reviewing the exoplanet detection techniques with particular consideration of the observational biases and then discuss the implications for planet formation with an eye toward how our solar system compares.

5.1 Overview of Keplerian orbits

The motion of planets around the Sun was famously deciphered by Johannes Kepler. Kepler's first law states that planets orbit in an ellipse with the star at the focus; the second law is a statement about conservation of angular momentum – planets sweep out equal areas in equal time intervals; the third law says that the square of the orbital period is proportional to the cube of the semi-major axis. Kepler's laws were later generalized by Isaac Newton in his universal law of gravitation.

The fundamental plane of an elliptical orbit is shown in Fig. 5.1 with the star at the focus of the ellipse. The ellipse is parametrized by the semi-major axis a, and the semi-minor axis b. The orbital eccentricity is defined as the ratio of c/a, and the planet sweeps out an angle ν, which is referenced to the point of periastron.

In practice, the orbital plane is randomly oriented in space. Figure 5.2 shows the angles that define the orientation of the fundamental elliptical orbit with respect to the plane of the sky. First, imagine that the plane of the sky (the reference plane) passes through the star-centered focus of the ellipse; the intersection of these two planes is called the line of nodes. The orientation of periastron passage is defined by ω; $\omega = 90°$ or $270°$ if we are looking along the long semi-major axis of the orbit. The inclination, i, references the tilt of the orbit and is defined so that $i = 90°$ when the orbit is viewed edge-on and $i = 0°$ when viewed face-on. For the special case of a circular orbit, $e = 0$ and there is no periastron point, so ω is undefined for circular orbits. The third angle, Ω, is a rotation perpendicular to the plane of the sky. This last angle is not relevant for Doppler observations (because it does not change the radial component of the velocity) or for transit observations; it can only be measured with direct imaging or astrometric techniques.

Although the planetary orbits have been shown in the reference frame of the star, the star and planet actually orbit the center of mass (COM).

5.2 Doppler surveys for exoplanets

Doppler surveys have detected more than 500 planets and this was the first successful technique for detecting planets outside our solar system. This technique is unique in providing masses for exoplanets, modulo the generally unknown orbital inclination. The first detected planets were gas giants that orbit close to their host stars. This turned out to be a bias of this technique: close-in gas giants exert the largest possible gravitational force on the host star and produce the most significant stellar reflex velocities. While improvements in this technique have permitted the detection of one planet with a mass similar to the Earth, this technique has severe incompleteness with decreasing mass and increasing orbital periods.

5.2.1 The Doppler effect

Owing to the Doppler effect, spectral lines from the stellar atmosphere are periodically blue-shifted and then red-shifted as the star orbits the COM over one orbital period. Only the velocity component along the line of sight (the radial velocity) between the observer and the source produces a Doppler shift.

An emitted photon of wavelength λ_0 will be Doppler shifted to a new wavelength λ as described by the theory of special relativity (Einstein, 1905). The reflex velocities that planets induce in their host stars are typically small. For example, the reflex solar velocity from Jupiter is $\sim 12\ \mathrm{m\,s^{-1}}$ and the tug of the Earth on the Sun is a mere $0.1\ \mathrm{m\,s^{-1}}$. The Doppler shift can safely be expressed in the non-relativistic form without incurring any measurable errors:

$$\lambda = \lambda_0 \left(1 + \frac{v}{c}\right). \tag{5.1}$$

When calculating the velocity of a source star, the velocity of the observer (the "barycentric velocity") must be subtracted from the measured radial velocity in order to recover the velocity of the source. The Jet Propulsion Lab's HORIZONS[1] ephemeris system provides the velocity of the Earth about the solar barycenter (including gravitational effects from planets and moons) with an impressive precision of about one millimeter per second.

The Doppler technique was the first method to detect planets around other stars. Latham *et al.* (1989) used Doppler velocity measurements to detect the first substellar mass object orbiting the star HD 114762. They interpreted this object as a likely brown dwarf since they derived $M \sin i \sim 12 M_{Jup}$ (Jupiter masses) for the companion. The object resides at the mass boundary between planets and brown dwarfs, however, the unknown inclination likely means that the true mass is in the brown dwarf regime. Planetary mass objects were found orbiting the neutron star PSR 1257+12 by Wolszczan and Frail (1992). This rapidly spinning neutron star was serendipitously oriented so that a narrow synchrotron beam swept across the solar system like a beam from a lighthouse. Careful monitoring of the pulsar timing permitted the detection of three planets that were just a few times the mass of the Earth. More than 20 years later, the precision of pulsar timing measurements still exceeds the precision that has been achieved with other Doppler techniques. A few years later, Mayor and Queloz (1995) detected the first planet around a Sun-like star. This was the beginning of an era of successful Doppler planet surveys (see the review by Fischer *et al.*, 2014).

The stellar radial velocity semi-amplitude K_*, can be expressed in units of $\mathrm{cm\,s^{-1}}$ as a function of the orbital eccentricity e, the orbital period P (in years), the combined stellar and planetary mass (in solar mass units), and the planet mass $M_P \sin i$ (in units of Earth masses, $M_\oplus$):

$$K_* = \frac{8.95\,\mathrm{cm\,s^{-1}}}{\sqrt{1-e^2}} \frac{M_P \sin\, i}{M_\oplus} \left(\frac{M_* + M_P}{M_\odot}\right)^{-2/3} \left(\frac{P}{\mathrm{yr}}\right)^{-1/3}. \tag{5.2}$$

Because only the projected line-of-sight velocity is measured, the inferred planet mass from Doppler measurements is $M_P \sin i$, the product of the true planet mass and the sine of the orbital inclination; the true mass of the planet cannot be determined. However, in a statistical sense the probability that the orbital inclination is within a particular range $i_1 < i < i_2$ is given by:

$$\mathcal{P}_{incl} = |\cos(i_2) - \cos(i_1)|. \tag{5.3}$$

Thus, there is an 87% probability that orbital inclinations lie between 30° and 90°, implying that the true planet mass is statistically within a factor of two of $M_P \sin i$ for the vast majority of Doppler-detected exoplanets.

[1] http://ssd.jpl.nasa.gov/?ephemerides

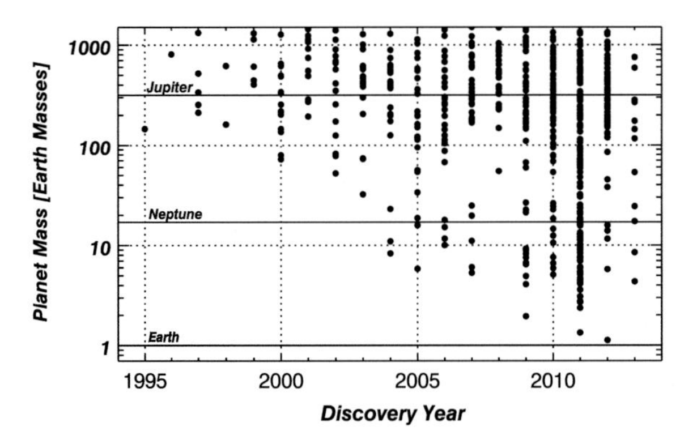

Fig. 5.3 Detection of exoplanets over time. Black dots indicate Doppler detections. Solid horizontal lines indicate the masses of Jupiter, Neptune and Earth.

In order to derive the orbital period and other orbital parameters, the radial velocity observations must span at least one complete orbit. Assuming circular orbits and a K_* corresponding to a given velocity precision (say, 1 $\mathrm{m\,s^{-1}}$), the minimum detectable mass can be calculated over a range of orbital periods to determine the threshold for the minimum detectable planet mass for the Doppler technique.

Figure 5.3 shows a time line of exoplanet detections; clearly the community has been addressing the technical challenges and improving the Doppler measurement precision so that the minimum detectable planet mass has been dropping over time. The data point close to one Earth mass indicates the companion to α Cen B (Dumusque *et al.*, 2012) with an orbital period of 3.24 days. The question now is whether we can further improve the Doppler precision so that Earth analogs in habitable zone orbits can be discovered around nearby stars.

5.2.2 Current limitations to Doppler precision

The measurement of Doppler shifts induced by orbiting planets is technically challenging. In order to reach a Doppler precision of 1 $\mathrm{m\,s^{-1}}$ the wavelength must be known with a relative precision of at least 10^{-9}, a nontrivial requirement. A Doppler shift corresponding to 1 $\mathrm{m\,s^{-1}}$ typically moves the stellar lines by less than 1/1000th of a pixel on a CCD detector. However, the spectrum will move on the detector for other reasons, too. Variations in the temperature, pressure, or mechanical flexures can shift the stellar spectrum by more than a pixel. Time-varying imperfections in the CCD can compromise the measurement of wavelength shifts. Contamination from a Moon-lit sky or a background star can induce spurious velocity signals. In short, everything matters and all of these issues must controlled or tracked.

5.2.3 Stellar noise for Doppler techniques

Even if the technical and engineering challenges are perfectly managed, another threat to precise Doppler measurements remains outside of our control: coherent velocity flows on the surface of the star. Any technique that relies on observations of the host star to detect the unseen planet (Doppler, transits, astrometry, but not direct imaging or microlensing) will be affected by signals arising from the stellar photosphere. Planet hunters often refer to these signals as stellar noise. Of course, the stellar noise is signal to our colleagues who study the Sun and other stars; these signals include p-mode oscillations and features that are correlated with time-variable magnetic fields: variability in granulation, starspots, meridional flows (see chapters on the Sun and its magnetic activity listed in Table 1.2). The magnitude of these variable velocities can be hundreds of meters per second, making them important even if they are diluted by the integrated stellar flux.

In the case of p-mode oscillations, the variability has a typical period of a few minutes for stars like the Sun. The amplitude of the radial velocity signal from p-modes depends on whether the pressure modes are in resonance and is also a function of the spectral type of the star. However, radial velocities from typical p-mode signals usually have an amplitude that is no more than a few meters per second. By taking long or multiple exposures, it is possible to average over p-mode oscillations. From the perspective of the planet hunters, this high frequency contribution to errors is the least serious of the potential stellar noise sources.

Convective granules are a more significant source of stellar noise. In principle, what goes up must come down; however, the intensity of the hot upward flows is greater than the cool downward flows. This produces asymmetry in the spectral line. If the granulation were in a steady state, that asymmetry would not matter. However, magnetic fields cause a local suppression of granulation. Because the magnetic fields are time-variable, the granulation flows are variable and the spectral line profile will be time-variable. With high enough signal-to-noise and high enough spectral resolution, it may one day be possible to distinguish between the effect of stellar noise and Doppler shifts associated with the bulk motion of stars. Current instruments do not have that resolution, so the Doppler analysis code interprets this as a shift in the line centroid over time – i.e., a spurious net stellar bodily Doppler shift. The time scale for convective flows ranges from several minutes to a couple of hours, but the magnitude of this effect is difficult to assess.

Cool spots in the stellar photosphere also cause spectral line profile variations. As a spot emerges from behind the stellar disk as the star rotates, it blocks out light from the approaching limb of the star and the Doppler-broadened spectral line has less light in the blue wing. Later, the spot moves across the rotating star, and blocks light from the receding edge of the star; now the spectral line profiles all have less intensity in the red wing. The Doppler code interprets these line profile variations as a net red shift followed by a net blue shift with a periodicity that matches the rotation period of the star. The rotation period of stars on Doppler planet surveys is uncomfortably close to the orbital periods of planets that we want to detect; this has led to confusion in the interpretation of data on more than one occasion. The spot signal is further complicated because it attenuates over a few rotation cycles and differential rotation and spot migration produces spots with slightly varying periods.

Longer-term magnetic activity variations, comparable to the solar cycle, have also been correlated with radial velocity variations. All of the above issues only represent “the devil that we know”. There are additional noise sources and velocity flows in stars that are less well understood, such as meridional flows. Current instruments do not have the ability to resolve most of the photospheric noise from Doppler shifts. Without new instrument designs and analysis techniques that have the ability to detect photospheric velocities, the Doppler technique will be limited to a precision of about 1 $\mathrm{m\,s^{-1}}$.

5.3 Transit technique

In the lucky case where the orbit of a planet takes it along a path that crosses our line of sight to the star, the planet will block out a fraction of the stellar flux. The decrease in brightness scales with the ratio of the cross sectional area of the planet to the star (see Fig 5.4, reproduced from a review by Winn, 2011). Thus, if we can measure or estimate the radius of the star, we can easily calculate the size of the transiting planet.

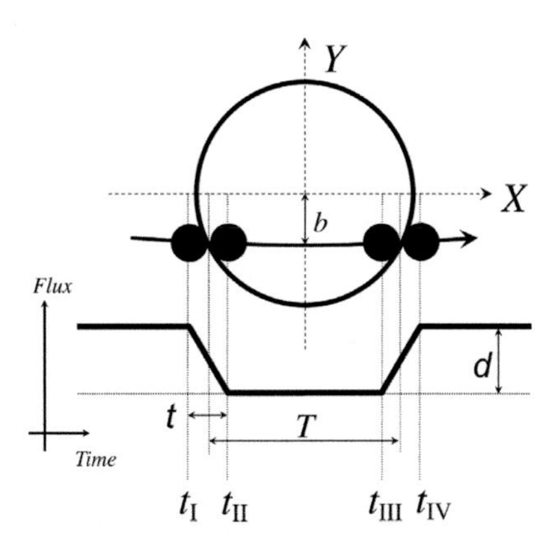

Fig. 5.4 The sketch of a transit light curve shows that the measured flux from the star begins to decrease during ingress. The flux is at a minimum after the planet has completed ingress and before the planet begins egress. After egress, the measured flux returns to the pre-transit value. Reproduced with permission from Winn (2011).

Transit observations uniquely provide a measurement of the radius of exoplanets for cases where the stellar radius is known. Ground-based transit observations only have the precision to detect gas giant planets; however, the spaced-based *Kepler* mission has detected thousands of planet candidates and confirmed planets with radii as small as the Earth. The real bonus comes when transit and radial velocity measurements can be combined to calculate an average density for exoplanets with masses comparable to or greater than Neptune. The technique is limited to a narrow range of essentially edge-on orbital configurations (inclinations close to 90°) and to relatively short orbital periods (up to about 1 year). However, geometrical correction can be made to deduce the statistics of these planets. Thanks to the *NASA Kepler* mission, we now know that small rocky planets are far more common than gas giants.

The first transiting planet was detected around the Sun-like star HD 209458 (Henry *et al.*, 2000; Charbonneau *et al.*, 2000). In this case, the photometric monitoring of the star began after the planet was first discovered by the Doppler technique. Although the inclination was unresolved by the radial velocity measurements, the short period of this planet meant that the transit probability was about 10% (see Eq. (5.4)) and the other orbital parameters derived with Doppler data were used to predict the putative transit time. Because both the size and the mass of the planet were known (in the case of transiting planets, we know the inclination so the Doppler measurements yield a true mass for the planet, not just $M \sin i$), the mean density of the planet was easily calculated. Density is a powerful characterization parameter that reveals information about the internal structure and atmospheres of exoplanets.

There are also programs that carry out nearly continuous photometric monitoring with the hope of a serendipitous transit observation. The HAT-NET (Bakos *et al.*, 2007), MEarth (Charbonneau *et al.*, 2009), and the XO Project (McCullough *et al.*, 2005) are examples of ground-based transit surveys. Examples of space-based missions that have been used to search for transiting planets include the Hubble Space Telescope (HST; Brown *et al.*, 2001), Spitzer (Knutson *et al.*, 2007a), CoRoT (Deleuil *et al.*, 2000) and *Kepler* (Borucki *et al.*, 2003).

In order for the planet to transit, the impact parameter (b in Fig. 5.4) must be less than unity, which corresponds to the angular radius of the star. In practice, this is a requirement for nearly edge-on inclinations; as shown in Fig. 5.5, most of the orbital inclinations for the planet candidates from the NASA *Kepler* mission are indeed between 87° and 90°. The ratio of the stellar radius to the semi-major axis a, or R_*/a, is also of fundamental importance in the geometry for transits. The probability that a given planet will transit is given by the following expression:

$$\mathcal{P}_{tr} = 0.0045 \left(\frac{\mathrm{AU}}{a}\right) \left(\frac{R_\star + R_\mathrm{p}}{R_\odot}\right) \left[\frac{1 + e\cos(\pi/2 - \omega)}{1 - e^2}\right]. \tag{5.4}$$

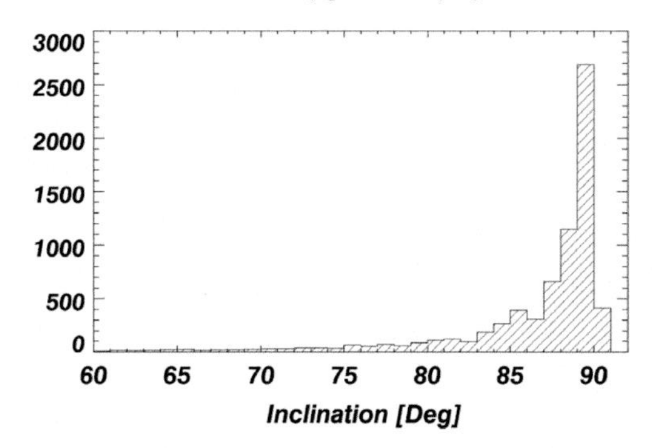

Fig. 5.5 The distribution of inclinations for the *Kepler* transiting planet candidates are highly biased toward edge-on ($i = 90°$) configurations (drawn from the list of KOIs in the NASA Exoplanet Archive).

This relation can be used to back out a geometrical correction for planet occurrence rates determined by transits. To appreciate the observational detection biases for the transit technique, it helps to assume circular orbits so that the last term in Eq. (5.4) reduces to unity. Then it is clear that for a given size of star, planets that have small semi-major axes and large radii are most easily detected.

Transiting gas giant planets uniquely permit studies of gas giant atmospheres. Although exoplanets cannot be spatially resolved from the star, it is possible to obtain a faint transmission spectrum of the exoplanet atmosphere, generally with low-resolution spectroscopy. Most transmission "spectra" are really spectrophotometric observations, obtained with broadband photometry (e.g., J, H, and K bands). The game plan is to obtain a transit light curve in each bandpass. Before the beginning of the transit, the only flux contribution is from the star. During transit, the flux in each bandpass is a combination of the flux from the star and the transmission spectrum of the planet. The light curve is then modeled in each bandpass, fitting for the ratio of the radius of the planet to the radius of the star and for limb-darkening (discussed below). This model of the planet radius provides three points for the planet's transmission spectrum, in the broad J, H, K bandpasses. Extracting spectral information based on three (very low resolution) points is challenging; as a result, the detection of molecules such as H_2O, CO_2, and CH_4 and assessments about the thermal structure of atmospheres in hot and warm Jupiters and Neptunes can be controversial (Tinetti *et al.*, 2010; Crouzet *et al.*, 2012; Grillmair *et al.*, 2008; Barman, 2008; Madhusudhan and Seager, 2009).

Planets that transit their host stars are also occulted when they pass behind the star. The planet occultation is sometimes called a secondary eclipse and it provides a unique opportunity to obtain an isolated spectrum of the star. The stellar spectrum obtained during occultation can be subtracted from the unresolved combined spectrum of the star plus planet to yield a spectrum of the planet alone.

5.3.1 Limb darkening

The shape of the transit light curve provides information about the stellar atmosphere. When we look at the spherical star, it appears brighter and bluer in the center and redder near the edges or limb of the star. Both the density and temperature are decreasing as a function of the stellar radius. When we look near the edge of the star, we see down to an optical depth of $\tau \sim 1$ but we are looking through a column of relatively cooler and lower density gas that is higher up in the stellar atmosphere. When we look at the center of the star, we also see down to an optical depth of $\tau \sim 1$; however, this column of gas extends to deeper physical depths in the star and is therefore hotter and higher in intensity. Because of limb-darkening, a transiting planet will block more flux from the bright center of the star than near the edges.

A beautiful example is shown in Fig. 5.6, which is reproduced from Knutson *et al.* (2007b). Each of the transit curves was taken with a different bandpass using the Space Telescope Imaging Spectrograph (STIS) on the Hubble Space Telescope

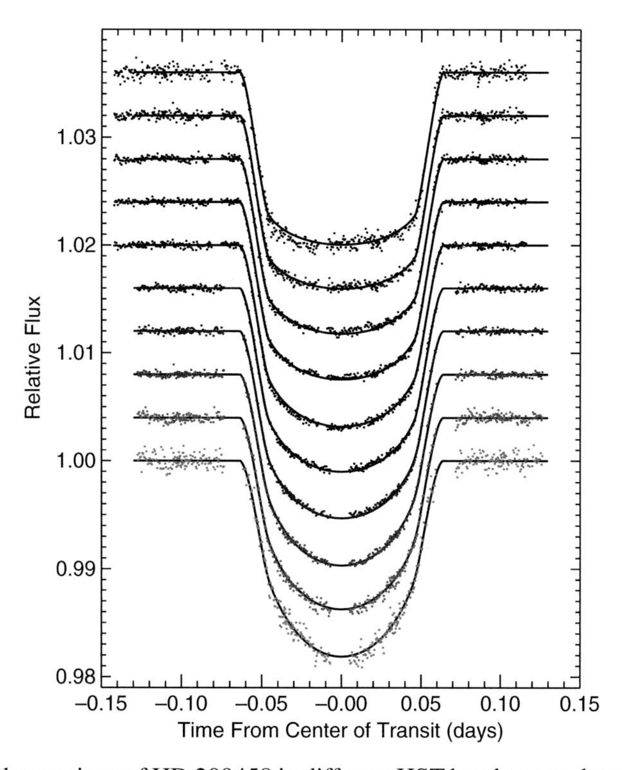

Fig. 5.6 Observations of HD 209458 in different HST bandpasses show the wavelength dependency of limb darkening (red wavelengths at top to blue at the bottom). (From Knutson *et al.*, 2007b).

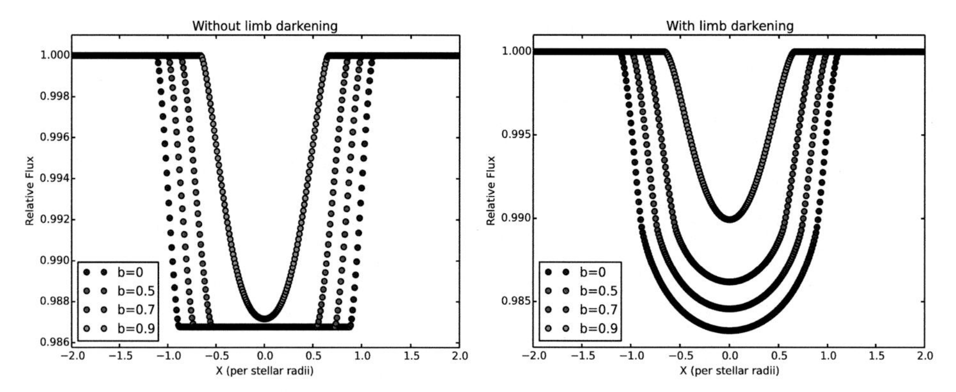

Fig. 5.7 Left: the shape of the transit light curve depends on the impact parameter *b* (from 0 to 0.9 from the outside inward in the diagram). Right: inclusion of limb darkening also affects the shape of the ingress and egress, making the curves more rounded. Figures courtesy of Meg Schwamb.

(HST). Limb darkening changes the shape of ingress and egress and the light curves are more rounded for the blue wavelengths of light than for the red wavelengths. Knutson *et al.* (2007b) used a nonlinear limb-darkening law to model the wavelength-dependent shapes of the transit curves in Fig 5.6.

Changing the impact parameter also affects the shape of the light curve because it changes the duration of the transit. At an impact parameter of zero, the planet is perfectly aligned with the diameter of the star and the maximum transit duration occurs. When the impact parameter is close to unity, only a grazing transit is observed. Figure 5.7 (left) shows the difference in the shape of the transit curve for different impact parameters without considering limb darkening. In this figure, the same wavelength bandpass was assumed for all four (synthetic) light curves.

If limb darkening is included, there is an additional change in the shape of ingress and egress for the curves, and Fig. 5.7 (right) shows a more realistic set of transit curves for impact parameters of 0, 0.5, 0.7, and 0.9. Although both the impact parameter and limb darkening will change the shape of the transit curve, these effects can be distinguished because the impact parameter is wavelength independent.

5.3.2 Stellar noise for transit techniques

As with Doppler observations, the flux for transit measurements comes from the host star. Therefore, starspots can be a source of additional noise. The cooler starspots result in a diminution of flux (especially at bluer wavelengths). When a planet crosses a starspot, the sum of the flux decrement from the transit plus the starspot is not as great and the star brightens slightly. The shape of the photometric

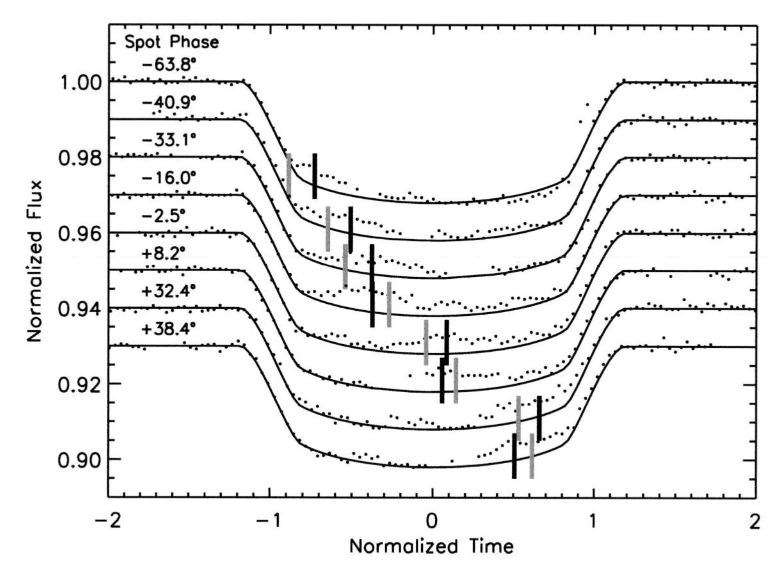

Fig. 5.8 As the planet in CoRoT-2 transits starspots on HD 189733, which are cooler than the rest of the star, less flux is blocked. The rotation period of the planet is different from the rotation period of the spots, and the spots advance in this time series of transit light curves. These data were cleverly used to determine the alignment of the planet with the equatorial plane of the star (reproduced from Nutzman *et al.*, 2011).

perturbation to the light curve depends on the relative size of the planet and spot and the relative spot temperature. This effect can be seen in Fig. 5.8 for the CoRoT-2 transiting planet (Nutzman *et al.*, 2011). For this star, the rotation period is roughly 4.5 days and the typical spot lifetime is about 55 days. The orbital period of the planet is only 1.74 days, so as the planet circles around, the spot cluster has advanced slightly on the star.

5.4 Direct imaging

The majority of exoplanets that have been detected are within 5 Sun–Earth distances (AU) of their host stars. This is due to the detection biases of the two predominant exoplanet detection techniques, the Doppler technique and the transiting method. The direct imaging technique offers the most promising prospect to detect gas giant exoplanets in wide orbits. The opportunity to directly image an exoplanet has enormous appeal. With enough photons from the planet, one day in the future it might be possible to see clouds rotating on the surface of the planet or to take a spectrum of the exoplanet atmosphere from a direct image.

A 10-m telescope imaging at H band has a 32-milliarcsecond diffraction limit. Such an instrument has sufficient spatial resolution to detect a planet on a 5-AU orbit around a star at 150 pc, approximately the distance to the Orion star-forming region (Oppenheimer and Hinkley, 2009). However, scattered light from the star generally prevents the detection of planets at small angular separations. The critical requirement for direct imaging is high contrast, the brightness ratio between a planet and its host star. This requirement is less severe for wider angular separations where scattered light from the star is less intense. While the contrast requirement for imaging a young and hot Jovian planet is 10^{-8}, the prototypical high-contrast observation of GL 229 B with a relatively wide angular separation was detected with a contrast of only 10^{-4} (Oppenheimer, 1999; Oppenheimer *et al.*, 2001).

Improvement of image reduction techniques will enable direct exoplanet detections in more systems, for example, HR 8799 (Marois *et al.*, 2008), Fomalhaut (Kalas *et al.*, 2008), and β Pictoris (Lagrange *et al.*, 2010). With the advent of next generation adaptive optics systems, instruments such as the Gemini Planet Imager (Macintosh *et al.*, 2006), the Project 1640 (Hinkley *et al.*, 2008), and SPHERE (Dohlen *et al.*, 2006) will deliver better than 10^{-10} contrast close to the diffraction limit of a telescope. These instruments will not only image young Jovian planets but also obtain low-resolution spectra to study their atmosphere.

5.5 Microlensing

The microlensing technique was developed to search for dark matter in the form of massive compact halo objects, or MACHOs. The method works in the following way: the light from a distant source brightens when a "lensing" star passes between the line of sight of the observer and the background source. The lensing star warps spacetime through the mathematical construct of an Einstein ring; light from the source bends around the lensing star and the observer detects more photons. The duration of the photometric brightening (i.e., the microlensing event) is a function of the mass of the lens star; however, the brightening amplification depends almost entirely on the impact parameter (the alignment of three objects: the observer, the lens, and the source).

The brightening amplification is remarkably insensitive to the mass of the lens. Even low mass planets (in orbit around the lensing star, or free-floating planets) can induce strong amplification of the source starlight if the alignment is good. This is what makes the technique useful for the detection of low mass exoplanets, when they orbit at angular separations near the Einstein Ring.

A historical challenge for microlensing detections has been the follow up observations required to search for the lens star. Because 70% of the stars in the galaxy are M dwarfs, the lensing star is likely to be faint and difficult to recover, making the detection more difficult to characterize. However, the microlensing community is tightly organized with observing stations at all latitudes on the Earth and rapid response follow-up. The recovery rate of the lensing stars has improved and clever new techniques are being developed to measure microlensing parallaxes, yielding

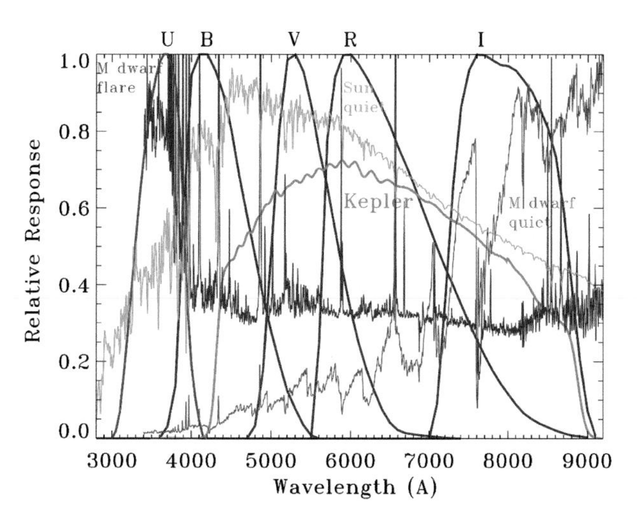

Fig. 2.5 Wavelength coverage of several standard filters used in optical astronomy, along with the wavelength coverage of the filter used in the Kepler mission. Overplotted are also spectral energy distributions of a quiescent and flaring M-dwarf atmosphere, taken from Kowalski *et al.* (2013). The solar spectrum is the 1985 Wehrli Standard Extraterrestrial Solar Irradiance Spectrum from http://rredc.nrel.gov/solar/spectra/am0/.

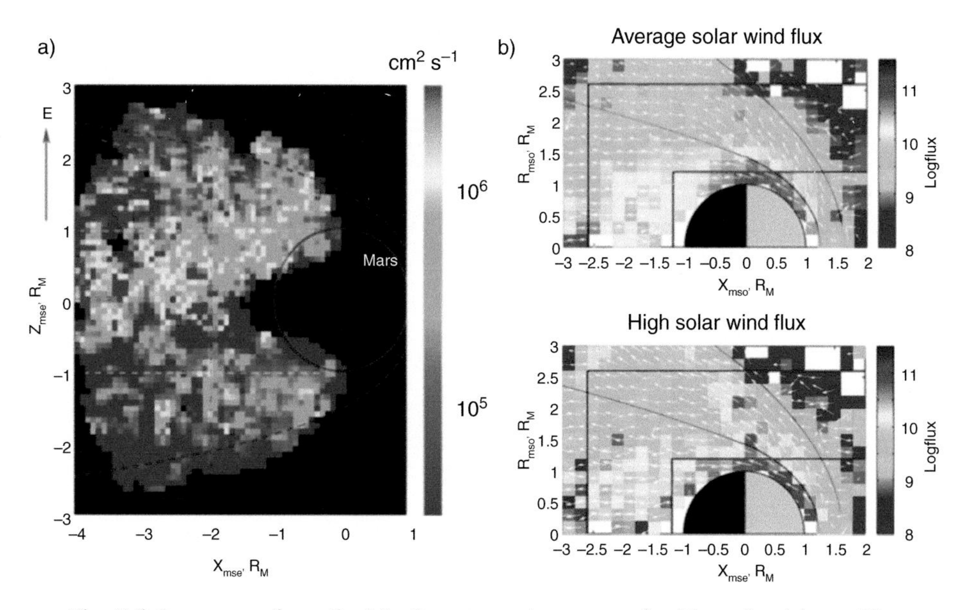

Fig. 7.6 Ion escape from the Martian atmosphere, organized by solar drivers. The Sun is to the right in both panels. (a) Escaping ion fluxes downstream from Mars are greater in the hemisphere of upward directed (with respect to the planet) solar wind electric field (Barabash *et al.*, 2007); (b) escaping ion fluxes downstream from Mars are greater during periods of high solar wind flux. (From Nilsson *et al.*, 2011.)

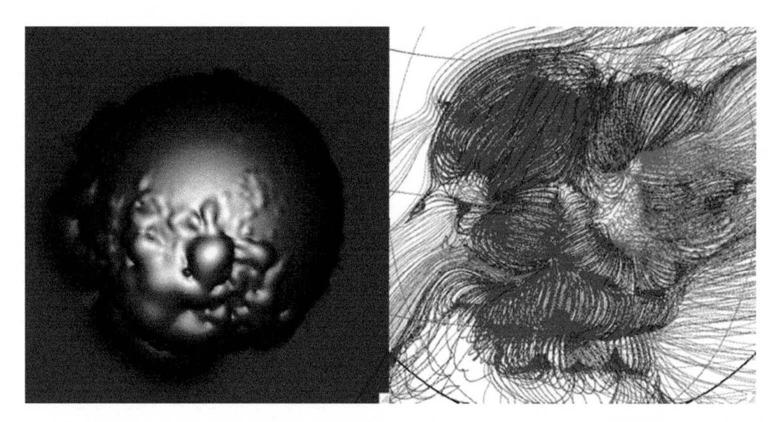

Fig. 7.8 Influence of magnetic fields on planetary near-space environments. Magnetic fields supply magnetic pressure (left: for Martian crustal magnetic fields) that deflect solar wind, but also modify magnetic topology (from Brain, 2006); (right: for the strong Martian crustal fields in the southern hemisphere, where red denotes closed field lines and blue denotes field lines open to the solar wind at one end) that enable exchange of particles and energy between the atmosphere and solar wind. Both renderings result from model calculations that include contributions from crustal fields and external drivers (solar wind or IMF).

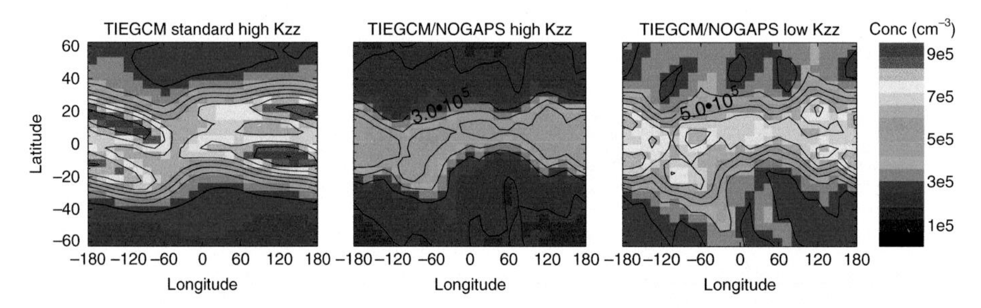

Fig. 9.11 Calculated averaged peak F_2 electron density for March, 1300 local for three TIEGCM simulations. Left column uses standard K_{zz} = 125 m^2 s^{-1} with vertical winds from the right column of Fig. 9.7. Middle uses NOGAPS-ALPHA vertical winds. Rightmost field is with NOGAPS-ALPHA vertical winds and K_{zz} divided by 5.

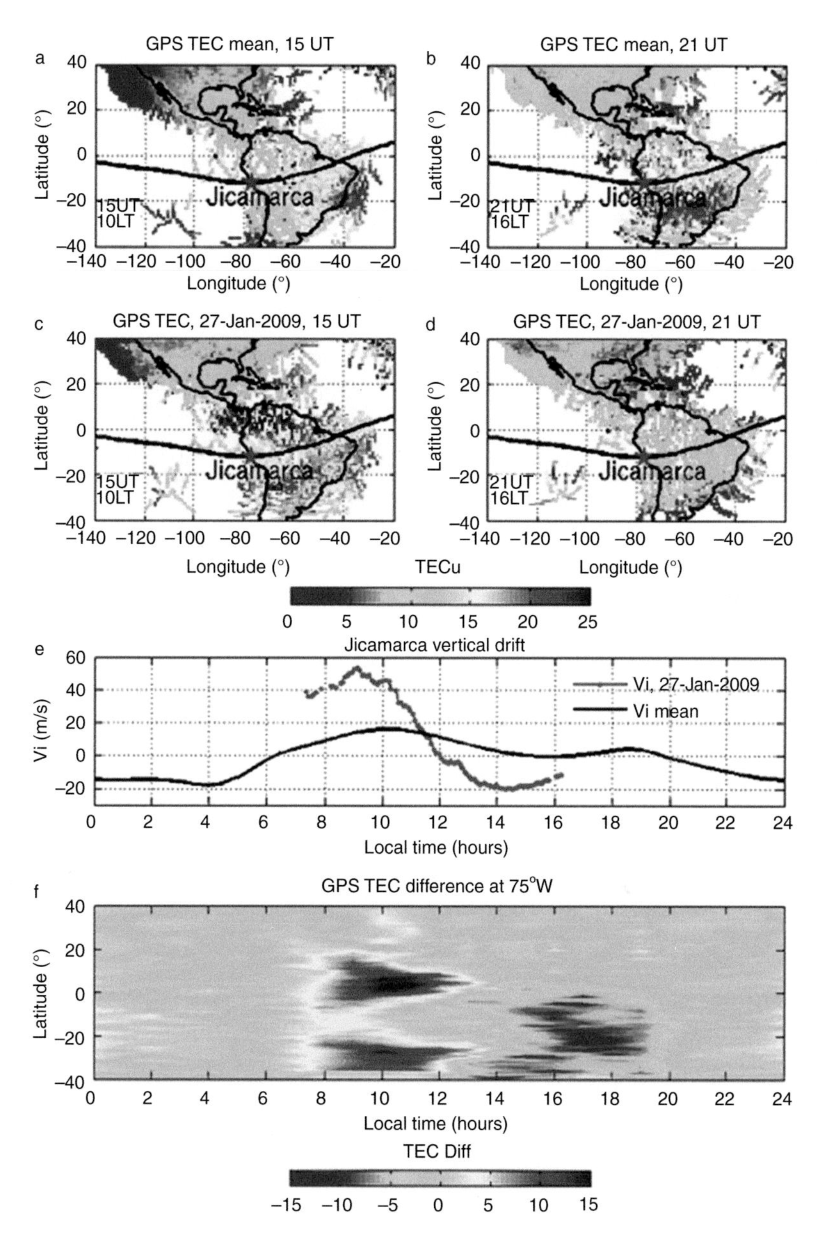

Fig. 9.12 Perturbations to the ionosphere, both total electron content (TEC) and vertical ion drift from the sudden stratospheric warming (SSW) of January 2009. The top row shows typical morning (15 UT = 10 local time at 75° W) and afternoon (21 UT) TEC fields over South America. The second row shows these fields after the SSW with a notable enhancement of TEC in the morning. The third panel shows the difference in the vertical ion drift as measured from Jicamarca Peru. The bottom panel shows difference fields between the SSW perturbation and the mean case as a function of local time, emphasizing the morning TEC enhancement and the afternoon depletion. (From Goncharenko *et al.*, 2010.)

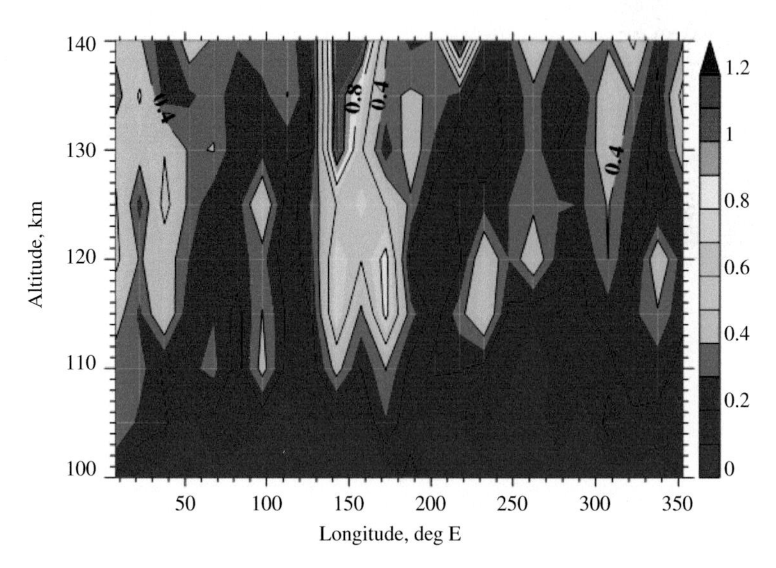

Fig. 9.14 Density variance data from the Mars Odyssey accelerometer in 15° and 5-km bins as function of longitude over about 127 orbits. (From Fritts *et al.*, 2006.)

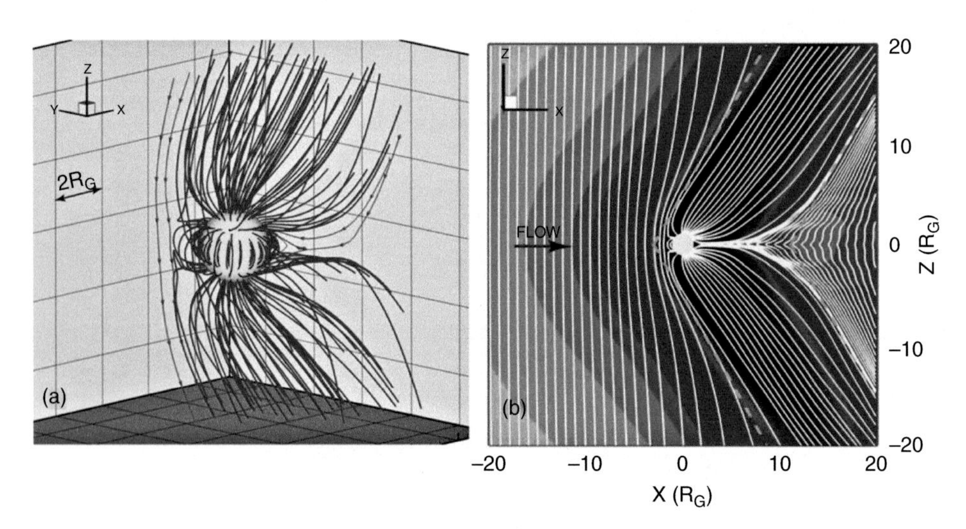

Fig. 10.5 (a) Selected magnetic field lines in Ganymede's magnetosphere from an MHD simulation. (b) Magnetic field lines projected onto the $x-z$ plane at $y = 0$. The x-component of the plasma flow velocity is shown in color. Orange dashed lines are tilted relative to the background field at the Alfvén angle and the flow is excluded from regions downstream of the left hand dashed lines, reappearing only in regions about 5 R_G further downstream. In the simulation, the sphere of radius 1.05 R_G is the inner boundary for plasma flow. (From Jia *et al.*, 2008.)

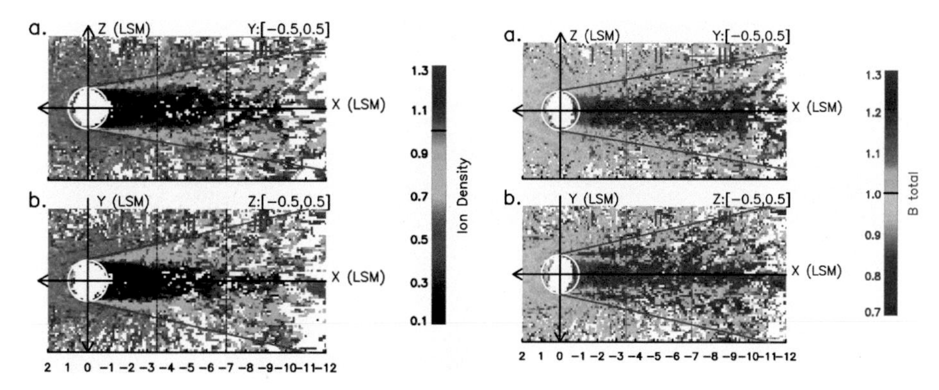

Fig. 10.7 Ion density (left) and magnetic field (right) in the vicinity of Earth's Moon from measurements by the *Artemis* spacecraft. The parameters represented by color are normalized by their values in the upstream solar wind. The x-axis is antiparallel to the solar wind flow. The data are plotted in the $x-z$ plane which is the plane of the solar wind field and the flow, and in the $x-y$ plane, perpendicular to this plane. The red lines diverging in the direction of negative x denote the wake boundary across which the density changes significantly. The divergence from the wake center is controlled by the propagation of fast mode waves. (From Zhang *et al.*, 2014.)

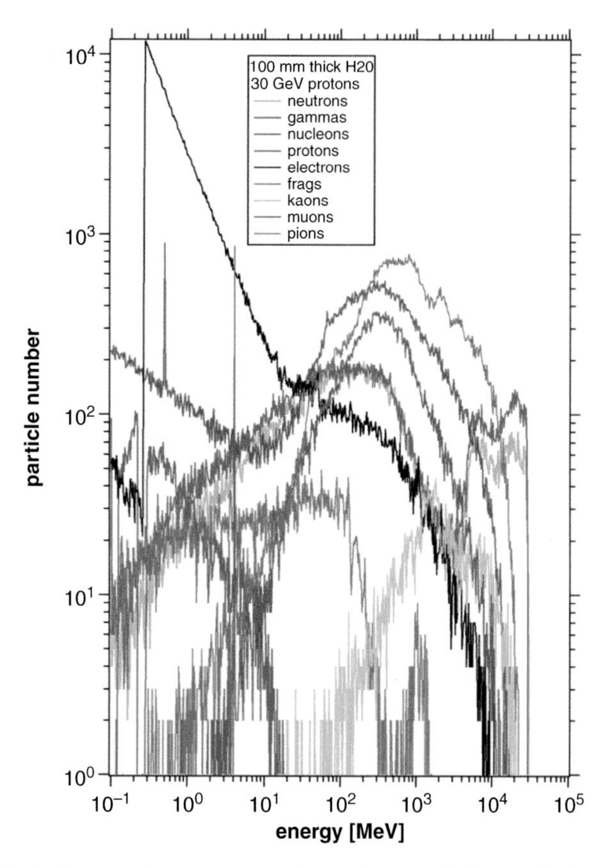

Fig. 12.11 Galactic cosmic-ray spectra for various particle populations. (Diagram provided by D. Haggerty, JHUAPL.)

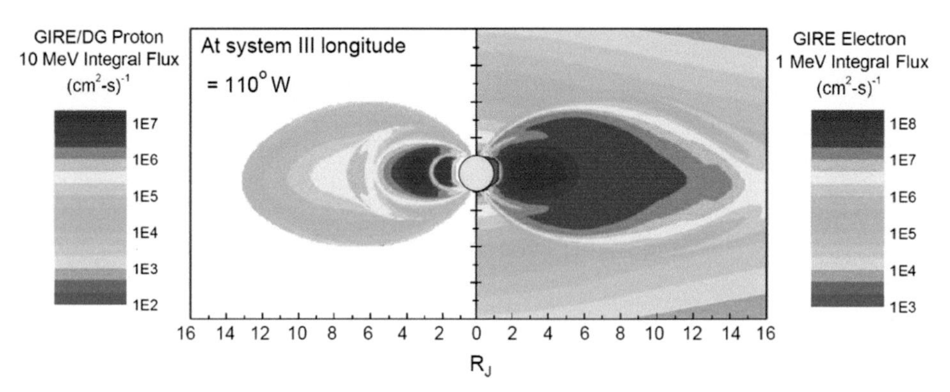

Fig. 12.12 Two-dimensional integral flux distributions for electrons (right section) and protons (left section) at Jupiter based on the Divine-Garrett/GIRE radiation models. (From Paranicas *et al.*, 2009.)

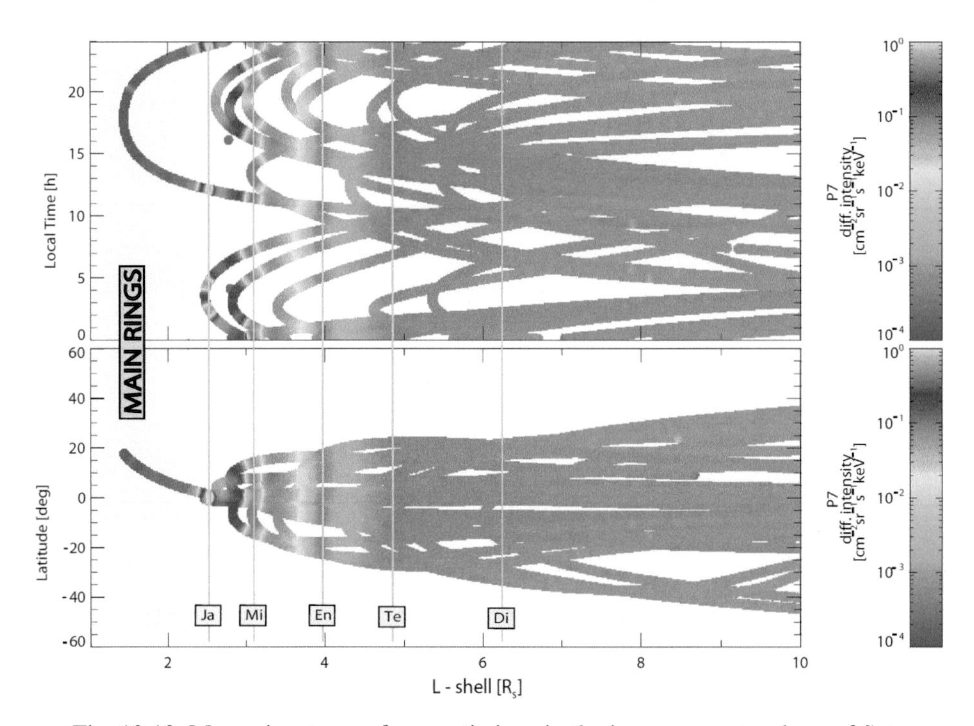

Fig. 12.13 Macrosignatures of energetic ions in the inner magnetosphere of Saturn as a function of L-shell and either local time (upper panel) or latitude (lower panel). Color-coded are the differential intensities of ions (> 10 MeV/nucleon) as measured between 2004 and 2007 by the Low Energy Magnetospheric Measurement System LEMMS onboard the Cassini spacecraft. (From Roussos, 2008.)

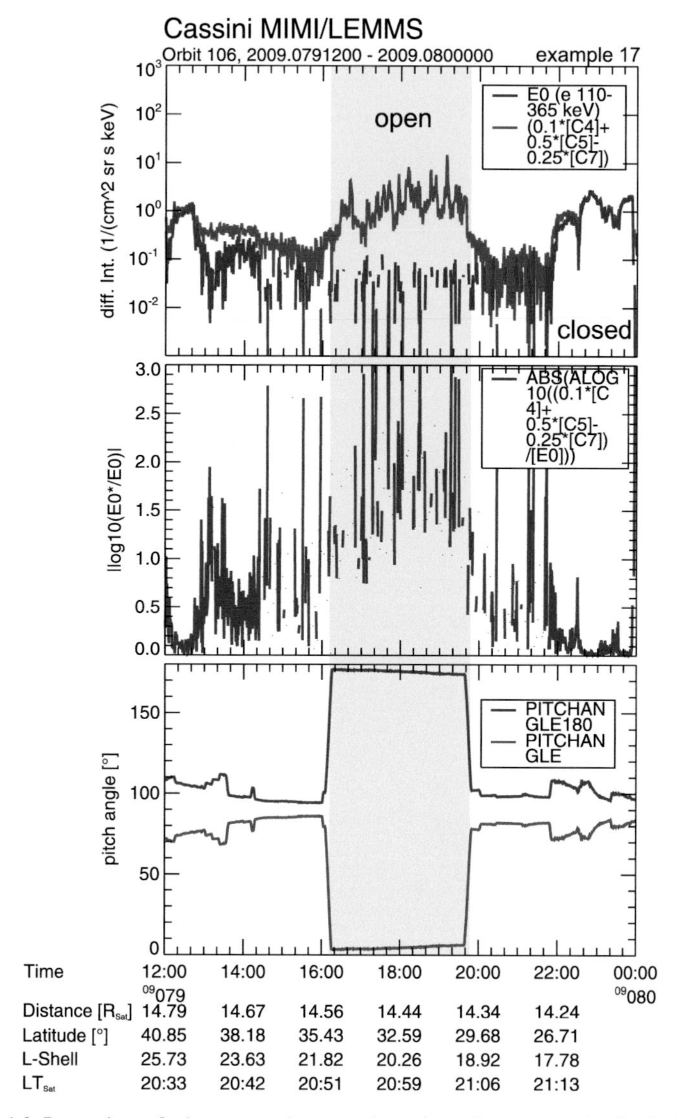

Fig. 12.16 Intensity of electrons along and against the magnetic-field direction inside Saturn's magnetosphere.

Caption for 13.10 (overleaf)

The 3D CAT reconstructed visualization of the distribution of solar-wind density upwards of $8\,e^{-}\,cm^{-3}$ (brighter colors toward yellow mean increasing density) on the left-hand side and high-velocity portions (blue) on the right-hand side showing the developing and changing reconstructed structure of the 13–15 May 2005 coronal mass ejection (CME) event sequence. The left-hand density images are highlighted with green cubes to encompass the reconstructed volume of the mass portion of the CME. This same highlighted volume is depicted on the

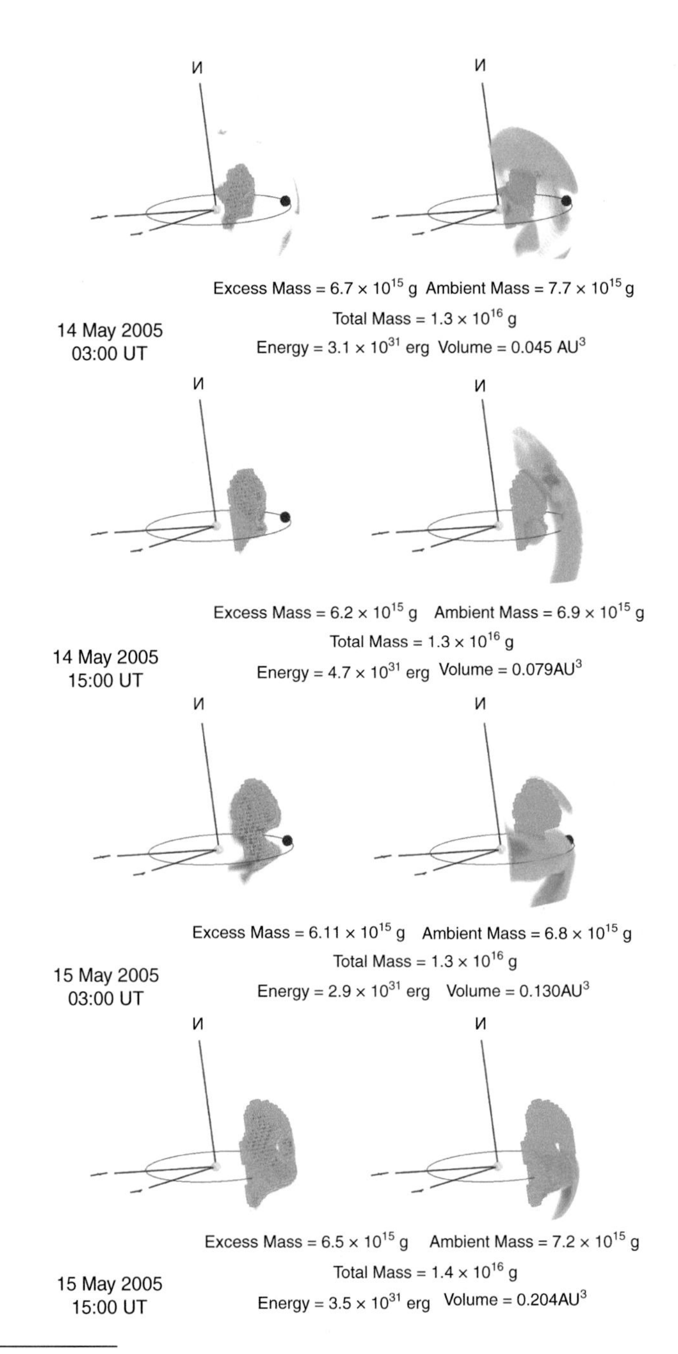

Caption for 13.10 (cont.)

right-hand velocity reconstructions for illustrative purposes. Each image is labeled with the masses, volume, and energy values on each date and time as shown. All non-CME-related features have been removed for clarity of viewing when displaying the 3D volume. The axes are heliographic coordinates with X-axis direction pointing toward the vernal equinox, and Z-axis directed toward solar heliographic North. An r^{-2} density increase has been added to better-show structures further out from the Sun (the central sphere) to the Earth (the blue sphere) along with the Earth orbit (ellipse). (From Bisi *et al.*, 2010a.)

distances and masses of the lensing object with a few astrometric and photometric ground-based observations.

This technique will be particularly powerful for determining exoplanet masses when WFIRST (the highest-ranked project in the 2010 Astronomy Decadal Survey) is launched. If parallaxes can be recovered for most of the WFIRST detections, it will make this technique a game-changer. Microlensing will become a powerful technique for detecting exoplanets at separations beyond the ice line and for understanding exoplanetary architectures.

5.6 Astrometry

Astrometry is one of the oldest techniques and has been used to measure stellar parallaxes, proper motions, and binary star orbits. With this technique, the changing position of the star in the plane of the sky is measured with respect to other objects – typically background stars. Like the Doppler technique, at least one full orbital period must be observed to map out exoplanet orbits. However, this technique recovers the full three-dimensional orbit, so there is not the $M \sin i$ degeneracy of the Doppler technique.

Attempts to detect exoplanets with ground-based astrometry have been challenging because the center of mass for a star–planet system is generally inside the radius of the star. As a result, the photometric centroid barely moves. Astrometry is better leveraged for planets at large separations because the center of mass moves outside of the star. However, these planets also have longer orbital periods and the astrometric precision must be maintained for years.

In all cases, the astrometric wobbles induced by orbiting planets are tiny and it is an enormous challenge to identify background reference stars that do not move. Some improvement in ground-based astrometric precision have been realized with the use of adaptive optics to shrink the twinkling star. However, astrometry is best carried out above the Earth's atmosphere. The Hipparcos mission operated from 1989 to 1992 with a measurement precision of 1 milliarcsecond. Astrometry is about to undergo a new revolution. The European Space Agency (ESA) launched the Gaia mission in 2013. This mission will make the largest and most precise three-dimensional map for a population of more than one billion stars in the Milky Way galaxy. The collecting area of the Gaia telescope is 30 times the size of Hipparcos and the positional accuracy and proper motion measurements for most stars will be improved by a factor of 200.

5.7 Comparative planetology

5.7.1 Exoplanet formation

How do all of the exoplanet detections fit with our understanding of the formation and evolution of the solar system? The solar nebula theory provides a theoretical description for the formation of the solar system. Indeed, it has been said that this model is so elegant, that it is hard to imagine that it could be wrong. The solar nebula theory neatly explains most observations: the planets closest to the Sun form in a hot environment and as a consequence these planets are small and comprised of refractory elements (i.e., elements whose solid state withstands high temperatures); the more massive gas giants form beyond the ice line (a distance where it is cold enough for dust grains to be coated with icy mantles) where the feeding ground is more voluminous; jovian planets have moons that were either captured or that form as mini-solar-systems; the planets all orbit in the same direction in the disk because they inherit the same angular momentum vector; the solar system is littered with leftover debris such as asteroids and comets. The theory supports the idea first suggested by Kant and Laplace that the proto-Sun was surrounded by a primordial spinning disk of dust and gas. All of the material that makes up the Sun drained through this disk.

Note that the primordial or protoplanetary disks are different from reprocessing disks or debris disks, which can be observed around older main sequence stars. Debris disks are caused by collisions of small bodies in the disk at later stages and can even be detected around old main sequence stars. Debris disks are dusty, gas-poor structures that evolve and dissipate with Poynting–Robertson drag (Wyatt, 2008) as stellar radiation causes dust grains to lose energy and spiral inwards.

The study of protoplanetary disks has made tremendous advances in parallel with the discovery of exoplanets. Lada and Wilking (1984) inferred the presence of dusty shells around young stellar objects (YSOs) in Ophiuchus based on an excess of infrared flux; light from the star that was trapped and scattered by dust particles. Their classification of three different types of YSOs suggested evolutionary stages. However, the geometry of the dust distribution was not actually observed until the refurbished Hubble Space Telescope (HST) resolved flattened pancake-shaped structures around young stars in the Orion Nebula (O'dell and Wen, 1994). However, observations cannot yet see into the protoplanetary disks because the disks are optically thick at most wavelengths. It is only the outer regions (beyond ~ 40 AU) where the disk becomes optically thin to millimeter wavelengths that observations are secure. Thus, theory currently outpaces observational evidence about the temperature and pressure structure and the evolution of protoplanetary disks, a situation which should improve with data from the Atacama Large Millimeter Array (ALMA).

The mass of the protoplanetary disk is a fraction of the stellar mass and evolves with the central star. Our understanding of the physics and chemistry of protoplanetary disks is distilled in Fig. 5.9. The temperature is about 1500 K near the inner part of the disk and along the flared outer layers. These high temperature are too hot for grain growth, but a few AU from the protostar the disk mid plane

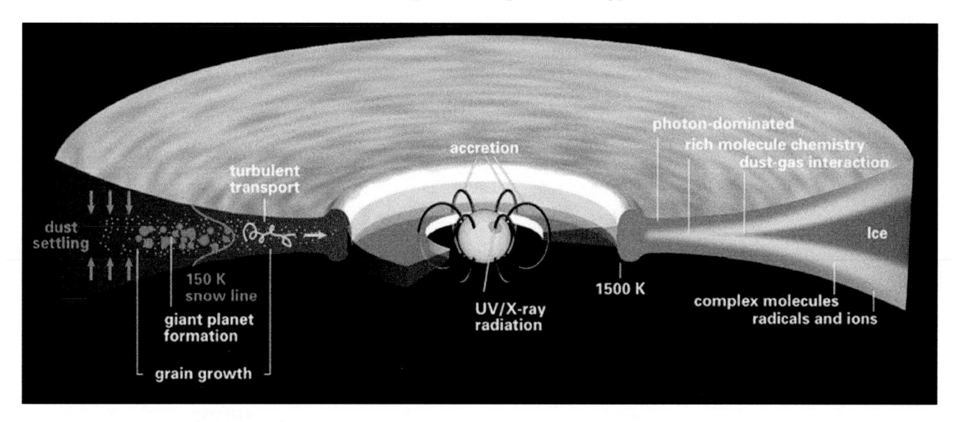

Fig. 5.9 A sketch of the structure and processes of protoplanetary disks. From a talk by Dmitry Semenov PPVI (Henning and Semenov, 2013; https://www.youtube.com/watch?v=F2IDOeeNy8c).

is cool enough for icy grains to stick and grow. The opacity of the disk is set by the dust, which gradually decouples from the gas and settles toward the mid plane, increasing transparency of the disk over time.

Protoplanetary disks provide the initial conditions for planet formation. The formation of gas-giant planets was described in a seminal paper by Pollack *et al.* (1996). In the first phase of planet formation, the planet grows by runaway accretion of solid material. The second phase of growth is very slow; both solid and gas accretion are nearly time independent and this phase sets the planet formation time scale. Once the planet core reaches a mass of about $10 M_\oplus$, the third phase of runaway gas accretion begins, growing the planet mass from ten to a few hundred $M_\oplus$. Pollack *et al.* (1996) estimated that gas-giant planet formation should take roughly 10 Myr. However, observations of protoplanetary disks in the 1990s presented a conundrum: the primordial disks appear to be nearly ubiquitous around stars that are 1 Myr; at 2 Myr only about half of young stars have disks and, by 10 Myr, the disks are essentially gone. Figure 5.10 shows the fraction of protoplanetary disks found in young cluster stars (Mamajek, 2009).

One triumph that emerged from the discovery of exoplanets was a solution to the disagreement between theory and observations for the formation time scale of gas-giant planets. The first detected gas-giant planets orbited close to their host stars providing evidence that exoplanets could undergo orbital migration. Thus, planets were not restricted to a planetesimal feeding ground at a fixed orbital radius; instead, the planet embryos are pushed around in the disk by planet–planet interactions and tidal torques. The access to a wider part of the disk suggests a wider feeding zone for more rapid accretion of planetesimals that would shorten the second phase of gas-giant planet formation described by Pollack *et al.* (1996).

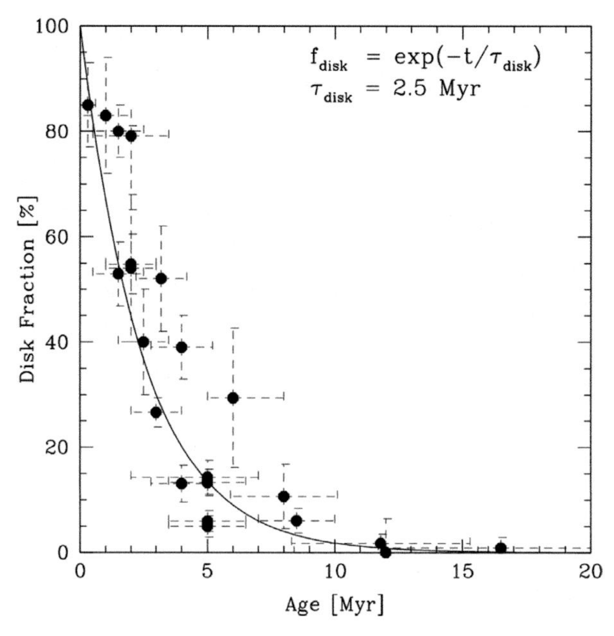

Fig. 5.10 Primordial disk fractions of stars in young clusters (Mamajek, 2009). These observations show that the dust disks last for only a few million years.

5.7.2 Exoplanet migration

The realization that exoplanets are mobile during the early stages of formation has led to many studies of dynamical interactions. The details of migration and the parking mechanisms that place gas-giant planets just a few stellar radii away from their host stars are an active area of research (Lin *et al.*, 1996; Batygin, 2012). In the younger primordial disk with significant gas and dust density, the planet embryos will clear gaps in the disk. In this case, material can pile up at both the inner and outer edges of the gap. When the disk mass at the edges of one of these gaps is comparable to the mass of the planet embryo the disk will exert a torque that causes the planet to migrate. The outer edge of the disk causes inward migration while the inner edge of the disk can produce outward migration. When multiple planet embryos exist in the disk it is possible for the outer embryo to become locked into a resonant orbit with the inner planet, a process called convergent migration. As the disk clears, convergent migration can leave planets in resonant orbits that persist stably over the lifetime of the star. This effect is especially powerful for resonances where the ratio of the orbital periods (P_{outer}/P_{inner}) is close to an integer number, N. Planets with small N are said to be in mean-motion resonance (MMR) and the exchange of angular momentum between MMR planets is flagged by oscillations in eccentricity and orbital periods.

Another way to push exoplanets inward is through gravitational encounters. There are several proposed mechanisms that excite orbital eccentricity including secular migration (Wu and Lithwick, 2011), planet–planet scattering (Ford and Rasio, 2008; Nagasawa *et al.*, 2008), and Kozai perturbation in which gravitational interactions result in coupled variations in orbital inclination and eccentricity (Wu, 2003; Fabrycky and Tremaine, 2007; Naoz *et al.*, 2011). High-eccentricity planets with a small enough periastron passage eventually experience tidal circularization and can end up in short-period orbits.

Different migration mechanisms predict distinct observables. A particularly interesting observable is stellar obliquity, the relative angle between the stellar rotation vector and the vector of planet orbital plane. The stellar obliquity can be measured by observing the Rossiter–McLaughlin (RM) effect (Rossiter, 1924; McLaughlin, 1924). The RM effect is caused by a transiting object blocking some of the light from a rotating star. First, the planet crosses the approaching limb of the rotating star, decreasing the contribution of blue-shifted light in the spectral line and a few hours later the planet crosses the receding limb of the rotating star, decreasing the contribution of red-shifted light. The systematic decrement of Doppler-shifted light in the composite spectral lines results in a distortion of line profile, which is (mis)interpreted as a change in the radial velocity of the star. The shape of the RM curve during transit is entirely dependent on the stellar obliquity. Consequently, the stellar obliquity is determined by modeling the anomalous radial velocity signals during a transiting event.

Disk-driven migration is expected to produce a small stellar obliquity whereas gravitational encounters that temporarily pump up the orbital eccentricity of gas-giant planets should result in a wide range of stellar obliquities including retrograde orbits. The latter has been observed for many transiting planets (Winn *et al.*, 2010; Albrecht *et al.*, 2012) suggesting that high-eccentricity mechanisms drive gas-giant planets inward. However, it has also been suggested (Batygin, 2012) that the observed stellar obliquity range may reflect a primordial stellar obliquity due to interactions between protoplanetary disk and a companion star. Interestingly, the small stellar obliquity of low-mass multi-planet systems suggests well-aligned vectors for the stellar spin and planetary orbits (Sanchis-Ojeda *et al.*, 2012; Albrecht *et al.*, 2013). It is certainly possible that gas-giant and low-mass planets migrate by different mechanisms.

In summary, the most important revisions to the solar-nebula model and our understanding of planet formation can be attributed to one source: the addition of dynamical interactions between planets and the primordial disk. These dynamical interactions speed up the accretion time scales, produce mean-motion resonances, scatter planets out of the disk into non-coplanar orbits that can be detected by the Rossiter–McLaughlin effect and even eject some planets.

Several other studies have also suggested an important transition at $\sim$1.5–1.7 Earth radii. Rogers (2014) applied a hierarchical Bayesian statistical method for a sample of *Kepler* planets with determined mass and identifies a transition radius above $1.6R_\oplus$. Lopez and Fortney (2014) model radii for planets with mass between $1-20M_\oplus$ considering different compositions and suggest a physically-motivated transition radius at $1.75R_\oplus$. Buchhave *et al.* (2014) study the metallicity distribution of 406 *Kepler* planet host stars. They find two characteristic planet radii (1.7 and $3.9R_\oplus$) that divide planets into three populations: terrestrial planets, gas-dwarf planets, and gas-giant planets.

Both the mass–radius relationship and the transition radius from rocky to non-rocky planet help us to better understand the formation history of small planets. Planets that form in-situ in the inner part of the disk would consist primarily of rocky materials and possibly a primordial H/He atmosphere (Chiang and Laughlin, 2013). In comparison, planets that have undergone significant migration should contain more volatile materials such as astrophysical ice (H_2O, CO, and NH_3). The debate of whether *Kepler* close-in planets form in-situ (Chiang and Laughlin, 2013; Hansen and Murray, 2013) or migrate (Swift *et al.*, 2013; Schlichting, 2014) should eventually gain evidence from studies of exoplanet atmospheres that add constraints on their chemical composition.

5.7.3 Exoplanet geology

Thousands of planet candidates were discovered by the *Kepler* mission, allowing for precise measurements of exoplanet radii. The combination of the radius and mass measurements (either from the Doppler technique or from transit timing variations) provide a mean density for hundreds of exoplanets and allow us to begin considering the bulk composition of unseen planets that orbit stars hundreds of light years away from us. The varying bulk composition of exoplanets results in different curves that cut through the mass–radius parameter space shown in Fig. 5.11.

Planets with radii smaller than 4 times that of the Earth can exhibit a remarkable diversity of compositions (Rogers and Seager, 2010). Weiss and Marcy (2014) considered the *Kepler*-detected planets with radii smaller than 4 times that of the Earth. Although their Doppler precision was not sufficient to measure reflex velocities from these small planets, they were able to place statistical limits on the exoplanet masses. They found that these small planets could be divided into two radius regions. Planets smaller than 1.5 Earth radii increase in density with increasing radius and seem to have a composition that is consistent with rock. Planets with radii between 1.5 and 4 times the radius of the Earth showed decreasing density with increasing radius, suggesting that the larger planet radius was a product of gaseous envelopes. Weiss and Marcy (2014) also concluded that the significant

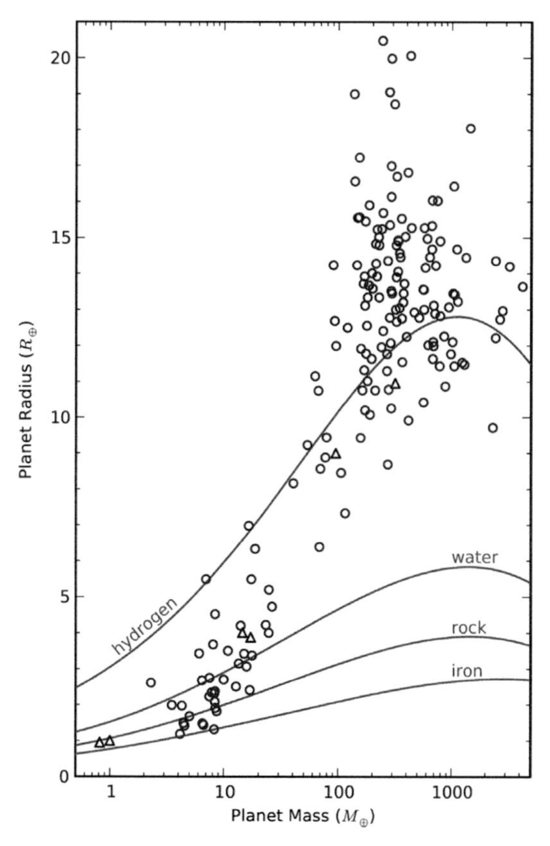

Fig. 5.11 Masses and radii of well-characterized exoplanets (circles) and solar-system planets (triangles). Curves show models for idealized planets consisting of pure hydrogen (Seager *et al.*, 2007), water, rock (Mg_2SiO_4), or iron. (From Howard *et al.*, 2013).

amount of scatter in the mass–radius parameter space suggested a large diversity in planet composition at a given radius.

5.7.4 Exoplanet statistics

With thousands of exoplanets and exoplanet candidates, it is possible to carry out statistically significant studies of the attributes of exoplanets. It is common to plot exoplanet mass as a function of orbital period when showing the distribution of exoplanets. However, that figure simply reflects the observational incompleteness and biases of the detection techniques and does not contain very much fundamental information about exoplanets.

However, there are other correlations that do reveal fundamental information. One of the first observed statistical correlations established that gas-giant planets

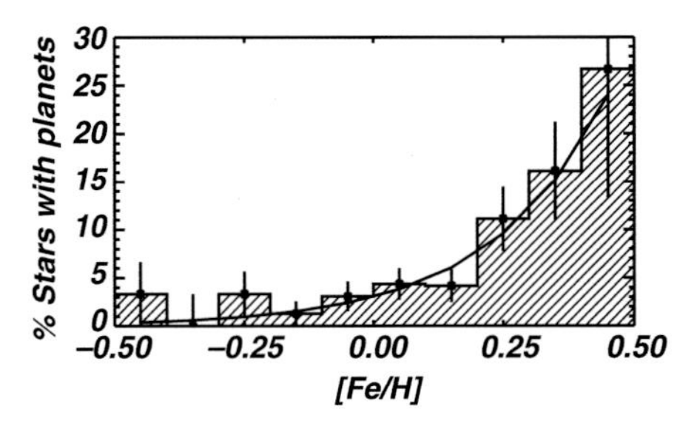

Fig. 5.12 High metallicity stars are more likely to host gas-giant planets than sub-solar metallicity stars. Figure from Fischer and Valenti (2005).

form more frequently around metal-rich stars (Gonzalez, 1997; Santos *et al.*, 2004; Fischer and Valenti, 2005, Johnson *et al.*, 2010). This planet-metallicity correlation was used as evidence for core accretion as the formation mechanism for gas-giant exoplanets that orbit closer than a few AU around their host main-sequence stars (see Fig. 5.12).

Interestingly, a similar correlation with host-star metallicity has not been identified for smaller Neptune-like or rocky planets (Sousa *et al.*, 2008; Neves *et al.*, 2013). The discovery of so many small planets with the *Kepler* mission has enabled a more thorough search. Buchhave *et al.* (2012) measured metallicity for a sample of 152 *Kepler* planet stars hosting planets with radii smaller than the radius of Neptune ($4R_\oplus$) and did not find a metallicity correlation. Everett *et al.* (2013) obtained spectra of 220 faint *Kepler* planet host stars and reached a similar conclusion. Buchhave *et al.* (2014) then expanded their metallicity measurements to include 406 *Kepler* planet host stars. In their recent data, the average metallicities for gas-giant planets ($R_P > 3.9R_\oplus$) and gas-dwarf planets ($1.7R_\oplus < R_P < 3.9R_\oplus$) are above the solar metallicity (0.18 ± 0.02 dex and 0.05 ± 0.01 dex), the average metallicity for terrestrial planets ($R_P < 1.7R_\oplus$) is consistent with the solar metallicity at $\sim 0.02 \pm 0.02$ dex. With their larger sample, it was clear that stars with either gas-giant planets or gas-dwarf planets were preferentially metal-rich, suggesting a planet-metallicity correlation for these two types of planets. However, it remains unclear whether such correlation exists for rocky planets. Wang and Fischer (2013) examined the same dataset as Buchhave *et al.* (2014). After accounting for systematic errors of stellar properties from the Kepler Input Catalog, Brown *et al.* (2011), they reported a modest planet-metallicity correlation for terrestrial planets at 4.2σ level.

Many stars in the solar neighborhood are components of multiple-star systems (Duquennoy and Mayor, 1991; Fischer and Marcy, 1992; Raghavan *et al.*, 2010; Duchêne and Kraus, 2013) and many planets have been detected in binary or multi-star systems. Initially, exoplanets were discovered orbiting one individual star in the binary star system (Cochran *et al.*, 1997; Eggenberger *et al.*, 2004). Recently, exoplanets have been discovered in difficult to detect circumbinary orbits, where the planet orbits both stars (Doyle *et al.*, 2011; Welsh *et al.*, 2012; Schwamb *et al.*, 2013).

Circumbinary planets can be detected via the timing variation of eclipsing binaries (Deeg *et al.*, 2008; Beuermann *et al.*, 2010). Since the launch of the *Kepler* mission, ten circumbinary exoplanets have been discovered around eight *Kepler* stars. The occurrence rate of circumbinary planets is estimated to be $\sim$10% (Welsh *et al.*, 2014; Armstrong *et al.*, 2014) assuming the orbital plane of circumbinary planets roughly align with the binary orbital plane. The occurrence rate could be much higher if the orientation of planet orbits is more isotropic.

It is expected that planet formation may be impeded in systems where the binary stars have small separations (e.g., $\sim$10–200 AU). This is supported both by simulations (Thébault *et al.*, 2006; Kley and Nelson, 2008; Thebault, 2011) and observations (Desidera and Barbieri, 2007; Kraus *et al.*, 2012; Wang *et al.*, 2014) that find a smaller fraction of exoplanets in binary star systems. It is not surprising that the dynamics of binary star systems stir things up and challenge planet formation. What is surprising is that the planets exist there at all.

Our view of exoplanets is still skewed by the observational sensitivities of the techniques that we use. However, the discoveries that have been made have helped us to revise our understanding of planet formation and the formation of the solar system. We see that planet formation is a chaotic process and that disks are sculpted by gravitational interactions to a greater extent than we appreciated by considering our own solar system. We now know that almost every star has planets and that planet formation is far more robust than astronomers expected.

6

Planetary dynamos: updates and new frontiers

SABINE STANLEY

In the study of heliophysics, planetary dynamos are important in understanding how various planetary bodies produce their magnetospheres which then so intricately interact with the solar wind. The dynamo mechanisms in planets are also very similar to those in the Sun, as well as in stars and other astrophysical bodies. Investigating planetary magnetic fields, therefore, provides data points in understanding magnetohydrodynamic processes in a broad range of astrophysical settings.

The investigation of planetary dynamos was predominantly focused on Earth's magnetic field until the mid-to-late twentieth century when planetary missions began to provide data on magnetic fields of other planets. Through our exploration of the solar system, we have discovered the diversity of planetary magnetic fields and realized the importance of magnetic fields in acting as probes of planetary interior structure, composition, and thermal evolution. As examples, magnetic field data were fundamental in discovering the global oceans of Europa, Ganymede, and Callisto, they demonstrated that Mercury and Ganymede each have a liquid iron outer core, and provided a main line of support for a helium-insolubility layer in Saturn.

Several aspects of planetary dynamos have been covered in previous chapters in the Heliophysics series. For a review of theoretical magnetohydrodynamics, applicable to planets as well as other astrophysical bodies, see Ch. 3 in Vol. I (Rempel, 2009a). In addition, an overview of planetary magnetic field properties can be found in Ch. 13 of that volume (Bagenal, 2009) and further details on the geomagnetic field and planetary dynamos can be found in Ch. 7 of Vol. III (Christensen, 2010).

This chapter serves two purposes. First, it provides an update on our understanding of planetary magnetic fields and dynamos from new mission data and dynamo models since the previous volumes of this series were written. Wherever possible, we refer the reader to specific chapters in previous volumes (see Table 1.2) rather than repeat too much information. However, we review the most important concepts and findings needed here so that this chapter is also self-contained. Second, this chapter delves into the frontier (or fringe, depending on your perspective) of planetary dynamo studies by reviewing our understanding of dynamos in small bodies and extrasolar planets.

6.1 Dynamo fundamentals

Dynamo action refers to the conversion of mechanical energy into electromagnetic energy through induction. In planets, the mechanical energy is supplied by fluid motions in electrically conducting regions inside the planets and the electromagnetic energy produces the observed planetary magnetic fields. A dynamo is referred to as *self-sustaining* if it does not require any external magnetic field contributions for regeneration (except initially for a starting seed field).

6.1.1 The magnetic induction equation

The fundamental equation governing this induction process is known as the *Magnetic Induction Equation*:

$$\frac{\partial \vec{B}}{\partial t} = \nabla \times (\vec{u} \times \vec{B}) + \lambda \nabla^2 \vec{B}, \tag{6.1}$$

where $\vec{B}$ is the magnetic field, t is time, $\vec{u}$ is the fluid velocity, and λ is the magnetic diffusivity, defined as:

$$\lambda = \frac{1}{\mu \sigma}, \tag{6.2}$$

where μ is the magnetic permeability and σ is the electrical conductivity.

The Magnetic Induction Equation can be derived from Maxwell's equations and Ohm's law in the magnetohydrodynamic limit (i.e., with fluid velocities much slower than the speed of light). The derivation is given in Sect. 3.3 of Vol. I (Rempel, 2009a).

6.1.2 Requirements for planetary dynamo action

At present, a complete minimal set of necessary and sufficient conditions for planetary dynamo action is not known. The definition of a planetary dynamo hints at some necessary conditions for this process. The planet must contain an electrically conducting fluid region. There must also be motions in this fluid region and hence a power source for the mechanical energy associated with these motions. Below we discuss other necessary conditions that are required for dynamo action.

6.1.2.1 The Magnetic Reynolds Number criterion

By inspecting the two terms on the right-hand side of Eq. (6.1) we see that magnetic field can grow or decay in time through two processes. The first term involves interactions of the velocity and magnetic fields through electromagnetic induction and acts as a source/sink term for field generation. The second term represents diffusion due to Ohmic dissipation. To ensure magnetic field does not decay away in time, field must be generated as fast or faster than its diffusion. A necessary condition for self-sustained dynamo action is therefore that the induction term be larger than the diffusion term in Eq. (6.1). By using characteristic scales for the variables in the Magnetic Induction Equation (i.e., B for the magnetic field scale, U for the velocity scale and L for a length scale) we derive a common measure of the ratio of field generation to field diffusion known as the *Magnetic Reynolds Number*:

$$Re_M = \frac{|\nabla \times (\vec{u} \times \vec{B})|}{|\lambda \nabla^2 \vec{B}|} \approx \frac{UB/L}{\lambda B/L^2} = \frac{UL}{\lambda}. \tag{6.3}$$

Upon first glance, it seems reasonable that the Magnetic Reynolds Number must be larger than unity in order for dynamo action to be possible. However, more rigorous theoretical analyses suggest that the lower bound for Re_M is instead closer to π^2 (Jones, 2008) and planetary numerical dynamo simulations typically find Re_M must be larger than $\sim 20-50$ for self-sustained dynamo action to occur. These higher values are due to the complexities in the velocity field morphologies that cannot be captured in the simple estimate given in Eq. (6.3).

6.1.2.2 Power source for fluid motions

In most planetary dynamo source regions, the fluid motions required for dynamo action result from convection. Thermal convection results if the heat output from the dynamo source region is higher than what can be transported down the conductive adiabatic gradient. This can be represented with the criterion:

$$q > q_{\text{ad}} \Rightarrow k|\nabla T| > k|(\nabla T)|_{\text{ad}} = \frac{k\alpha T g}{C_{\text{P}}}, \tag{6.4}$$

where q is the heat flux, subscript "ad" refers to the adiabatic value, k is the thermal conductivity, T is temperature, g is gravitational acceleration, α is the thermal expansion coefficient, and C_{P} is the heat capacity at constant pressure. The propensity for dynamo action is therefore strongly linked to the thermal evolution and heat-transport properties of the planetary interior.

Compositional convection may also be important for driving motions. These motions result when there are density differences in a multi-component fluid. For example, in the Earth's core, the solidification of the inner core releases a fluid enriched in light elements compared to the bulk core and hence, is buoyant and will rise, thereby driving motions.

In the absence of convection, the most feasible mechanism to generate flows in dynamo regions is fluid instabilities resulting from mechanical boundary forcings such as those due to precession, tides, or impacts.

6.1.2.3 Morphology of fluid motions

The intensity of fluid motions is also not a sufficient criterion for dynamo action. The morphology of the flow is also crucial. For example, in the spherical geometry of dynamo regions, basic flows such as solid body rotation or flows without radial components cannot sustain dynamo action. Motions must be fairly complex and three-dimensional (Jones, 2008).

Another concern is how to ensure generation of a large-scale magnetic field (i.e., with wavelengths much larger than the turbulent motions generating the field). To do so, the flow must contain a net helicity. Turbulent motions alone do not guarantee this as some symmetry-breaking mechanism is required. In planetary cores, the flow constraint due to rapid rotation acts as an excellent mechanism for generating this net helicity and large-scale field. Further information on dynamo generation mechanisms and constraints can be found in Ch. 3 in Vol. I (Rempel, 2009a).

6.1.3 Dynamo scaling laws

A major goal of dynamo studies is to develop predictions of dynamo characteristics based on the physical parameters (e.g., size, rotation rate, available energy) of the system. Researchers have been working to refine such scaling laws for planetary dynamos; see Sect. 7.6.5 of Vol. III (Christensen, 2010) for details. Aside from (hopefully) minor tunings, scaling laws seem to be effective in numerical dynamo simulations for predicting magnetic field strength, the degree of dipolarity of the magnetic field, and heat transport. Some of these scalings also seem to work well for actual planets (but not all!).

6.2 Planetary dynamos: updates

Since the previous volume in this series, advances have been made on several fronts which have improved our knowledge of planetary dynamos.

- New mission data have been gathered on planetary magnetic fields and interior properties.
- Computational resources and numerical methods have improved, allowing new regions of parameter space to be explored with numerical dynamo simulations.

- Significant theoretical and experimental work has been carried out on the properties of materials at high pressure and temperature, most notable for dynamo studies is the work on iron alloys and water.
- Paleomagnetic instrumentation advances have resulted in exciting new data on meteorite magnetism.

In the sections below we briefly review the main results discussed in Ch. 11 of Vol. III (Christensen, 2010) on planetary dynamos and discuss subsequent findings in these areas.

6.2.1 Terrestrial planets

6.2.1.1 Earth

Earth's dynamo is generated in the liquid Fe-rich core. The resulting surface field is predominantly axially dipolar (see Fig. 6.1) and experiences variability on a range of time scales. Paleomagnetic studies suggest that the geomagnetic field has maintained a similar field strength as today for at least the past three billion years.

The source of motions in the core is convection, both thermal and compositional in origin. It is believed that compositional convection is the dominant source in recent times. Seismic data demonstrate that, in addition to iron and nickel, the core must contain roughly 10% lighter elements such as Si, S, O, or H. Compositional convection results from the release of light-element-rich fluid at the inner core boundary upon solidification of the inner core as the planet cools. Thermodynamic estimates suggest that the inner core began solidifying as late as a billion years ago which implies that thermal convection must have been the dominant source of convection before then.

Chapter 11 in Vol. III (Christensen, 2010) provides more detailed information on the geodynamo. Below, I highlight two fundamental discoveries that have occurred since that chapter was written with profound implications for the geodynamo.

(1) New estimates for Fe conductivities Recent ab-initio density functional theory computations by Pozzo *et al.* (2012) have revised the thermal and electrical conductivity of Fe alloys to be 2–3 times higher than previously thought at core pressures and temperatures. This has two implications for the geodynamo. First, the higher electrical conductivity means that the diffusive time scales of the dynamo are longer than previously thought. Second, the higher thermal conductivity means that more heat can be transported down the core adiabat than previously thought.

With this revised adiabatic core heat flux, estimates for how much of Earth's surface heat flow comes from the core imply that the outer portion of the outer

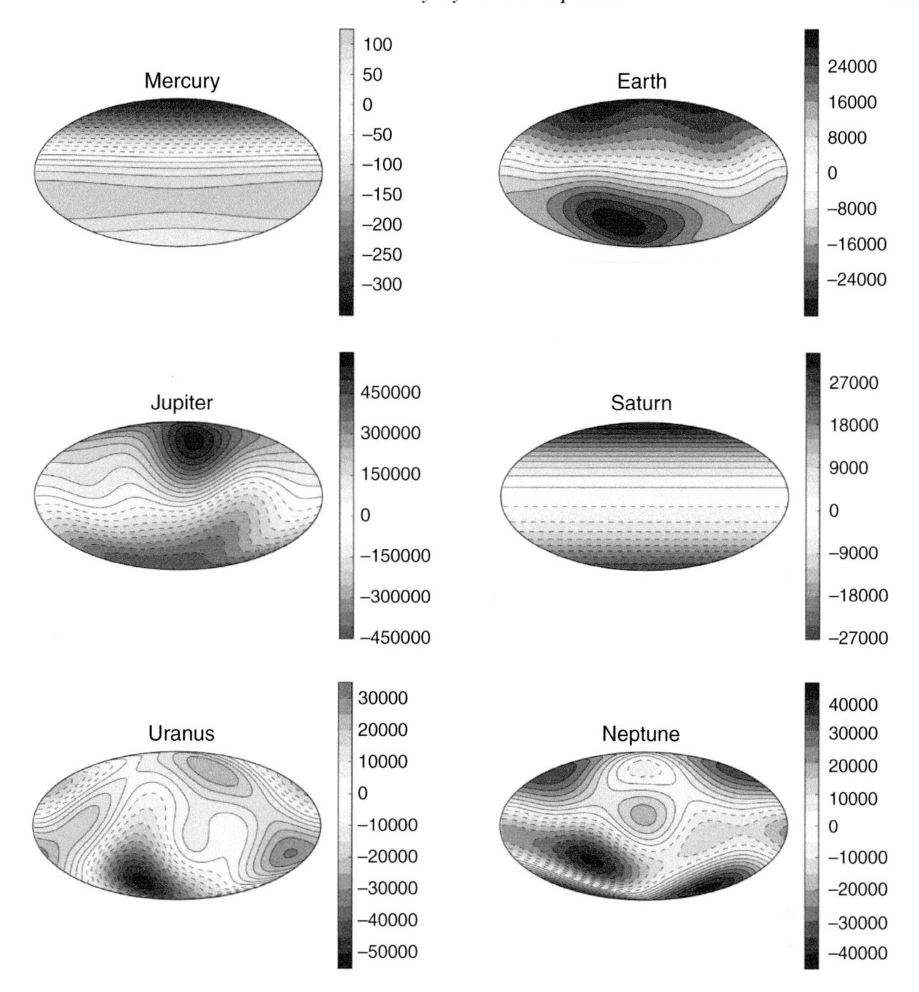

Fig. 6.1 Filled contours of the radial component of the surface magnetic field for planets in our solar system with active dynamos. Dashed contours represent negative values. Units are nT.

core is most likely thermally stably stratified. This would imply that convection is concentrated in the deeper portion of the core, that the smaller and faster scales of magnetic-field variability may be somewhat screened from observation by the stable layer, and that waves in the stable layer may contribute to the observed geomagnetic secular variation (Buffett, 2014).

(2) The translating inner core In the simplest prescription, the inner core is a spherical phase boundary where an iron-rich fluid is crystallizing as the planet cools. However, seismologists have known for some time that there are anomalies within the inner core. First, seismic waves travel faster in certain directions through the inner core. This is known as *seismic anisotropy*. Recent

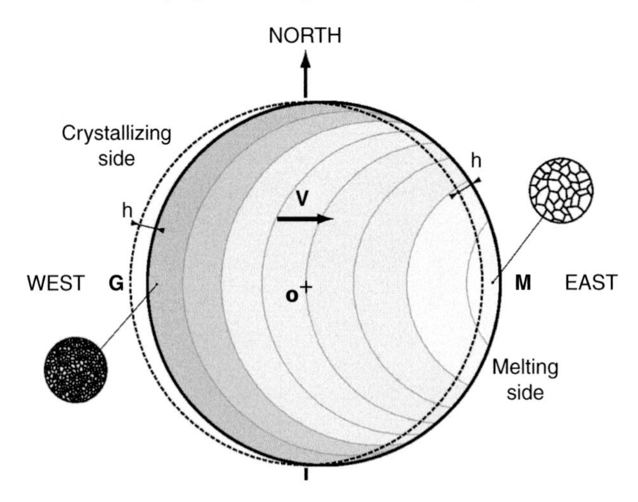

Fig. 6.2 Schematic of a translating inner planetary core due to inner-core convection. The dashed circle is the equilibrium position of the inner core. Thermal perturbations from degree-one inner-core convection cause the inner core to shift to the right inducing melting on the warm side and crystallization on the cold side. (From Monnereau *et al.*, 2010.)

work has also demonstrated that this anisotropy is different in the western and eastern hemispheres of the inner core. What could be responsible for such a lateral difference? Studies suggest that the inner core might evolve in a very interesting manner. If the inner core experiences large-scale solid-state convection (Buffett, 2009), then the thermal perturbations (resulting in density perturbations) in the inner core could shift it from the center of the planet due to gravitational forces. This would displace the inner core boundary from the geocentric freezing boundary (see Fig. 6.2). Ultimately, this process leads to crystallization at the inner core boundary predominantly occurring in one hemisphere and melting occurring in the other hemisphere (Monnereau *et al.*, 2010; Alboussiere *et al.*, 2010). This would make the age of the material in the inner core younger on the crystallizing side and older on the melting side explaining the hemispheric seismic differences. What this means for the dynamo is that the buoyancy sources driving the dynamo (the release of light elements and latent heat at the solidifying boundary) may not be homogeneous at the inner core boundary. Dynamo simulations by Aubert *et al.* (2013) have demonstrated that such variations may explain the inhomogeneity in secular variation rates in the Pacific and Atlantic hemispheres.

6.2.1.2 Mercury

The fact that Mercury possesses an intrinsic magnetic field has been known since the Mariner 10 mission in the mid 1970s. From Mariner 10 it was determined that Mercury's observed magnetic field is predominantly dipolar but much weaker than expected from standard scaling studies. Previous dynamo studies and simulations have worked to explain this weakness in the field strength (see Stanley and Glatzmaier, 2010, for a review).

The more recent MESSENGER mission has provided exciting new data on Mercury, including its magnetic field. In addition to the weakness of Mercury's field, the dipolar field is offset by approximately 480 km northward from the equator (equivalent to having a magnetic field with a $\sim$40% magnetic quadrupole component relative to the dipole). In addition, present data suggest the field is fairly axisymmetric (see Fig. 6.1) with a dipole tilt smaller than 0.8° (Anderson *et al.*, 2012).

New dynamo simulations are working towards producing all three of Mercury's field characteristics: (1) weak intensity, (2) predominantly dipolar but with a large quadrupole component, and (3) large-scale axisymmetry. To achieve (2) and (3) simultaneously is challenging due to dynamo selection rules (Bullard and Gellman, 1954). Models suggest that lateral thermal heterogeneities at the core-mantle boundary may be able to explain the strong quadrupole component (Cao *et al.*, 2014).

6.2.1.3 Mars

Mars does not have an active dynamo today, although it does have a crustal magnetic field indicating Mars did possess a dynamo in its early history that subsequently died. There may be a connection between the death of the Martian dynamo and the loss of its early thick atmosphere due to solar wind erosion in the absence of a global magnetosphere, although this is contentious (see Ch. 7). At first glance the Martian dynamo may be easily explained as a brief-lived version of the geodynamo with the explanation for the brevity lying in the smaller size (and hence faster cooling) of Mars' core. However, the Martian crustal field is extremely intense and concentrated in the southern hemisphere. If the Martian dynamo had produced an Earth-like axially dipolar dominated magnetic field, and the Martian crust is similar in age and composition in both hemispheres, then one would not expect this asymmetry in the crustal magnetic field.

To explain this feature, researchers have suggested that either crustal reworking after magnetic-field emplacement removed the magnetization in the northern hemisphere (Nimmo and Gilmore, 2001; Solomon *et al.*, 2005), or that the dynamo on Mars produced a very asymmetric surface magnetic field, where surface fields were strongest in the southern hemisphere (Stanley *et al.*, 2008; Amit *et al.*, 2011; Dietrich and Wicht, 2011). The reason for this single-hemisphere dynamo relies on hemispheric thermal variations on Mars' core–mantle boundary due to the same mechanism that generated the Martian crustal dichotomy during crust formation.

This may have been a spherical harmonic degree-one mantle circulation pattern or a giant impact in the northern hemisphere of Mars.

Researchers are also investigating the potential for giant impacts to kill the Martian dynamo. If an impact can transfer enough heat to the core, the core can stratify and hence convection and the dynamo will shut down (Arkani-Hamed and Olson, 2010).

6.2.1.4 Ganymede

Ganymede is the only moon in our solar system with evidence of a present-day dynamo. The Galileo mission to the Jupiter system detected magnetic field signatures at the other Galilean satellites, but they are due to influences from Jupiter's magnetic field interacting with electrically conducting layers in these bodies, rather than due to an internal dynamo. The Cassini mission at Saturn detected no intrinsic magnetic fields in Titan (Saturn's largest moon) or any of the smaller moons that it has visited. As discussed in the next section, Earth's Moon most likely had a dynamo in its past, but it has since decayed.

The challenge with Ganymede is to explain the longevity of the dynamo because, as a smaller body, it should have cooled fairly quickly and convection should have ceased by now. Present thinking is therefore that novel convection sources may exist in Ganymede. If sulfur is the main light element in Ganymede's core, then at the modest planetary pressures in Ganymede the core may actually begin solidification at its outer boundary resulting in Fe snow, or solid FeS may precipitate deeper in the core (Hauck *et al.*, 2006; Zhan and Schubert, 2012). These methods of convection are not well studied and it is likely that Ganymede's ability to maintain a dynamo is rooted in the details.

6.2.1.5 Moon

Evidence for crustal magnetism on Earth's Moon comes from Lunar Prospector's electron-reflectometry and fluxgate magnetometer instruments (Mitchell *et al.*, 2008; Purucker and Nicholas, 2010) as well as from paleomagnetic analyses of Apollo samples (Wieczorek *et al.*, 2006). This remanent crustal magnetic field is most likely due to dynamo action. Recent advances in seismology and paleomagnetic techniques have led to new insights regarding the past lunar dynamo:

(1) **Lunar seismology** The Apollo Passive Seismic Experiment recorded seismic activity from lunar quakes in the 1970s. Since then, seismologists have made significant advances in processing of seismic data for Earth-based studies and recently Weber *et al.* (2011) applied such methods to the old lunar seismic data. This unraveled much information on the lunar interior structure including the size of the lunar core ($\sim$400 km) and the fact that there is likely

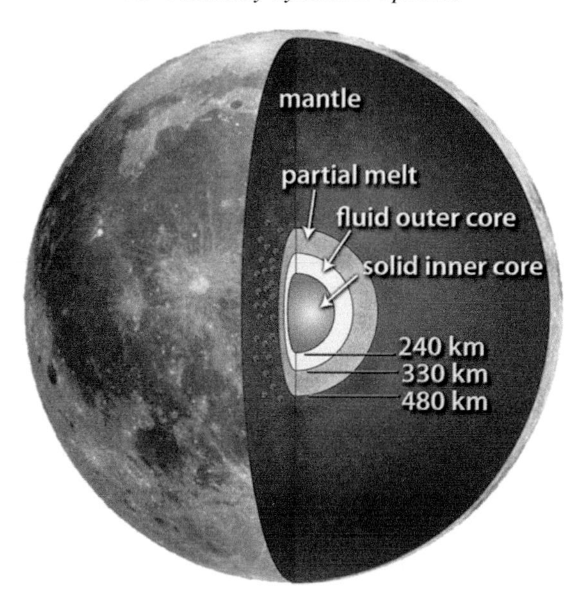

Fig. 6.3 Schematic of the lunar interior from lunar seismic data. (From Weber *et al.*, 2011.)

a solid inner core and a partially molten layer above the lunar core–mantle boundary (Fig. 6.3). This has allowed for more accurate modeling of lunar thermal evolution and dynamo simulations.

(2) Lunar paleomagnetism Advances in paleomagnetic techniques have resulted in re-analyses of the magnetic fields in lunar samples from the Apollo missions (Garrick-Bethell *et al.*, 2009; Shea *et al.*, 2012; Suavet *et al.*, 2013). This has allowed for estimates of the intensity of the magnetizing field present on the lunar surface as a function of time. These studies suggest that a lunar dynamo produced a fairly intense surface field (10–100 microtesla) from 4.25 to 3.56 billion years ago. This implies that the lunar dynamo was long-lived which is challenging to explain if it is driven by convective motions.

To explain the longevity of the lunar dynamo, alternative driving mechanisms have been proposed. Repetitive torques on the lunar mantle by large impacts have been shown to produce core–mantle boundary forcings that are energetic enough to sustain a dynamo and produce flows that are dynamo-capable (Le Bars *et al.*, 2011), but due to the interim nature of impacts, it is difficult to understand how the longevity of the field can be explained. Alternatively, precessionally forced flows in the core early in lunar history may provide the answer (Dwyer *et al.*, 2011). Numerical simulations have demonstrated that this mechanism can explain both the necessary intensity and longevity of the lunar surface field (Tian *et al.*, 2014). It is also possible that a relatively wet and compositionally stratified lunar mantle can keep convection going in the core for long enough to explain the paleomagnetic data (Evans *et al.*, 2014).

6.2.2 Giant planets

All four giant planets in our solar system have dynamo-generated magnetic fields. Sections 7.3 and 7.7 of Vol. III (Christensen, 2010) provide a nice discussion of the morphology of the giant-planet magnetic fields as well as their interior dynamo-region structure. Below, we highlight some recent results from numerical simulations and mission data on these bodies.

6.2.2.1 Jupiter

Jupiter's magnetic field is very similar in morphology to Earth's field in that it is dominated by an axial dipole component (Fig. 6.1) and has a similar dipole tilt ($\sim 10^\circ$). Jupiter's surface field is about ten times stronger than Earth's, which is expected based on simple scaling laws using Jupiter's size and rotation rate.

The jovian dynamo is generated in the metallized hydrogen region of the planet which extends from deep in the planet out to a radius of about $0.85 R_{\mathrm{Jup}}$. Dynamo scaling studies for Jupiter are capable of predicting the similarity of their field morphologies if you take into account that the dynamo region in Jupiter is very thick, like in Earth, and the planet is a rapid rotator.

However, there is one major difference between Jupiter and Earth that might be important for dynamo action. As a gas giant planet, there is no sharp boundary in physical properties between the dynamo region and the surrounding layers. Instead, Jupiter's physical parameters can depend strongly on pressure and temperature (and hence depth) in the planet. For example, density, as well as the electrical and thermal conductivities vary by orders of magnitude from the atmosphere of the planet to its deep interior. Recent Jupiter dynamo simulations have attempted to recreate the interior dynamics in a body with these varying properties (Stanley and Glatzmaier, 2010; Duarte *et al.*, 2013). The key goal is to produce a simulation that simultaneously generates the famous observed surface zonal jets on Jupiter while also generating a dynamo that produces a surface magnetic field similar to Jupiter observations.

6.2.2.2 Saturn

Saturn's magnetic field is of similar intensity to Earth's, but is unique in its level of axisymmetry (Fig. 6.1). No non-axisymmetric spectral components are required to explain present Saturn data, even with re-analysis of Cassini data providing field models resolved to spherical harmonic degree $L_{max} = 5$ (Cao *et al.*, 2012).

The lack of non-axisymmetry in the observed data has been discussed as problematic since Voyager observations. This is due to Cowling's theorem (Cowling, 1933) which states that a dynamo cannot generate a perfectly axisymmetric magnetic field. The most likely explanation for Saturn's field is therefore that the helium insolubility layer which surrounds the dynamo region in Saturn is responsible for attenuating non-axisymmetric field components so they are not visible at the planetary surface (Stevenson, 1980).

Numerical dynamo simulations by Christensen and Wicht (2008) and Stanley (2010) have implemented stably stratified layers surrounding dynamo regions in an effort to demonstrate the feasibility of this mechanism. Although the studies use substantially different thicknesses for the helium insolubility layer, both are able to produce more axisymmetrized fields.

A data analysis by Cao *et al.* (2012) suggest another unique feature of Saturn's field geometry. There appears to be a preference for odd harmonics (i.e., modes with equatorial anti-symmetry) in the surface magnetic field spectrum. The signs of the largest odd modes (dipole and octupole) result in concentrations of field in the polar regions. This is opposite to what is observed on Earth where the dynamo-generated magnetic fields are weaker near the poles. Recent dynamo models attempt to explain this feature in addition to the field's axisymmetry (e.g., Cao *et al.*, 2012).

6.2.2.3 Uranus and Neptune

Data from the Voyager 2 mission demonstrated that the ice giants have multipolar magnetic fields rather than the axial-dipolar magnetic fields of other solar-system bodies (Fig. 6.1). Previous numerical models involving stably stratified layers interior to the dynamo-generating water-rich layers were used to explain this field morphology (Stanley and Bloxham, 2004, 2006). Recent models involving 3D-turbulence dynamos (Soderlund *et al.*, 2013) or the lower electrical conductivity of ionic water (Gómez-Pérez and Heimpel, 2007) have also been proposed as solutions.

The most significant insight into the ice giant dynamos has, arguably, resulted from new ab-initio studies of the properties of water at high pressure and temperature. Redmer *et al.* (2011) demonstrate that a new phase of water, known as *superionic water* may occur in the deeper regions of Uranus and Neptune (see Fig. 6.4). The physical properties of this new phase, such as its electrical conductivity, thermal conductivity, and viscosity may strongly impact the dynamo processes in these bodies. As Redmer *et al.* (2011) demonstrate, it may not be a coincidence that the radius of the stably stratified layers required in the models by Stanley and Bloxham (2004, 2006) occur at approximately the same depth as this new water

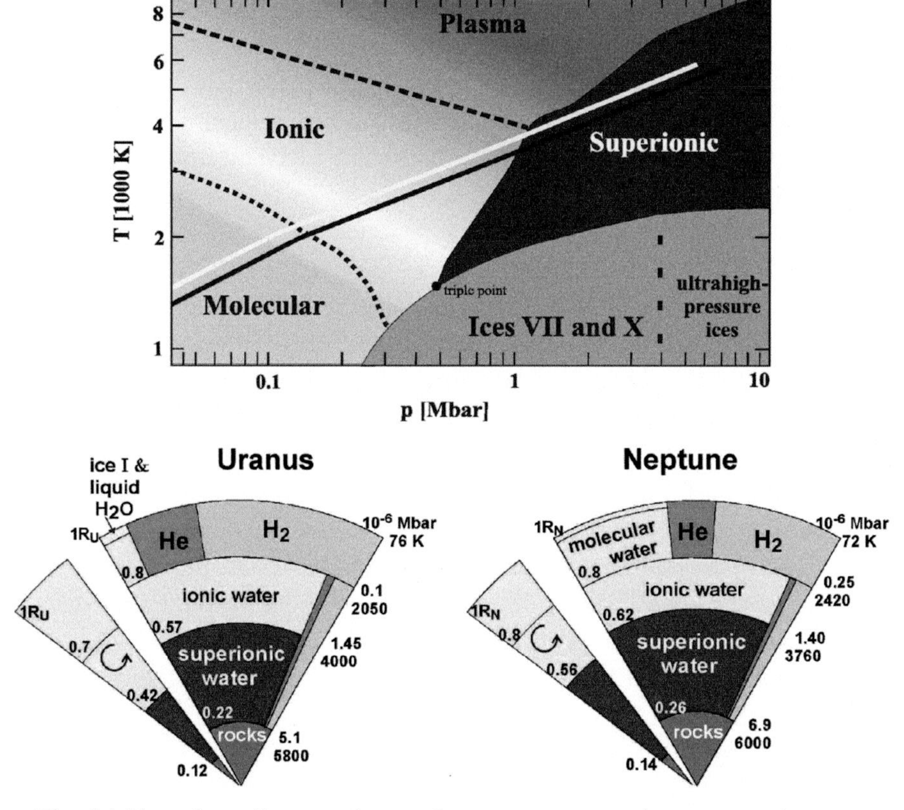

Fig. 6.4 Top: phase diagram of water for temperatures and pressures relevant to the ice giant planet interiors. Isentropes for Uranus (white) and Neptune (black) are also shown. Bottom: three-layer interior composition models for Uranus and Neptune that reproduce the gravity field data. The thin slice on the left of each figure is the structure of the dynamo source region used in Stanley and Bloxham (2006) for dynamo models. Figures from Redmer *et al.* (2011).

phase. Further work on the properties of superionic water will likely produce the biggest advances in understanding the ice giant dynamos.

6.3 Planetary dynamos: new frontiers

6.3.1 Small body dynamos

The Magnetic Reynolds Number criterion in Eq. (6.3) makes dynamo action in smaller bodies problematic due to the inherent smaller length scales. Exacerbating this problem is that thermal conduction is relatively more efficient at cooling small bodies and therefore driving fluid motions through convection for long times in smaller bodies is also problematic. This means that the lifetimes of small-body dynamos are typically much shorter than in larger bodies, all other things being equal.

Here, we consider two groups of small bodies for which some evidence of past or present dynamo action exists: planetesimals and asteroids.

6.3.1.1 Planetesimals

Planetesimals were the large (tens to hundreds of km) building blocks of planets that were present in the early solar system during planet formation. Although no planetesimals currently exist (unless you count asteroids and comets), there are remnants of planetesimals in the form of meteorites that have been found on Earth. Some very old meteorites, such as a group of basaltic achondrites called the Angrites (Weiss *et al.*, 2008) and the CV chondrite, Allende (Carporzen *et al.*, 2011), demonstrate strong magnetization for which the best explanation is that they formed on parent bodies which had differentiated to form cores early in solar system history and sustained an active dynamo for millions of years.

Planetesimals can differentiate fairly early in solar system history due to the formation of magma oceans on these bodies which result from radiogenic heating by ^{26}Al in the early solar system. These magma oceans also aid in cooling the planetesimal cores rapidly enough to generate core convection (i.e., the core heat flows are super-adiabatic). Thermal modeling by Weiss *et al.* (2008) demonstrates that super-adiabatic heat flows can be maintained until the magma ocean solidifies and this process can last for tens of millions of years (Fig. 6.5a).

Using scaling laws to estimate Magnetic Reynolds Numbers (Fig. 6.5b) and surface magnetic field strengths, Weiss *et al.* (2008, 2010) show that these super-adiabatic core heat fluxes can result in dynamos with appropriate duration and field intensities to explain the Angrite magnetism (Fig. 6.5c).

The magnetism of the Allende meteorite is a bit of a puzzle because, along with other CV chondrites, its texture suggests it has not experienced significant melting, which one would expect if it formed on a differentiated planetesimal with a core. However, the magnetism in Allende can be explained if the magma ocean on the CV chondrites' parent body was not global, but instead, only occurred at depth (Fig. 6.5d). This would leave an unmelted shell surrounding the magma ocean and core which would be producing the dynamo (Elkins-Tanton *et al.*, 2011; Weiss and Elkins-Tanton, 2013).

6.3.1.2 Asteroids

No presently active dynamos have been found on asteroids. This is not surprising because, although these bodies are of similar sizes to planetesimals, they are now far too old to presently have convecting cores. However, it is possible that some differentiated asteroids possessed dynamos in their pasts. For example, paleomagnetic

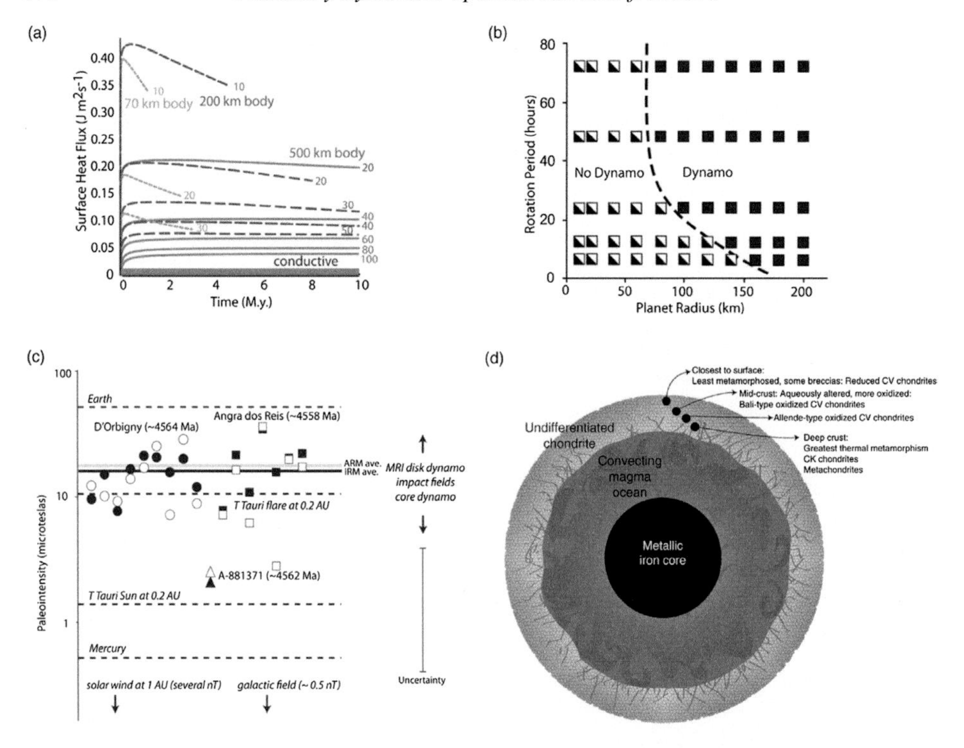

Fig. 6.5 Planetesimal dynamos. (a) Estimates of lifetimes of convection from thermal modeling. (b) Regions of phase space where dynamos have supercritical Magnetic Reynolds Numbers. (c) Paleointensity measurements for various Angrite meteorites as well as thresholds for magnetizing field strengths from different sources. (d) Interior structure of a chondrite parent body. Panels (a), (b), and (c) from Weiss *et al.* (2008). Panel (d) from Elkins-Tanton *et al.* (2011).

studies of the eucrite meteorite Allan Hills A81001, believed to have formed on the differentiated asteroid Vesta (with a core radius of $\approx$ 110 km); Russell *et al.*, (2012) suggest that Vesta possessed a surface field strength of at least 2 microteslas 3.69 billion years ago (Fu *et al.*, 2012). It is therefore possible that Vesta currently has a crustal magnetic field that could be measured from spacecraft magnetometers.

6.3.2 Extrasolar planets

The study of exoplanet dynamos is interesting for several reasons. First, the existence of a dynamo-generated magnetic field that can produce a large magnetosphere may have implications for habitability. Second, radio emissions from stellar wind–magnetosphere interactions provide a potential new detection mechanism for extrasolar planets.

Possible mechanisms to detect magnetic fields from extrasolar planets include detection of stellar spectral features indicative of magnetic interactions between close-in planets and their parent stars (Shkolnik *et al.*, 2003; Cohen *et al.*, 2009), and the observation of synchrotron emission from stellar wind interactions with planetary magnetospheres (Griessmeier *et al.*, 2011). At present, no radio emissions have been detected from extrasolar planets, although campaigns are searching. This is most likely due to the fact that the emissions need to be extremely strong and in an appropriate bandwidth to be detected from the Earth's surface. I shall make a bold prediction that it is only a matter of time for such emissions to be detected.

Observational campaigns have demonstrated that both a large range of masses and compositions are possible in planets (cf., Ch. 5). These necessarily have implications for the structure, composition, and evolution of the dynamo source regions in these bodies. In addition, planets can form in quite different environments than seen in our solar system. For example, many planets have been found that are extremely close to parent stars. This means they exist in a much hotter environment, that tidal effects can influence their orbits, that the stellar wind can be much stronger near the planets and that magnetic fields may even connect planets and parent stars. Below, we consider several classes of exoplanets that exhibit properties not seen in our solar system and discuss what this might mean for their dynamo-generated magnetic fields.

6.3.2.1 *Rocky planets*

The rocky (terrestrial) planets in our solar system all have similar structure and composition. Namely, an iron-rich core is surrounded by a rocky mantle made up predominantly of magnesium silicates. For example, Earth's mantle is approximately 80% $(Mg,Fe)SiO_3$ and 20% (Mg,Fe)O. The terrestrial planets differ slightly in the bulk Fe/Si ratio (higher for Mercury, possible lower for Mars) and the amount of Fe in the mantle, but to a large extent are quite similar to Earth. The rocky planets in our solar system are also relatively small, with Earth being the largest. Because the internal pressures in the planets depend on the planet size, this means that the highest mantle pressures seen in our solar system are in Earth, and are about 135 GPa near the core–mantle boundary. At these pressures, magnesium silicates are good electrical insulators and do not affect the magnetic field generation process in the core.

Exoplanet studies have demonstrated that terrestrial planets can form that are much larger than Earth. Dubbed *super-Earths*, these exoplanets naturally experience much higher pressures in their mantles. Pressures in deep mantles of super-Earths can reach the TPa ($= 10^{12}$ Pa) level. It is also possible that rocky exoplanets have mantles with significantly different composition than solar system planets. For example, they may have larger fractions of iron oxides like FeO or more exotic compositions.

There have been a few studies that use existing dynamo scaling laws to estimate surface field strengths in rocky exoplanets. For example, Driscoll and Olson (2011) consider optimal scenarios for super-Earth core evolution to derive magnetic dipole moments as a function of planetary mass and Zuluaga and Cuartas (2012) consider the influence of rotation on super-Earth magnetic fields. However, different exoplanet compositions and environments may result in other factors that need to be considered when predicting dynamo properties.

Theoretical and experimental work on different rocky compositions have found the regions of pressure–temperature phase space where these compositions become metallic (Ohta *et al.*, 2012; Nellis, 2010; Tsuchiya and Tsuchiya, 2011). Interesting results include that FeO metallizes at about 60 GPa (notice these pressures exist at quite shallow depth even in Earth's mantle), Al_2O_3 metallizes at 300 GPa (pressures not seen in solar system terrestrial mantles), and $CaSiO_3$ metallizes at 600 GPa (a pressure which would occur near the core–mantle boundary in a super-Earth with mass 5 times that of Earth).

If a rocky exoplanet contains a significant fraction of a mantle composition that is metallic at some depth (Fig. 6.6c), then there are implications for the dynamos in these bodies.

- Thermal and mechanical effects.
  - The metallic phase should decrease the mantle viscosity compared to the insulating phase. This may either make mantle convection easier resulting in faster core cooling or result in layered mantle convection making core cooling slower.
  - The metallic phase would also have a larger thermal conductivity than its insulating counterpart implying that heat could be removed faster from the core.
- Electromagnetic effects.
  - An electromagnetic screening effect would attenuate the rapidly time-varying magnetic fields in the dynamo source region from reaching the planetary surface.
  - The electromagnetic boundary condition at the core–mantle boundary (CMB) would be different if the mantle-side of the CMB were a good conductor.
  - Magnetic fields that penetrate (and effectively anchor in) the metallic mantle layer could experience significant stretching due to shearing motions on the core-side of the CMB resulting in new field generation mechanisms.
  - If the temperatures are high enough in the planet such that the metallic mantle layer is liquid, then it is also possible that a dynamo may operate in this mantle layer.

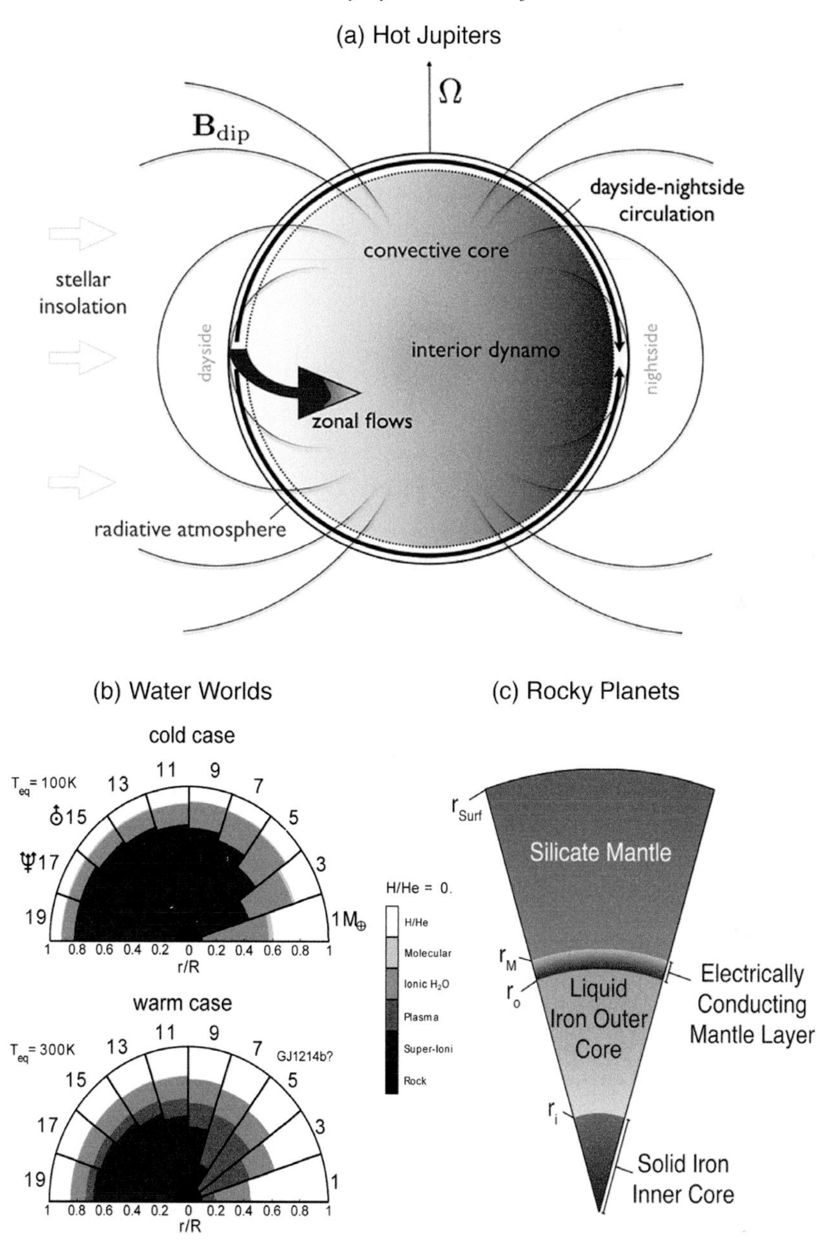

Fig. 6.6 Interior structure diagrams for various categories of exoplanets. (a) Schematic of a Hot Jupiter indicating dayside–nightside atmospheric flows which can interact with a dynamo-generated magnetic field (from Batygin *et al.*, 2013), (b) Potential interior structures for water-rich planets of varying mass and equilibrium temperature (from Tian and Stanley, 2013). Possible interior structures corresponding to Uranus (⛢), Neptune (♆) and GJ1214b are marked. (c) Interior structure schematic for a rocky exoplanet with an electrically conducting mantle layer (from Vilim *et al.*, 2013).

Vilim *et al.* (2013) investigated the electromagnetic effects of an electrically conducting mantle on the dynamo and the observable surface magnetic field. They demonstrated that metallized mantle layers result in stronger magnetic fields inside the dynamo source region, but that the field can be somewhat weaker at the surface, especially for planets with thinner convective fluid cores. This implies that for rocky exoplanets, those with smaller masses (i.e., less likely to have metallic mantles) may exhibit stronger surface magnetic fields than planets with larger masses.

6.3.2.2 Water-rich planets

The water-rich bodies in our solar system include comets ($\sim$1–10 km radii), the solid, icy-surfaced moons ($\sim$10–1000 km radii), and the fluid ice giant planets, Uranus and Neptune ($\sim$25 000 km radii). Here, we concentrate on ice giant exoplanets as these objects may actually have dynamo action occurring in their water-rich layers (as opposed to an iron-rich core). They may differ in size from our ice giants as well as be located at different orbital distances resulting in hotter or colder surface temperatures. In addition, the amount of hydrogen and helium atmosphere surrounding the water-rich layer may vary. All three of these aspects can affect the dynamo source region in these bodies.

As discussed in Sect. 6.2.2.3, the phase diagram of water has been recently revised to include a transition to the superionic water phase at high pressures and moderate temperatures. This phase may exist in exoplanets and if so, is likely to have implications for the dynamos in these bodies. Tian and Stanley (2013) created 1D interior structure models for water-rich bodies with a range of masses, H/He envelope mass fraction, and equilibrium temperatures. By calculating the temperature–pressure profiles for these bodies and comparing them to the phase diagram of water by Redmer *et al.* (2011), they found that small changes in planetary mass and moderate changes in H/He content can significantly affect which water phases are present in the planet (Fig. 6.6b). This may result in large differences for the dynamo source regions if, for example, the viscosity, electrical conductivity, or stability of the superionic water layers differ from ionic and plasma phases. One prediction from these models is that GJ1214b (a possible water-rich exoplanet), does not have a significant superionic water layer, and instead, may have a thick plasma phase of water in its deep interior where the dynamo is generated.

6.3.2.3 Hot Jupiters

Hot Jupiters are gas giant planets with close-in orbits (i.e., within $\sim$0.1 AU of their host star). Owing to this proximity to the parent star, these planets are highly irradiated and are most likely tidally locked such that they have permanent daysides and nightsides. The locations of the dynamo regions in these bodies are, at first glance, expected to be similar to that in Jupiter, namely at depths such that hydrogen sufficiently metallizes to produce a supercritical Magnetic Reynolds Number. However at least a couple of complications need to be considered.

- Because of the highly irradiating environment, the radiative-convective boundary in the atmospheres may be quite deep ($\sim$100–1000 bar, compared to a depth of $\sim$0.01–1 bars in Jupiter).
- Owing to the permanent dayside/nightside divide, there may be significant lateral thermal variations deep in these bodies. This means that the physical properties that depend on temperature (such as the thermal and electrical conductivity) may vary significantly with longitude.

Both of these speculations depend on how efficiently the outer atmospheric layers in these bodies can redistribute heat laterally (i.e., from the dayside to the nightside as shown in Fig. 6.6a). Several global circulation models (GCMs) of the atmospheric dynamics in Hot Jupiters attempt to answer this question (Showman *et al.*, 2011).

Another scenario that arises in Hot Jupiters is the partial ionization of certain alkali metal species, such as Na and K, in atmospheric layers due to the high temperatures. This can result in atmospheric layers with significant electrical conductivity. Batygin and Stevenson (2010) demonstrate that electrical currents in these atmospheric layers driven by the dayside–nightside flows may generate enough ohmic dissipation to explain the inflated radii of these planets. Therefore, there is a coupling between the internal dynamo-generated field and the flows in these ionized atmospheric layers. Several groups have worked on producing more sophisticated models of the magnetic interactions between the dynamo and these atmospheric layers (Perna *et al.*, 2010; Menou and Rauscher, 2010; Rauscher and Menou, 2013; Batygin *et al.*, 2013; Rogers and Showman, 2014). These studies demonstrate that zonal jets in the atmospheric layers are likely damped by Lorentz forces.

6.4 Outlook

For studies of planetary dynamos, there is much to look forward to. Advanced numerics and experiments will bring new insights from numerical dynamo simulations, paleomagnetism, and high-pressure material physics. Complementary information about planets (e.g., gravity fields, compositional studies, thermal evolution) will also aid in answering the fundamental questions in this area. There will also be new mission data on planetary magnetic fields. Upcoming or active magnetic missions that will provide new data include the following.

- **Juno** En route to Jupiter, Juno will provide the best magnetic field data for Jupiter to date. This year-long polar orbiter is predicted to resolve the surface magnetic field up to spherical harmonic degree $L_{max} \approx 20$ and there is a possibility of detecting secular variation in the field. This will be the first time such time variability has been observed from a dynamo-generated field in a planet other than Earth.
- **Cassini** Cassini's final orbits will be at higher latitudes and closer to Saturn than has been possible to date. This will provide new magnetic field data in polar regions that may help to answer outstanding questions about Saturn's magnetic field.
- **Swarm** The Swarm constellation of satellites was launched in 2013 and will provide new global geomagnetic data to answer outstanding questions regarding the geodynamo.
- **BepiColombo** Expected to reach Mercury in 2024, the BepiColombo mission will provide new magnetic data for Mercury. With the time between MESSENGER and BepiColombo, it will be interesting to investigate possible changes in the magnetic field.
- **Juice** The Jupiter Icy moons Explorer (slated for possible arrival at the Jupiter system in the 2030s) will provide the first new data on Ganymede's magnetic field as well as explore the magnetic environment and internal oceans of the Galilean satellites.

Looking beyond our solar system will also be crucial, both by providing further data points from exoplanet magnetic field studies and by refining our understanding of MHD processes in other astrophysical bodies.

7

Climates of terrestrial planets

DAVID BRAIN

Suppose we detect a planet half the size of Venus orbiting a 5 billion year old M-type star at 0.5 AU. To our surprise the planet has detectable radiation belts. How might the planet's climate and surface habitability differ from that of Venus?

The prospect that the scientific community might be faced in the next decade or two with questions like the one above is exciting. It is also daunting, because we will likely be required to make inferences about distant planets based on partial information about their environment, orbit, and characteristics. Fortunately, we have at our disposal abundant information about the planets in our own solar system, and more than a half century of practice studying how very complex climate systems function.

This chapter provides the interested reader with an overview of the likely links between heliophysics and climate. Climate is typically defined as the long-term (multi-decade or longer) average of weather. For example, Merriam Webster defines climate as "the average course or condition of the weather at a place usually over a period of years as exhibited by temperature, wind velocity, and precipitation", while Wikipedia currently describes climate as "a measure of the average pattern of variation in temperature, humidity, atmospheric pressure, wind, precipitation", atmospheric particle count, and other meteorological variables in a given region over long periods of time. Climate studies therefore investigate properties of planetary atmospheres – properties that are influenced by both intrinsic characteristics of the planet and by interactions with the host star. To determine whether a planet is or has been habitable at its surface, therefore, it is helpful to understand how a variety of processes act together to influence climate over time. Heliophysical processes are important components of this understanding.

7.1 Current climates of terrestrial planets

This chapter focuses on the global climates of terrestrial (rocky) planets Venus, Earth, and Mars. Because they have atmospheres, gas and ice giant planets such

Table 7.1 *Present characteristics and climates of the terrestrial planets*

	Venus	Earth	Mars
Radius	6050 km	6400 km	3400 km
Heliocentr. dist.	0.72 AU	1 AU	1.52 AU
Rot. period	243 days	24 hours	24.6 hours
Surface temp.	740 K	288 K	210 K
Surface press.	92 bar	1 bar	7 mbar
Composition	96% CO_2; 3.5% N_2	78% N_2; 21% O_2	95% CO_2; 2.7% N_2
H_2O content	20 ppm	10,000 ppm	210 ppm
Precipitation	None at surface	Rain, frost, snow	Frost
Circulation	1 cell/hemisph., quiet at surface but very active aloft	3 cells/hemisph., local and regional storms	1 cell/hemisph. or patchy circulation, global dust storms
Max. surf. wind	~3 m/s	>100 m/s	~30 m/s
Seasons	None	Comparable north. and south. seasons	Southern summer more extreme

as Jupiter, Saturn, Uranus, and Neptune have climates, as do planetary moons with gravitationally bound atmospheres ranging from very thick (e.g., Saturn's moon Titan) to considerably more tenuous (e.g., Jupiter's moons Io and Europa). Dwarf planets (e.g., Pluto) also have climates, though we know relatively little about them at present. The terrestrial planets are of special interest because they are thought to have been habitable at their surfaces at some point during solar system history. They formed under similar conditions (see Ch. 4 in Vol. III), with early atmospheres that were more similar than they are today. The present day climates of Venus and Mars provide a useful contrast to that of Earth, and exploration of the root causes for differences in the present climates of all three planets allows us to better understand the processes that control climate everywhere. Their current climates are summarized in Table 7.1, and discussed briefly below.

The surface pressures of the three planets differ by more than four orders of magnitude, with a Martian pressure less than 1% that of Earth, and Earth's pressure a little more than 1% that of Venus. The atmospheric density near the surface of Venus is approximately 8% that of liquid water, while the atmospheric density near the surface of Mars is comparable to the density at altitudes higher than ~35 km on Earth, more than three times the altitude of Mt. Everest. Despite their large differences in mass, the atmospheres of Venus and Mars have similar bulk compositions, with carbon dioxide (CO_2) comprising ~95% by volume, followed by molecular nitrogen (~3%), and argon (~1%). Earth's atmosphere, by contrast, is composed mainly of nitrogen and oxygen, followed by argon. Earth's atmospheric composition likely mirrored that of Venus and Mars early on, but much of Earth's atmospheric CO_2 now resides in carbonates on the ocean floors, leaving nitrogen as the most common constituent. Earth's abundant atmospheric oxygen is believed to have been contributed by photosynthetic bacteria.

The surface temperatures of the three planets also differ widely, in part due to the distance of each planet from the Sun and in part due to the quantity of greenhouse gases in each atmosphere. Earth is the only of the three planets with a surface temperature (and pressure) appropriate for liquid water to be stable for long periods of time, thanks to $\sim$30 K of greenhouse warming. The atmosphere of Venus is too hot for water to exist as liquid at the surface, while the Martian atmosphere has too low a surface pressure (liquid water would sublime, except at the lowest elevations). The atmosphere of Venus is very dry, indicating that any surface water driven into the atmosphere by the high temperatures no longer resides there. The relative atmospheric water content at Mars is an order of magnitude larger than at Venus and, given the low atmospheric pressure, is often nearly saturated. Despite the near 100% Martian relative humidity, Earth still has roughly 50 times more water molecules (per particle of atmosphere) than Mars. The composition, temperature, and water content lead to different forms of precipitation on the three planets. Earth has a variety of forms of water precipitation, while Mars has carbon dioxide and water frost. Venus has no precipitation at the surface due to its high temperatures; any precipitation that forms higher in the atmosphere would turn to vapor before reaching the ground.

Circulation patterns on the three planets also differ. Earth possesses three circulation cells in each hemisphere, leading to prevailing winds organized by latitude. The circulation results, in a simplified sense, from an equator-to-pole temperature gradient that causes warm air to rise at the equator and fall at the poles. Earth's rotation provides a Coriolis influence that breaks the circulation cells into three regions, keeping the warmest air relatively confined at low latitudes. Venus, by contrast, rotates very slowly. Thus, heat is transferred efficiently from the equator to polar regions, leading to uniform surface temperatures as a function of latitude and local time (see Bullock and Grinspoon, 2013). Mars rotates at nearly the same rate as Earth but has only one circulation cell per hemisphere, though there are some arguments to suggest that while there is a net circulation, air tends to move in localized regional cells (see Rafkin *et al.*, 2013). Air at the surface of Mars moves sufficiently quickly to drive dust devil activity, while the surface of Venus is very still. At higher altitudes on Venus, however, the atmosphere superrotates on time scales of days (e.g., Kouyama *et al.*, 2013).

While Earth's seasonal variations, caused by a 23.5° tilt relative to its orbital plane, will be well known to the reader, seasonal variations on Venus and Mars are substantially different. Venus has nearly no seasonal variation due to a very small ($\sim$3°) axis tilt. Mars has a tilt of 25°, similar to that of Earth, but the planet's greater orbital eccentricity (a 21% difference between the perihelion and aphelion distances compared to 1.4% and 3.3% for Venus and Earth, respectively) leads to shorter and more intense summers in the southern hemisphere compared to the north. Strong heating during southern summer drives enhanced dust devil activity, which can couple across circulation cell boundaries and grow into planet-encompassing dust storms that last several weeks.

7.2 Evidence for climate change

Abundant evidence points to changes in the climate of all three terrestrial planets on a variety of time scales. Here, we focus on evidence for climate change over tens of thousands of years or longer. The reader is also directed to discussions in Chs. 4, 11, and 12 in Vol. III (cf., Table 1.2).

The most compelling evidence for climate change on Venus comes from measurements of the isotopes deuterium and hydrogen in the atmosphere today (Fig. 7.1a). Deuterium is far scarcer than hydrogen in the atmospheres of all planets. However, the ratio of deuterium to hydrogen (D/H) in the Venus atmosphere – about two deuterium atoms for every 100 hydrogen atoms – is more

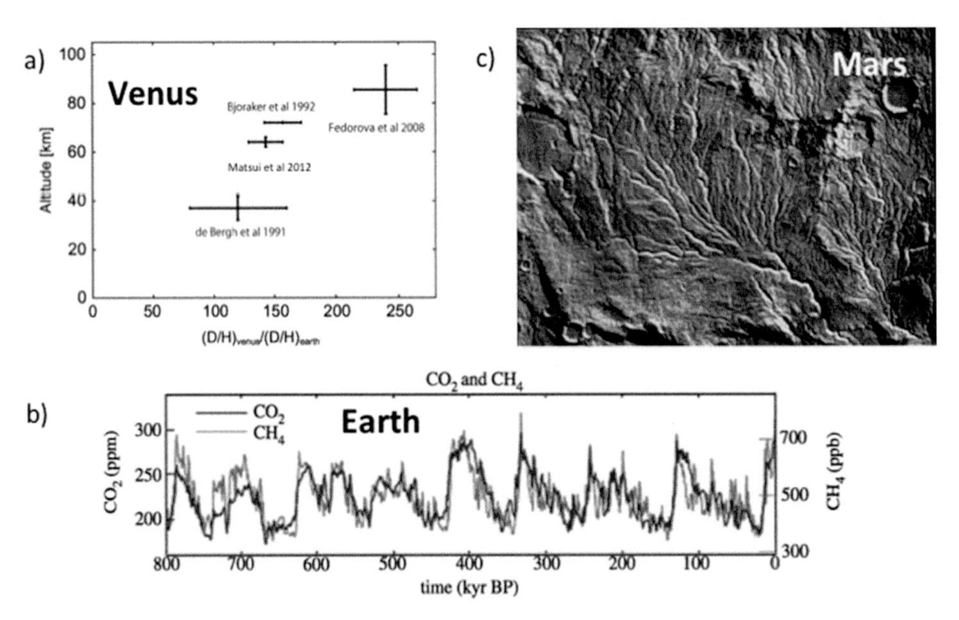

Fig. 7.1 Evidence for climate change on the terrestrial planets. (a) Determinations of D/H in the Venus atmosphere relative to terrestrial atmospheric D/H (Matsui *et al.*, 2012); (b) Earth's atmospheric carbon dioxide and methane concentrations as a function of time, as determined from ice cores (Hansen *et al.*, 2013); (c) a dendritic river valley network in the Warrego Valles region of Mars (courtesy NASA Viking).

than 100 times the same ratio calculated for Earth (Donahue *et al.*, 1982) and most other solar-system objects. There is little reason to expect that Venus formed with a D/H ratio significantly different from that of Earth, so we infer that the D/H ratio on Venus increased after the planet formed. Specifically, it is thought that hydrogen atoms (possibly from a primordial ocean) preferentially escaped the planet's gravity compared to deuterium and were lost to space. Section 7.5 explains this physical process in detail. The loss of hydrogen to space at Venus simultaneously explains the measured D/H ratio and the relative scarcity of water in the Venus atmosphere today: water was dissociated in the atmosphere and the hydrogen removed to space.

One might wonder whether the surface of Venus holds any clues about its past climate, in the same way that the geologic records of Earth or Mars can teach us about time periods billions of years ago. Unfortunately, the low-impact crater abundance at Venus suggests that the planet has been entirely resurfaced within the past several hundred million years. Whether this occurred in a global event or gradually over time is debated (Phillips *et al.*, 1992; Schaber *et al.*, 1992), but the implications for inferring climate change are identical in either case: it is not straightforward to use surface features as indicators of ancient climatic conditions at Venus.

In contrast to Venus, evidence for climate change on Earth is abundant and comes in many different forms. On time scales of hundreds to thousands of years, the measured growth rate of tree rings and coral as a function of time are used to infer various aspects of climate such as atmospheric and ocean temperature, precipitation, and ocean salinity. On time scales extending back hundreds of thousands of years, layered ice cores are used to infer atmospheric temperatures (through isotope ratios in the ice and layer thickness), atmospheric composition (through gases trapped in air pockets in the ice), and even which plants were present (through trapped pollen). Beyond a million years ago we are reliant on rock geochemistry, fossils, and sediment to provide information about atmospheric temperature, composition, climate shifts, and even surface pressure.

The terrestrial climate record from all of these sources suggests that Earth's climate varies on many time scales, with departures in temperature of as much as 10–15 °C over Earth's history (e.g., Hansen *et al.*, 2013). There are many inferred cold (glaciation) and warm periods that have been tied with changes in atmospheric conditions and diversity of life. Similarly, there are a few major changes in atmospheric composition, the most notable of which is the oxygenation of the terrestrial atmosphere more than two billion years ago (Bekker *et al.*, 2004; Ch. 4 in Vol. III), likely caused by the rise of oxygen-producing bacteria and the subsequent depletion of sinks for oxygen at Earth's surface (e.g., Kaufman *et al.*,

2007). Analysis of the size and depth of fossilized raindrop imprints in sedimentary rock even suggests that Earth's surface pressure has varied by as much as a factor of two over 2.7 billion years (Som *et al.*, 2012). Taken together, the evidence provides a caution against interpreting the present day climates of other terrestrial planets too finely, and assuming only monotonic changes in planetary climates over billion year time scales. At the same time, one of the most notable aspects of the terrestrial record is the fact that water has existed as liquid at the surface for most of the planet's history, suggesting that despite short-term deviations Earth's climate has been relatively stable over its history, in likely contrast to Venus and Mars.

Mars also provides several lines of evidence suggesting past climate that differs from today. This evidence can be broadly classified as geomorphologic, geochemical, or atmospheric (Jakosky and Phillips, 2001). Geomorphologic evidence includes surface features that are unlikely to have formed in today's environment. These include dry dendritic (branching) river valley networks (Fig. 7.1c), river delta deposits, possible regions of sedimentary rock, smoothed and rounded rocks imaged by Mars rovers, and possible ancient ocean shorelines. These features all suggest an ancient Mars where liquid water was abundant and active in shaping the surface of the planet. Further, highly eroded crater rims and a paucity of small craters relative to what might be expected from the abundance of large craters suggest that the ancient atmosphere was much more efficient at eroding surface features (i.e., thicker) than today – perhaps as thick as 0.5–3 bar, or even more.

Geochemical evidence on Mars demonstrates that water was the liquid responsible for creating the observed surface features, and not some other chemical species. Observations made from both orbiting spacecraft and surface rovers show the presence on both regional and local scales of minerals that require water to form and/or incorporate water as part of their crystal structure. These minerals include sheet silicates (phyllosilicates, including clays), sulfates, and carbonates.

Finally, a number of Martian atmospheric isotope ratios (D/H, $^{38}Ar/^{36}Ar$, $^{13}C/^{12}C$, $^{15}N/^{14}N$, $^{18}O/^{16}O$) point to the stripping of atmospheric particles to space over billions of years, similar to the inference drawn from D/H measurements at Venus (Jakosky and Phillips, 2001). Each of these species is enriched in the more massive isotope, suggesting that the lighter isotope has been preferentially removed to space from the upper atmosphere. In some instances (e.g., argon), no other processes are known that are capable of altering the isotope ratio. Together, the isotope ratios suggest that 50%–90% of the total atmospheric content has been removed to space from stripping processes alone.

7.3 How do climates change?

With abundant evidence that planetary climates are not static, we next turn our attention to the planetary characteristics and processes that can be responsible for changing climate. For this discussion we focus primarily on surface temperature, which directly or indirectly influences many other aspects of climate. The reader is also directed to Vol. III, Ch. 16 (Brasseur *et al.*, 2010).

Surface temperature is determined by the global energy budget for the atmosphere. An expression for the energy budget is given by

$$\frac{S}{d^2}(1 - A)\pi R_{\mathrm{p}}^2 = \sigma T_{\mathrm{eff}}^4 4\pi R_{\mathrm{p}}^2, \tag{7.1}$$

where S is the stellar irradiance at 1 AU ($\sim$1360 W/m^2 for our Sun, sometimes called the solar constant), d is the distance in units of AU from the star to the planet, A is the planet's Bond albedo (0 for perfectly absorbing and 1 for perfectly reflective), R_{p} is the radius of the planet, σ is the Stefan–Boltzmann constant, and T_{eff} is the effective radiating temperature of the planet. This equation assumes (quite reasonably) that incident and outgoing radiation at a planet are balanced, and that the planet is rotating (i.e., effectively redistributes energy). On the left-hand side, the incident flux of energy from the star at the location of the planet (S/d^2) strikes the disk cross section of the planet facing the star (πR_{p}^2) and is absorbed at the surface of planet according to its albedo ($1 - A$). On the right-hand side, the planet radiates energy away ($\sigma T_{\mathrm{eff}}^4$) over its entire surface area ($4\pi R_{\mathrm{p}}^2$). For the solar system Eq. (7.1) can be reduced to $T_{\mathrm{eff}} = 280(1 - A)^{1/4} d^{-1/2}$ K.

The temperature in Eq. (7.1) is effective temperature, which is the temperature of a black body emitting the same amount of radiation to space as the planet. Effective temperature can be related to surface temperature of a planet with an atmosphere under a few assumptions. Here, it is sufficient to write

$$T_s^4 = (1 + \tau) T_{\mathrm{eff}}^4, \tag{7.2}$$

where T_s is the surface temperature and τ is the optical depth of the atmosphere. This expression is derived under the assumption of a plane-parallel gray atmosphere (i.e., assumed to be wavelength independent) under radiative equilibrium. The atmosphere consists of τ slabs of unit optical depth, each of which absorbs all radiation emitted from adjacent layers, and emits black-body radiation only to adjacent layers.

There are a few things to note from Eqs. (7.1) and (7.2). First, the size of a planet plays no role in the global energy balance. On average, each portion of a planetary surface absorbs and radiates its fair share of energy, so that the size of a planet is unimportant. Second, the interior heat from a planet plays essentially no role in radiative equilibrium, as evidenced by the fact that only effective and surface temperature appear in Eqs. (7.1) and (7.2), and not the temperature of the planetary interior. This is true for terrestrial planets today in contrast to Jovian planets, which emit more heat than they receive from the Sun. Third, the surface temperature of a planet with an atmosphere is always greater than the effective temperature - atmospheres that absorb radiation act to warm a planetary surface. Finally, the equations apply to global averages, and not to local variations in the radiative energy budget.

Equations (7.1) and (7.2) are provided because they nicely illustrate four main ways in which planetary climate can be altered. First, the amount of radiation from the star (S) can change. The solar constant at Earth varies by only ~0.1% over the course of a solar cycle. But studies of Sun-type stars suggest that the Sun is ~30% brighter today than it was when the terrestrial planet atmospheres first formed (Fig. 7.2a; Sagan and Mullen, 1972; see also Ch. 2 in Vol. III). This makes the stability of Earth's climate all the more remarkable, because the amount of energy encountering the top of the atmosphere has changed considerably over time.

Second, changes in the albedo (A) of a planet will change the amount of incident energy absorbed by the surface (and atmosphere). Variation in cloud cover, the

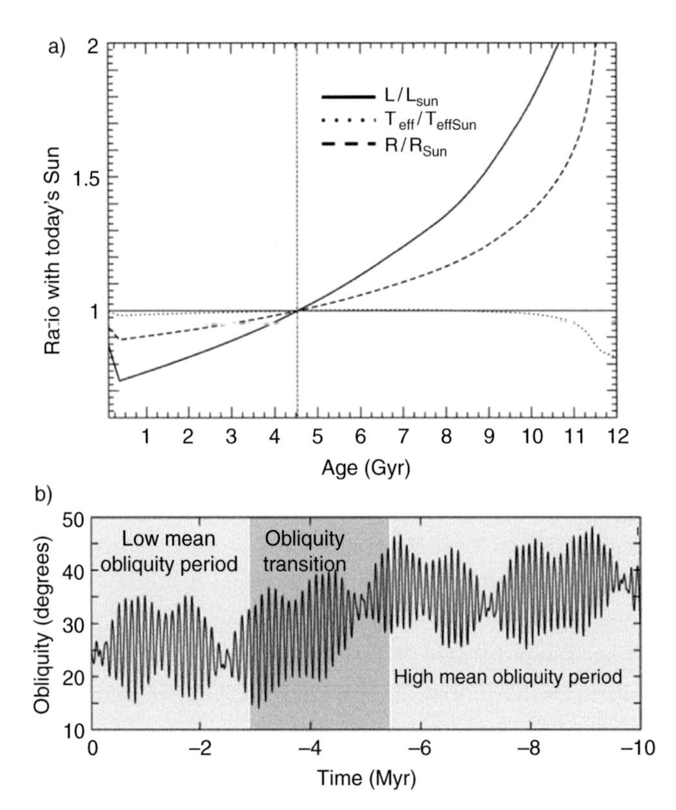

Fig. 7.2 Variation in climate drivers at terrestrial planets. (a) Modeled solar properties (luminosity, effective temperature, radius) as a function of time, relative to today's Sun (Ribas *et al.*, 2005); (b) Martian obliquity (i.e., tilt) as a function of time. (From Levrard *et al.*, 2004.)

extent of polar ices, vegetation, or wind blown dust, for example, can all change the albedos of the terrestrial planets, and will have an influence on the atmospheric energy budget. Venus has an albedo of ~0.9, while the albedos of Earth (~0.3) and Mars (~0.25) are considerably lower. Thus Earth and Mars absorb a larger fraction of incident sunlight than Venus. There are of course significant local variations in albedo and in the corresponding absorbed energy at different regions on the surface – especially at Earth.

Third, characteristics of a planet's orbit and rotation influence its energy budget. The amount of solar radiation encountering a planet varies with average orbital distance (d), with the result that Venus encounters roughly double the energy that Earth does, while Mars encounters ~45%. Ellipticity of the orbit (not captured explicitly in Eqs. (7.1) or (7.2)) influences variations in incident energy over a given orbit. For example, while the average energy incident at the top of the Martian atmosphere is ~45% that of Earth, it varies between 36% and 52% over a Martian year due to Mars' relatively high orbital ellipticity. This explains why the southern summer at Mars (near perihelion) is more extreme than northern summer. Tilt also influences the amount of sunlight that reaches each part of a planet's surface, making some portions of the planet cold and other portions warm. This effect influences where ices form at the surface, removing some gases from the atmosphere and changing albedo in some locations. Chaotic changes in the eccentricity, obliquity, and spin precession of Mars (Fig. 7.2b) and Earth over periods of tens to hundreds of thousands of years are thought to contribute to climate variations (see Ch. 11 in Vol. III), though the range of variation in both orbital properties (especially tilt) and climate is estimated to be larger at Mars due to the lack of a large moon (e.g., Laskar *et al.*, 2004).

Fourth, the amount of radiation-absorbing atmosphere (i.e., greenhouse gases) influences surface temperatures. The importance of the atmosphere is evident when we compare the amount of energy actually absorbed by Venus, Earth, and Mars. Considering the solar irradiance, average orbital distance, and globally averaged albedo, Earth absorbs the most incident sunlight of the terrestrial planets (~410 W/m^2), followed by a highly reflective Venus (~270 W/m^2), and then the more-distant Mars (~150 W/m^2). It may seem surprising at first that the absorbed energy at Earth's surface exceeds that of the much hotter Venus. However, a planet not only absorbs (and reflects) incident energy, it also radiates energy away. Greenhouse gases such as CO_2 and H_2O are efficient at absorbing the infrared energy emitted by the planet, keeping the lower atmosphere warmer than it would be in their absence. The more greenhouse gases are present, the more the surface is warmed, as shown in Eq. (7.2). The thick CO_2 atmosphere of Venus provides more than 500 K of greenhouse warming compared to the theoretical surface temperature in the absence of an atmosphere. Earth's atmosphere provides approximately 30 K of greenhouse warming. This warming, while much smaller than at Venus, is crucial to keeping our average surface temperature above the freezing point of water, making life and many aspects of our climate possible. The atmosphere of Mars, while dominated by CO_2, is too thin to provide substantial greenhouse warming today. The temperature is warmed only $\sim$5 K due to greenhouse gases. When we take into account the amount of greenhouse gases present in the atmospheres, we see that Venus retains a larger fraction of its radiated heat than the other planets, keeping its surface warmer despite its high albedo. Note also that the high albedo of Venus is due to its extensive cloud cover, which is a product of its climate. Thus albedo and greenhouse gas abundance are linked.

With this context in mind, we can examine how heliophysical processes may contribute to climate variation. They certainly should not influence the orbital characteristics of a planet in our solar system, though one can imagine the strength and timing of a stellar wind influencing planetary migration, and thus orbital distances, in other systems. They may indirectly influence albedo, by altering atmospheric chemistry and promoting the formation of clouds. Certainly, stellar properties directly influence the radiation from a star. And, perhaps surprisingly, interactions between a planet and its host star may change the abundance of climatically important atmospheric gases. This last idea involves an especially rich array of heliophysical processes, and is a focus for the rest of this chapter.

7.4 Atmospheric source and loss processes

As described in Sect. 7.2, surface temperature and climate are strongly affected by the amount of greenhouse gases in an atmosphere, which can be viewed as a combination of the total number of particles in an atmosphere (surface pressure) and its composition. A number of mechanisms are capable of changing atmospheric abundance and composition (Fig. 7.3; e.g., Hunten, 1992), only a few of which are heliophysical. We briefly describe them here.

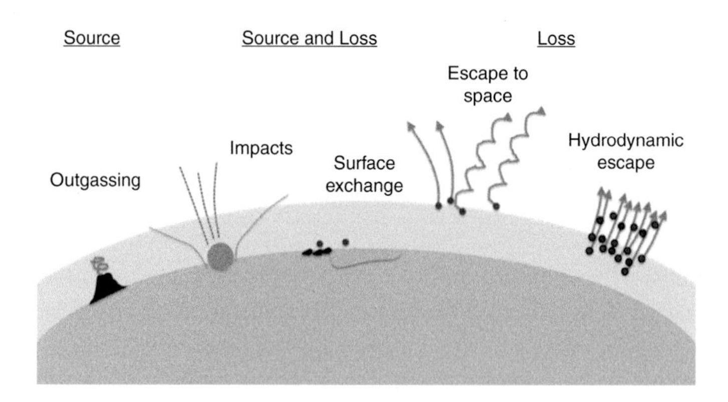

Fig. 7.3 Source and loss mechanisms for planetary atmospheres.

Volcanic outgassing from planetary interiors is thought to be the primary source for the terrestrial planet atmospheres we observe today. Water vapor is the most common gas released in terrestrial eruptions, followed by CO_2. Other commonly released gases include sulfur dioxide, nitrogen, argon, methane, and hydrogen. Outgassing should be a declining source of atmospheric particles over solar-system history, as the interior heat required to generate volcanic activity declines. Earth, as the largest terrestrial planet and therefore the one with the most interior heat, has the most evidence for volcanic activity today. Changes in both high-altitude atmospheric sulfur content and thermal emission from the surface suggest that Venus may still be active as well (Smrekar *et al.*, 2010; Marcq *et al.*, 2013). There is no direct evidence that Mars is active today, though relatively young lava flows suggest it could have been active as recently as a few tens of millions of years ago (Hauber *et al.*, 2011).

Atoms and molecules can be exchanged between a planet's surface layers and its atmosphere via a variety of processes and over many time scales. For example, changes in temperature can increase condensation rates to the surface, forming surface liquids or ices (evident on Earth and Mars). Chemical reactions (weathering) can also remove particles from the atmosphere, and is typically most effective in warm or wet environments (evident on Venus, Earth, and Mars). Adsorption removes atmospheric particles that stick to surface materials. Most or all of these processes can be considered to be reversible. Release of particles back to the atmosphere can involve changes in temperature, chemical reactions (including reactions with sunlight), and geologic events that allow subsurface reservoirs access to the atmosphere.

All planetary atmospheres are subject to impact from asteroids, comets, dust, and even atoms and molecules. Impactors of all sizes can deliver volatile species to an atmosphere (e.g., impact delivery is responsible for at least part of Earth's water inventory as well as meteoritic layers observed in terrestrial planet ionospheres). Impacts can also remove atmospheric particles via collisions, and sufficiently large impactors can additionally accelerate atmospheric particles via impact vapor plumes and lofted surface material (ejecta). Because the details of each impact determine whether there is a net gain or loss of atmospheric particles, it is not entirely clear how impacts have contributed to changes in atmospheric abundance and composition over time. It is certain that the importance of impacts has declined with time as the impactor flux decreases. Monte Carlo simulations suggest impacts have resulted in a net gain of atmospheric gases for Earth and Mars over solar-system history, and a net loss for Venus (Heath and Brain, 2014).

Hydrodynamic escape occurs when a light species escapes (thermally – see Sect. 7.6) in sufficient abundance that it becomes equivalent to a net upward wind, and drags heavier species with it through collisions. This process is usually enabled by high solar EUV flux or another form of heating. It should have been significant for all of the terrestrial planets during the first few hundred million years after formation, stripping away most of their primordial atmospheres. The present atmospheres, then, did not form in place on the terrestrial planets; instead they are the product of outgassing and impact delivery, along with subsequent evolution. Hydrodynamic escape may be ongoing at several observed exoplanets today (e.g., Vidal-Madjar *et al.*, 2003; Tian *et al.*, 2005), and perhaps even a few objects in our outer solar system (Tian and Toon, 2005; Strobel, 2009).

The removal of atmospheric particles to space from the upper layers of the atmosphere is commonly referred to as escape to space. This term typically excludes impacts by asteroids, meteoroids, and comets, and hydrodynamic escape is also often listed as a distinct process. Here, escape to space encompasses a set of approximately six processes, all of which provide escape energy to atmospheric particles. The energy is ultimately provided (sometimes directly, and sometimes indirectly) through interaction with the parent star and stellar wind. Escape to space is the most directly related to contemporary heliophysical processes of any of the atmospheric source and loss mechanisms, and is described in more detail in Sect. 7.6. It is currently thought that atmospheric escape has played an important role in the evolution of the climates of both Venus and Mars by altering atmospheric pressure and trace gas abundance.

7.5 Requirements and reservoirs for atmospheric escape to space

The removal of atmospheric particles via interactions with the Sun and solar wind is a scientific topic of much debate at present. There is little question that the processes occur; areas of investigation instead focus on whether escape to space is important for planetary evolution. Here, we describe the requirements for escape and reservoirs for escape in planetary atmospheres, and follow in Sect. 7.6 with a discussion of the mechanisms for removing particles.

All particles escaping from a planetary atmosphere share three characteristics. The first is that they have sufficient energy to escape the gravity of the planet. One can easily compute the necessary energy by assuming that a particle can escape when its kinetic energy exceeds its gravitational potential energy. Thus

$$v_{\text{esc}} = \sqrt{2GM/r}, \tag{7.3}$$

where v_{esc} is the escape speed, G is the universal gravitational constant, M is the mass of the planet, and r is the radial distance from which the escape occurs (typically the exobase, discussed below). Note that all terrestrial planets have similar bulk densities, so that we can replace M with density times volume, ρV. Volume is proportional to r^3, so that v_{esc} scales with radius. Table 7.2 shows typical escape

Table 7.2 *Escape velocities v_{esc} and escape energies E for protons and atomic oxygen*

	Venus	Earth	Mars
v_{esc}	10 km/s	11 km/s	5 km/s
$E(H^+)$	0.5 eV	0.6 eV	0.1 eV
E(O)	9 eV	10 eV	2 eV

speeds for Venus, Earth, and Mars. Mars has a much lower escape speed because the planet is smaller than Earth or Venus.

Because escape speed is independent of a particle's mass, escape energy must therefore be mass dependent. The table shows typical escape speeds for protons and atomic oxygen (both frequently considered in studies of escape). Escape energies range from fractions of an eV to $\sim$10 eV. Less-massive species require less energy to be removed from an atmosphere.

A second characteristic of an escaping particle is that it is unlikely to collide with other particles after acquiring sufficient escape energy. In planetary atmospheres, the region above which collisions are unlikely is termed the exobase, and is loosely defined as the location where the mean free path of a particle is equal to an atmospheric scale height

$$\frac{1}{n\sigma} = \frac{kT}{mg}, \tag{7.4}$$

where n is the number density of atmospheric particles, σ is the cross section for collisions, k is the Boltzmann constant, T is the atmospheric temperature, m is the average mass of atmospheric particles, and g is the local gravitational acceleration. An atmospheric particle must also not be directed downward, so that one can assume that approximately half of all particles given escape energy at or above the exobase location will eventually escape. In reality the exobase is not a sharp boundary; collisions still occur above the exobase, and particles below the exobase can still escape.

Finally, any escaping particles must not be confined to the planet by planetary magnetic fields. This requires either that an escaping particle be neutral, that the planet lack a magnetic field, or that any magnetic fields are weak enough that energized charged particles are able to easily traverse magnetic field lines. Venus lacks a measurable global magnetic field like that of Earth. Mars also lacks a global magnetic field but possesses localized regions of strongly magnetized crust that may locally trap energized atmospheric ions.

Owing to the highly collisional nature of planetary lower atmospheres, escape is generally limited to three regions of the upper atmosphere: the thermosphere,

Table 7.3 *Vertical extent and important species for upper atmospheric regions of terrestrial planets*

	Venus	Earth	Mars
Thermosphere	~120–250 km	~85–500 km	~80–200 km
	CO_2, CO, O, N_2	O_2, He, N_2	CO_2, N_2, CO
Ionosphere	~150–300 km	~75–1000 km	~80–450 km
	O_2^+, O^+, H^+	NO^+, O^+, H^+	O_2^+, O^+, H^+
Exosphere	~250–8000 km	~500–10 000 km	~200–30 000 km
	H	H, (He, CO_2, O)	H, (O)

the exosphere, and the ionosphere. The altitude and composition of these regions are summarized for each planet in Table 7.3, and the regions are described more generically below.

The thermosphere is a region of neutral particles extending from the mesopause up to the exobase, with temperature that increases as a function of altitude due to X-ray and EUV input from the Sun. The homopause separates the lower, well-mixed thermosphere from a large region where each species is in a separate diffusive equilibrium. Above the homopause, diffusive mixing is slower than gravitational separation, so each gas will take on its own independent scale height based on its mass. Thus, density falls off less quickly with altitude for less massive species (and one can now begin to see why hydrogen was preferentially removed from the top of Venus' atmosphere compared to deuterium, as discussed in Sect. 7.2). The thermosphere is a collisional region, and escape from deep within the thermosphere is unlikely. However, any neutral particle that reaches the top of thermosphere with escape energy should be removed from the atmosphere if it has an outward trajectory.

Thermospheric particles reaching an altitude where collisions are rare and having energy less than the escape energy will populate an extended exosphere or corona on ballistic trajectories. The exosphere is a region of neutral particles extending from the exobase upward to altitudes as high as tens of thousands of km. It is a non-collisional region (by definition), and so any upward-directed particle within it that has velocity greater than the escape velocity is very likely to be removed from the atmosphere. Particles with insufficient energy to escape will return to the exobase on ballistic trajectories unless they are first ionized.

The ionosphere is a region of ionized atmospheric particles extending from the exosphere down into the lower atmosphere (Ch. 13 in Vol. III). Ionospheres have a main density peak produced primarily by photoionization of thermospheric and exospheric neutrals by solar EUV. Other sources of ionization include photoionization by solar X-rays, impact by precipitating particles, and charge exchange with precipitating ions. Unlike neutrals, ionospheric particles can be accelerated by magnetic and electric fields. For planets lacking global magnetic fields, some ions reaching the exobase from below may be accelerated away from the planet by the passing solar wind, even if they initially lack escape energy. Others may re-impact the collisional atmosphere at the exobase.

7.6 Atmospheric escape processes and rates

A number of mechanisms are capable of giving atmospheric particles sufficient energy to escape from a planet. They are illustrated schematically in Fig. 7.4 and described below.

Neutral particles can escape an atmosphere in one of three ways: Jeans escape, photochemical escape, and sputtering. Jeans (or thermal) escape occurs because some fraction of neutral particles near the exobase will have sufficient energy to escape simply because the particles have a thermal distribution. Neutral temperatures near the exobase of all three planets are sufficiently low (~250–1000 K) that only species with small mass (H, D, and He) can escape via this mechanism in significant quantity. The process should be more efficient for Mars (due to its low gravity) and for Earth (due to its higher exobase temperature) than for Venus.

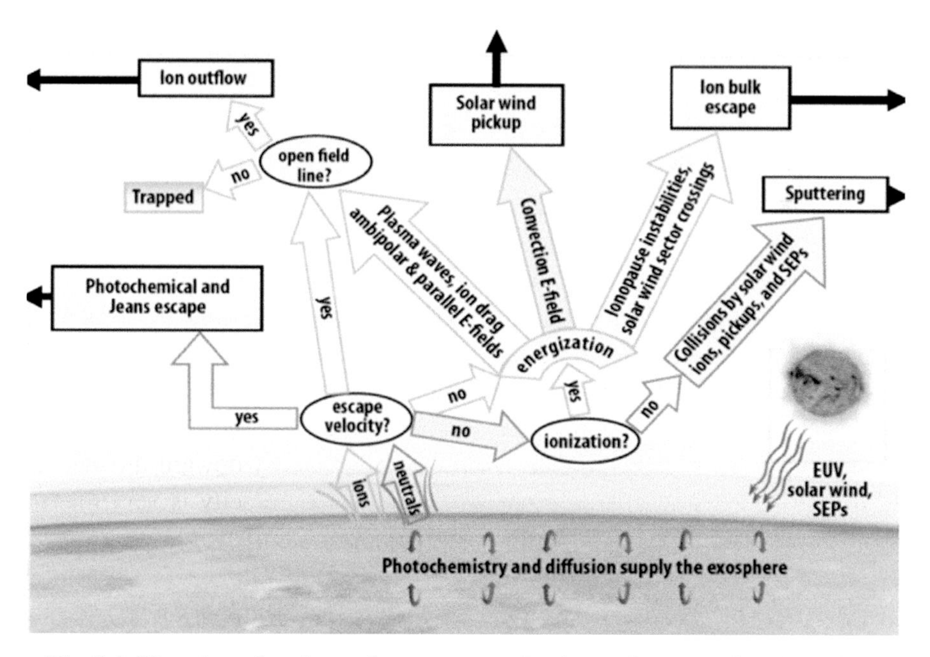

Fig. 7.4 Flowchart showing pathways to energization and escape of particles from a planetary atmosphere. Though Mars is depicted in the diagram, the six processes shown in the rectangular boxes are generic, and can apply to any planet (courtesy NASA MAVEN).

Photochemical escape refers to the escape of fast neutral particles energized by sunlight-driven chemical reactions. These reactions typically involve dissociative recombination of an ionized molecule with a nearby electron, resulting in two fast neutral atoms. Photochemical escape fluxes depend upon ionospheric molecular densities near the exobase, as well as electron density and temperature. Photochemistry is thought to be the dominant loss process for neutral species more massive than hydrogen and helium at Mars. Fast atoms produced photochemically at Venus and Earth are typically not energetic enough to escape the larger gravity.

Atmospheric sputtering occurs when atmospheric particles near the exobase receive sufficient energy from collisions to escape. Collisions occur when energetic incident particles (often ionospheric particles accelerated by electric fields near the planet) encounter the exobase. There are no unambiguous observations that sputtering is actively occurring at any of the terrestrial planets, or contributes significantly to the present-day atmospheric escape rate. However, estimates of loss rates due to sputtering for more extreme conditions suggest it may have been important earlier in solar-system history, especially for unmagnetized planets (Johnson, 1994).

Escaping ions have been directly measured at all three planets, and simulations of the solar-wind interaction with the plasma environments of all three planets are capable of predicting ion escape rates under various input conditions. A number of processes have been identified by which ions can escape an atmosphere, and different authors classify these processes in different ways. Here, we classify ion-loss processes into three categories: ion outflow, ion pickup, and bulk plasma escape.

Ion outflow refers to the acceleration of low-energy particles out of the ionosphere via plasma heating and outward directed charge separation (ambipolar) electric fields. In this case the ion acceleration can occur below the exobase, where collisions maintain a more fluid-like behavior. Ion outflow is the only significant ion loss process for the terrestrial atmosphere, and encompasses a number of processes referred to in the terrestrial literature, including wave heating, polar wind, and auroral outflow (Moore and Khazanov, 2010). Many of these processes should have analogs in the ionospheres of Venus and Mars, and in the localized crustal magnetic fields at Mars.

Ion pickup refers to the situation where a neutral particle is ionized (via photons, electron impact, or charge exchange) and accelerated away from the planet by a motional electric field ($\mathbf{E} = -\mathbf{v} \times \mathbf{B}$). Ion pickup occurs primarily for ionized exospheric neutrals (though some ionized thermospheric neutrals near the exobase region may escape via pickup as well). The motional electric field is usually supplied by the solar wind, so that the process is most relevant for compact magnetospheres unshielded by strong planetary magnetic fields (Venus and Mars) where the solar wind has access to exospheric regions with non-negligible density.

Bulk escape refers to any process which removes spatially localized regions of the ionosphere en masse. Bulk escape is relevant for unmagnetized planets, where the external plasma flow can create magnetic and/or velocity shear with the ionosphere. A popular example involves the Kelvin–Helmholtz (K-H) instability, which may form at the ionopause of Venus or Mars and steepen into waves that eventually detach from the ionosphere (Elphic and Ershkovich, 1984; Penz *et al.*, 2004). Other bulk escape processes are possible as well, such as transport via plasmoid-style flux ropes that may remove ionospheric plasma from Martian crustal magnetic-field regions (Brain *et al.*, 2010a).

It is convenient to think of ion-escape processes in terms of the type of electric field responsible for their removal. A simplified version of Ohm's Law describes the most important electric field terms that influence ion motion

$$\mathbf{E} = -(\mathbf{v} \times \mathbf{B}) + \frac{1}{ne}\mathbf{J} \times \mathbf{B} - \frac{1}{ne}\nabla\mathbf{P}_e, \tag{7.5}$$

where $\mathbf{E}$ is the total electric field, $\mathbf{v}$ is the plasma bulk velocity, $\mathbf{B}$ is the magnetic field, $\mathbf{J}$ is the current density, $\mathbf{P}_e$ is the electron pressure tensor, n is the plasma number density, and e is the electron charge. The three terms on the right-hand side of the equation are the motional electric field, the Hall electric field, and the electron pressure gradient. The Ohmic and electron inertial terms have been neglected. There is varying overlap of these three terms with the ion-escape processes described above. This alternate classification scheme has the advantage that each term can be evaluated unambiguously in global simulations, and compared with observations. It also highlights that a combination of processes can be responsible for removing ions from planetary atmospheres.

A number of estimates of escape rates have been produced for all three terrestrial planets, based both on observations and simulations. In broad terms, the present day global escape rate for Venus is estimated to be 10^{24}–10^{26} s^{-1} (Lammer *et al.*, 2009). The escape rate for Earth is 10^{25}–10^{27} s^{-1}, and for Mars is 10^{24}–10^{26} s^{-1}.

Two aspects of these estimates should strike the reader. First, the numbers appear to be very large. However, when we consider that the surface areas of the terrestrial planets are on the order of 10^{18} cm^2, we see that the escape rates are on the order of 10^6–10^9 cm^{-2} s^{-1}. In contrast, atmospheric densities near the surface range from 10^{17}–10^{20} cm^{-3}, and column densities range from 10^{23}–10^{27} cm^{-2}. So escape rates are a very small fraction of the number of particles in the present day atmospheres, though accumulated over ~4 billion years (~10^{17} s) they may be substantial. For this latter point the two orders of magnitude uncertainty in escape rates are crucial; they are the difference between heliophysical drivers being the main loss mechanism for planetary atmospheres or merely an afterthought in determining present-day atmospheric abundances.

Second, the escape rates for Venus, Earth, and Mars are all similar, within the admittedly large uncertainties. Given their similarity, we are forced to question a number of common assumptions. Does planetary size or distance from the Sun play a significant role in the removal of atmospheric particles to space? Does the presence of a global magnetic field significantly inhibit escape? And are differences in the present-day escape rates (long after the formation of secondary atmospheres) at all indicative of the amount of escape that has occurred at each body over solar system history? Each of the above questions is ripe for investigation.

Finally, it is important to keep in mind that escape to space not only influences atmospheric abundance but also atmospheric composition, which can be important in planetary evolution. One example is the aridity of the Venus atmosphere. The loss of atmospheric water is attributed to dissociation of the water in the atmosphere by sunlight, and the subsequent escape to space of oxygen. Water is only a trace gas in planetary atmospheres, but is an important greenhouse gas and is extremely important for habitability. So even if escape to space does not appreciably change atmospheric thickness, it may contribute in important ways to climate. Interestingly, the escape rates listed above, when converted to precipitable microns of water, amount to global layers of water only centimeters thick. More than this is assumed to have been lost from Venus, suggesting either that escape rates have changed over time (and are low today) or that other processes (such as impacts) have been important for removing water.

7.7 External drivers of escape

Observations, simulations, and common sense all tell us that atmospheric escape rates are not constant, and are influenced by a number of heliophysical drivers that vary on both short and long time scales. The reader is also referred to Chs. 2 and 11 in Vol. III.

The three main drivers are photons, charged particles, and electromagnetic fields. Photons deposit energy in atmospheres when they are absorbed by atmospheric particles. Extreme UltraViolet (EUV) and soft X-ray wavelengths (generated in the solar corona and chromosphere, and not to be confused with solar luminosity) provide the dominant energy source in upper-atmospheric regions. Charged particles in the solar wind also supply energy to planetary upper atmospheres and plasma environments. Table 7.4 summarizes some of the relevant quantities of the solar wind at each terrestrial planet. While density and velocity can each vary independently, studies of solar wind influences on atmospheric escape (especially the induced magnetospheres of Venus and Mars) typically use solar wind pressure (ρv^2) as the organizing quantity. Finally, the solar wind carries a magnetic field, which creates a convection electric field ($\mathbf{E}_{\mathrm{SW}}$) in the frame of the planet that

Table 7.4 *Typical properties of the solar wind (SW) and interplanetary magnetic field (IMF) at terrestrial planets*

	Venus	Earth	Mars
IMF strength	10–12 nT	6 nT	3 nT
Solar-wind speed	400 km/s	400 km/s	400 km/s
Solar-wind density	10–15 cm^{-3}	6 cm^{-3}	1–3 cm^{-3}
Alfvén speed	70 km/s	55 km/s	45 km/s
Mach number	5–7	6–8	8–10
SW H^+ gyroradius	1500 km	2500 km	5000 km
SW H^+ gyroradius/R_P	0.5	0.4	3

depends upon solar wind velocity and interplanetary magnetic field (IMF) strength and orientation (see Eq. (7.5)). Magnetic and electric fields organize charged particle motion, and electric fields accelerate charged particles; both effects influence the ability of charged particles to escape a planet's atmosphere.

The external drivers of atmospheric escape vary on four main time scales. Billion year time scales are associated with the age of the Sun, and both theoretical calculations and observations of Sun-type stars suggest that all three drivers should have declined in intensity with age (Fig. 7.5; see Chs. 2 and 11 in Vol. III). EUV flux varies by factors of several over a solar cycle (from solar minimum to solar maximum), and solar wind pressure varies by factors of 2–10. The IMF, in particular, varies with the solar rotation period, and all three drivers also vary on more rapid time scales of minutes to hours.

Variability in the heliophysical drivers should influence atmospheric escape rates. In general, an increase in solar EUV fluxes (e.g., a transition from solar minimum to solar maximum) is expected to result in an increase in loss rates of neutral particles. Energy from solar photons heats the upper atmospheric neutrals, so that Jeans escape rates should increase with solar EUV. This is likely to be true at Mars, but not at Earth where hydrogen escape from the exobase is limited not by the available energy, but by the supply (via diffusion) of particles from lower altitudes (see discussion in Tian *et al.*, 2013). Jeans escape should be negligible at Venus today, but may have been significant in the past if either exobase temperatures or solar EUV fluxes were much higher. Energy from solar photons is also used to drive the chemical reactions necessary for photochemical escape, so that contemporary Martian photochemical escape should vary with EUV flux. Neutral escape rates should be largely insensitive to changes in both the solar wind and the IMF, except for sputtering rates from Venus and Mars, which are thought to be dominated by re-impacting atmospheric pickup ions and will therefore increase as the pickup-ion population increases in response to changes in solar EUV.

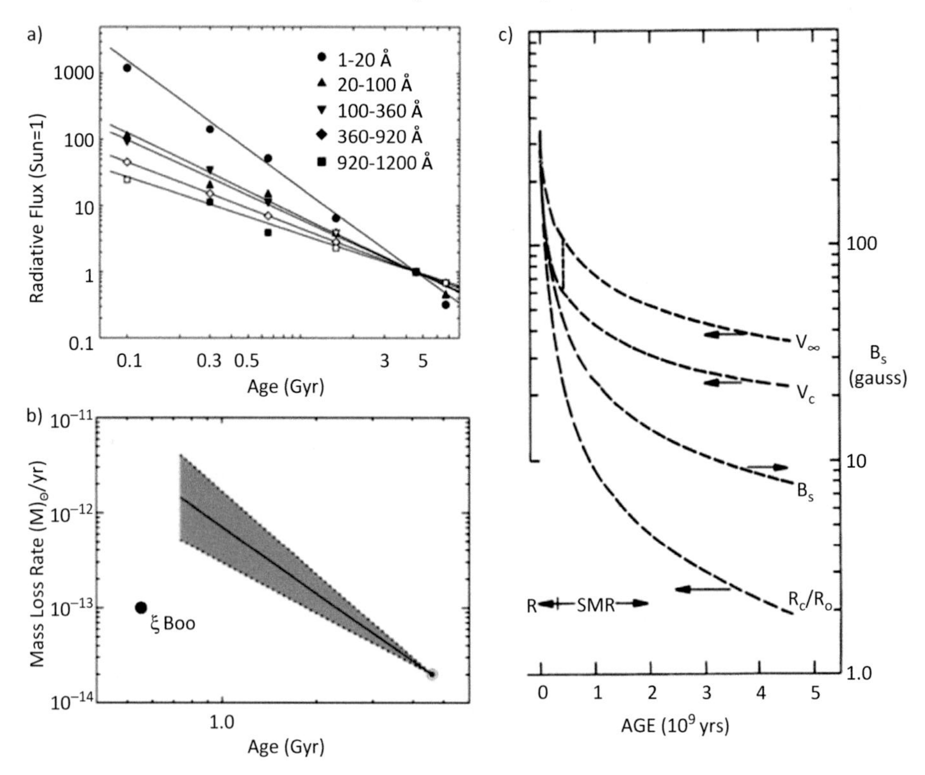

Fig. 7.5 Evolution of solar drivers of atmospheric escape. (a) Solar EUV photon flux, relative to today (from Ribas *et al.*, 2005); (b) solar mass loss rate (i.e., solar wind flux) (from Wood *et al.*, 2005); (c) interplanetary magnetic field (curve labeled B_S; from Newkirk, 1980).

Ion escape rates should also vary with the three drivers. An increase in solar-wind pressure will cause a corresponding decrease in the size of the magnetospheric cavity at all terrestrial planets, effectively lowering the pressure balance altitude between the solar wind and planetary obstacle to the flow. For Mars, with an extended neutral corona, an increase in solar-wind pressure exposes significant additional high-altitude neutrals to ionization and stripping by the solar wind (via electron impact and charge exchange). The IMF, by contrast, chiefly organizes the trajectories of escaping particles at Venus and Mars; large gyroradius pickup ions are preferentially accelerated away from the planet in regions where $\mathbf{E}_{SW}$ points away from the planet. At Earth, the orientation of the IMF affects the location and extent of cusp regions, from which outflowing ions escape. EUV fluxes have a more indirect effect. In total, one might expect the ion escape rate to increase at solar maximum due to the additional energy input from EUV. At unmagnetized Venus and Mars, however, the increased ionospheric content deflects the solar wind around the planet at higher altitudes and can prevent the interplanetary magnetic

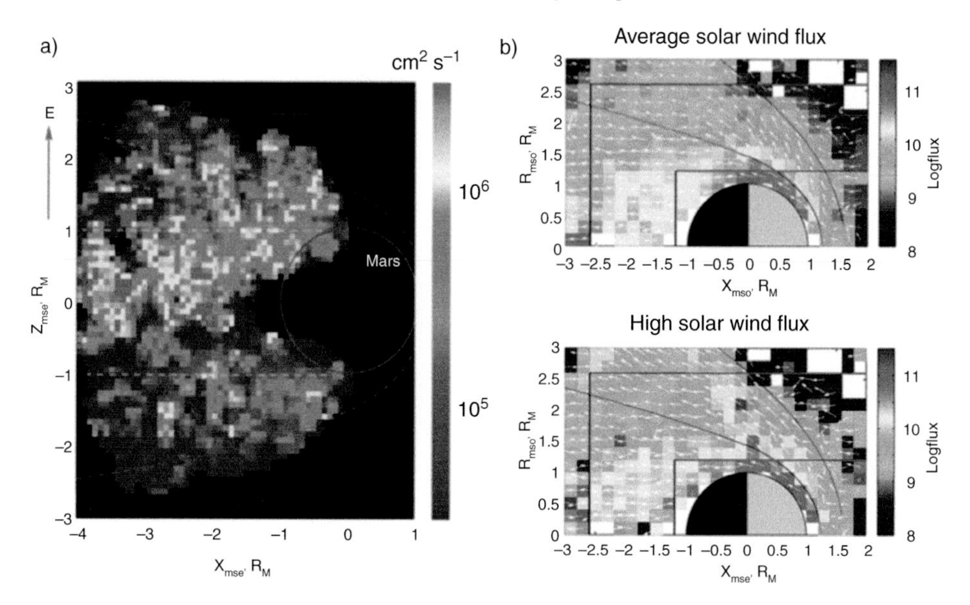

Fig. 7.6 Ion escape from the Martian atmosphere, organized by solar drivers. The Sun is to the right in both panels. (a) Escaping ion fluxes downstream from Mars are greater in the hemisphere of upward directed (with respect to the planet) solar wind electric field (Barabash *et al.*, 2007); (b) escaping ion fluxes downstream from Mars are greater during periods of high solar wind flux. (From Nilsson *et al.*, 2011.) A black and white version of this figure will appear in some formats. For the color version, please refer to the plate section.

field from entering the ionosphere. The escape of heavy ion species (which are concentrated at lower ionospheric altitudes) via pickup and bulk escape may therefore remain roughly constant, or even decrease during solar maximum periods, even as lighter ion species escape more efficiently.

Observations support the above assertions in a general sense, though quantification of many of the effects is still being teased out of the available data. Here, we take ion escape from Mars as an example. First, the organization of escaping ions by the IMF is borne out by observations (Fig. 7.6a). Next, the flux of escaping planetary ions has been correlated with the solar wind intensity (Fig. 7.6b). Finally, the fluxes of escaping ions measured at solar minimum and solar maximum differ by approximately an order of magnitude (Lundin *et al.*, 1990; Barabash *et al.*, 2007). These results should be cautiously interpreted, however, because the measurements were made by two spacecraft with different orbits and instruments.

Models also support these trends, and provide a useful complement to spacecraft observations because they are able to perform controlled experiments on how a planet responds when only one external driver is changed. Further, models are not limited by a spacecraft orbit or instrument observing geometry; they provide

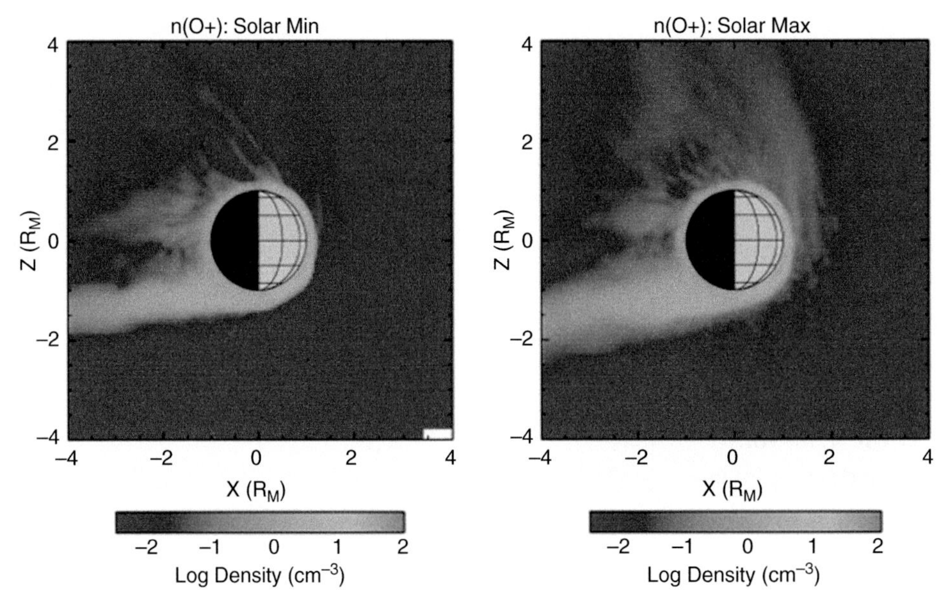

Fig. 7.7 Density of escaping atomic oxygen ions from Mars at solar minimum (left) and solar maximum (right) as predicted by a global hybrid plasma simulation. The Sun is to the right in each panel, and the solar wind electric field points toward $+y$. Courtesy E. Kallio and R. Jarvinen.

information about how the entire system responds. Again using ion escape from Mars as an example, a variety of models have been used to simulate the near-space environment and predict atmospheric ion escape rates. The models predict that escape rates increase from solar minimum to solar maximum (Fig. 7.7), and with solar-wind pressure (see discussion in Brain *et al.*, 2015). And the IMF orientation clearly controls the trajectories of escaping ions. The models employ different physical assumptions, boundary conditions, and implementation schemes, and so it is not surprising that the models disagree on the magnitude of each of the above-described effects. A major challenge facing the community at present is determining which models produce results that best match the observations. The answer is likely to depend on location, type of observation, and external conditions.

Solar storm periods provide extreme cases for each of the three drivers mentioned above, and in a sense provide a window into conditions earlier in solar-system history. Solar flares, solar energetic particle (SEP) events, and the enhanced magnetic field associated with coronal mass ejections (CMEs) have all been measured at the three terrestrial planets. Initial efforts at quantifying the effect of solar storm periods suggest that ion escape rates can increase by an order of magnitude or more at Mars and Venus (Futaana *et al.*, 2008), and suggest that Earth's escape rates increase less during similar periods (Wei *et al.*, 2012). However, much more work remains to be done on this topic, especially with regard to comparing the responses of the different planets to solar storms.

7.8 Internal drivers of escape

A number of characteristics of a terrestrial planet itself influence the properties and energetics of upper atmospheric reservoirs for escape, including transient events such as dust storms (e.g., for Mars), or longer-lived phenomena such as gravity waves that couple the lower and upper atmospheres. In the context of heliophysics, the nature of a planet's intrinsic magnetic field is of the greatest relevance.

Earth possesses a global dynamo magnetic field today (see Ch. 6 in this volume and Ch. 7 in Vol. III), while Venus lacks a measurable dynamo field. Mars also lacks a dynamo magnetic field but has crustal magnetic fields (Acuña *et al.*, 1998) that have significant influence on the upper atmosphere and plasma environment. The strength of the crustal fields and their higher concentration in the more ancient southern hemisphere suggest that they formed in the presence of an ancient global dynamo magnetic field, which shut off as many as 4.1 billion years ago (Lillis *et al.*, 2008). Temperature gradients in planetary interiors are important for driving the convection necessary to sustain a dynamo; in simple terms the smaller Mars cooled more quickly than Earth and became incapable of supporting an internal dynamo early in the planet's history. Venus may have hosted a dynamo at one time, but the much hotter surface makes it highly unlikely that any remanent crustal magnetism has been preserved (and orbiting spacecraft have not detected any). Thus we are left today with an intrinsic magnetosphere at Earth that deflects the solar wind at large distances from the planet ($\sim 10\ R_E$), and induced magnetosphere at Venus that deflects the solar wind at much closer distances ($\sim 1.3 R_V$), and a similarly sized (with respect to the planet) induced magnetosphere at Mars punctuated by mini-magnetospheres tied to specific regions of the crust and that rotate with the planet.

How do the different magnetic fields influence ion-escape processes and rates? In general terms, planetary magnetic fields influence escape processes in two main ways (Fig. 7.8). First, they add magnetic pressure from the planet that contributes to balancing the pressure in the solar wind, causing the solar wind to be deflected farther from the planet and thus farther from its atmosphere. Second, magnetic fields can alter the topology of magnetic field lines near the planet, reconnecting with the IMF. The presence of intrinsic magnetic field creates closed field lines that shield an atmosphere from the solar wind and trap ions, and open field lines in cusp regions that enable exchange of energy and particles between the atmosphere and solar wind.

It is reasonable to consider that the three ion-escape processes described in Sect. 7.6 should all be influenced by the presence of a planetary magnetic field.

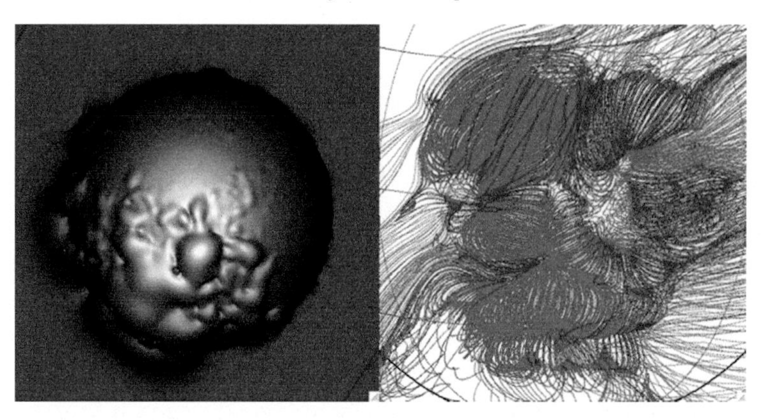

Fig. 7.8 Influence of magnetic fields on planetary near-space environments. Magnetic fields supply magnetic pressure (left: for Martian crustal magnetic fields) that deflect solar wind, but also modify magnetic topology (from Brain, 2006); (right: for the strong Martian crustal fields in the southern hemisphere, where red denotes closed field lines and blue denotes field lines open to the solar wind at one end) that enable exchange of particles and energy between the atmosphere and solar wind. Both renderings result from model calculations that include contributions from crustal fields and external drivers (solar wind or IMF). A black and white version of this figure will appear in some formats. For the color version, please refer to the plate section.

Both ion pickup and bulk plasma escape should be reduced in the presence of a magnetic field, because the solar wind is more effectively shielded from the ionosphere. In addition, ionization of atmospheric neutrals by the solar wind should be reduced in the presence of a magnetic field, decreasing the population of particles available for escape. Ion outflow occurs when cold atmospheric ions are accelerated by electric fields. The electric fields result from a variety of processes, and it is unclear whether outflow should decrease or increase in the presence of a magnetic field. However, intrinsic fields create vertically oriented flux tubes, which should facilitate vertical ionospheric transport. Several outflow acceleration mechanisms also rely on vertical field lines to form or maintain electric fields that accelerate particles, suggesting that outflow is likely to be more effective when a planet is magnetized.

Of the neutral loss processes, sputtering is most likely to be influenced significantly by planetary magnetic fields. Sputtering is primarily caused by planetary ions re-encountering the exobase. Magnetic fields influence both the trajectories of charged particles, and their formation (by preventing the solar wind from accessing high altitude regions of neutral atmosphere). Thus one would expect sputtering loss to be inhibited in the presence of planetary fields. Jeans escape should not be affected by magnetic fields because it primarily involves solar photons and neutral particles. Photochemical escape may be indirectly affected because the fast neutral particles lost via the process are produced when a planetary ion and electron recombine. Like sputtering, the influence of a planetary magnetic field on ion production and motion may influence the loss rate. However, unlike sputtering, most of the processes involved in photochemical loss typically take place deeper in a planet's atmosphere (at or below the exobase) than sputtering, so influences of a magnetic field may be negated by particle collisions.

When considering the total atmospheric loss from a planet, it has often been assumed that the presence of a magnetic field results in lower escape rates. The argument is that intrinsic magnetic fields prevent the solar wind from directly accessing the atmosphere, shielding it from solar wind-related ion loss. Above, we argued that three of six loss processes (ion pickup, bulk plasma escape, and sputtering) should be less efficient in the presence of planetary magnetic fields, and two more should be unaffected or only weakly changed (Jeans escape and photochemical escape). Further, evidence discussed in Sect. 7.2 suggests that Mars lost substantial atmosphere, and that both Venus and Mars lost atmospheric particles to space more efficiently than Earth. Though there are few planets to compare and comparisons are complicated by the lack of controls (e.g., size, distance from the Sun, etc.), it may be telling that Earth possesses both a global magnetic field and a habitable atmosphere.

However, we mentioned in Sect. 7.6 that the measured atmospheric escape rates for Venus, Earth, and Mars are comparable within the current uncertainties. It has recently been proposed that magnetic fields, rather than shielding a planetary atmosphere from stripping by the solar wind, actually collect solar-wind energy and transfer it to the ionosphere along field lines (Strangeway *et al.*, 2010). Global magnetic field lines converge near the cusps, so that the energy is more spatially concentrated than for unmagnetized planets. The escape rate for a given planet may be comparable when it is magnetized, or even greater because planetary magnetic fields extend much farther than the planet's atmosphere, giving it a larger energy collecting cross section in the solar wind. One key difference with magnetized planets is that the concentrated energy in cusp regions is likely to lead to more efficient removal of heavy species.

There are a few caveats. First, there are large error bars at present on planetary atmospheric escape rates at all three terrestrial planets, so that we are still unsure whether the escape rates are similar. Second, not all solar-wind energy collected by a planet need go into removing atmospheric particles. It could instead drive chemistry or upper-atmospheric dynamics and heating. However, measurements from Earth suggest that the solar-wind Poynting flux scales exponentially with the upward directed ion flux in cusp regions (Fig. 7.9). Finally, accelerated ions in Earth's cusps need not escape the planet at all. It is currently uncertain what

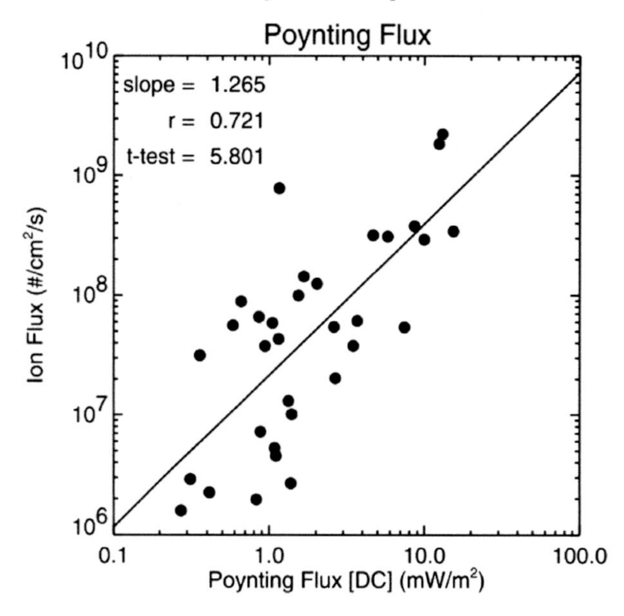

Fig. 7.9 Measurements of outflowing oxygen ions from Earth's cusp regions against solar wind Poynting flux. (From Strangeway *et al.*, 2005.)

fraction of ions leave Earth's magnetosphere, and what fraction is returned to the planet along magnetic field lines. A number of issues must be investigated and resolved before we can determine whether magnetic fields protect an atmosphere from being lost.

7.9 Frontiers

It should be apparent that there are multiple likely connections between heliophysics and the climates of terrestrial planets. Yet we are still uncertain whether Mars lost a bar of atmosphere to space, or whether Earth's magnetic field is a shield for the atmosphere. There are clearly exciting and important questions to be tackled in the coming years. Where and how is the community likely to make progress in the next 5–10 years? Below, we discuss three frontier research areas.

First, global simulations of planetary plasma environments have become more and more capable over the past two decades. These models are capable of simulating global magnetospheric interactions with increasingly accurate physical assumptions and spatial resolution as computers become ever faster. Still, very few models are capable of treating a global dipole magnetic field as a knob to be turned on or off in order to study the influence of magnetic fields on atmospheric escape processes. Fluid models for Earth's solar-wind interaction could turn off the planetary dipole, but would not be capable of accurately capturing several of the kinetic ion-escape processes (which may dominate escape from Earth). At least one hybrid (kinetic ions and fluid electrons) model has simulated Mars with and without a global dipole, but the simulated planetary field was made very weak so that the model was computationally tractable, and the model is not optimized for studying escape from low-altitude regions (Kallio and Barabash, 2012). Within the next several years, however, it seems likely that a model will become capable of simulating magnetized and unmagnetized versions of Earth, Venus, and Mars using physical assumptions that allow investigation of all ion-escape processes. Further, current efforts to couple global plasma simulations with models for the exosphere and thermosphere hold promise for capturing all ion and neutral escape processes in a single self-consistent simulation (e.g., Dong *et al.*, 2014).

Next, analyses of existing and ongoing observations should provide useful constraints for models on the importance of individual loss processes at each planet under varying conditions, and the role of magnetic fields. Existing observations of atmospheric escape from Venus, Earth, and Mars are typically presented for a single object. More detailed comparisons between objects are in order. Spacecraft missions such as MAVEN, which arrived at Mars in 2014, will allow investigation of the physical processes that result in escape (Jakosky *et al.*, 2015). MAVEN measures the drivers of escape (solar photons, particles, and fields), the atmospheric reservoirs for escape, and the escaping particles. Earlier missions, while very productive, have been hampered by incomplete observations. Further, MAVEN measures atmospheric escape processes from both magnetized and unmagnetized regions of Mars, which holds promise for comparisons between atmospheric regions that differ only in their magnetic field. These results are likely to be useful in assessing whether magnetic fields reduce escape rates.

Finally, the lessons from models and observations of solar-system objects can be applied to exoplanets. Since 1995, the number of known exoplanets in our Galaxy has grown, from zero into the thousands, as detection methods have improved substantially and dedicated spacecraft missions and telescopes have been commissioned. Now, we find ourselves at a point where we infer that most stars in our Galaxy have planets, and that there may be as many as 40–100 billion habitable exoplanets in the Milky Way (Petigura *et al.*, 2013). The word habitable is tricky, though. The studies that calculated 40–100 billion habitable planets assumed that Venus and Mars were habitable. This is a fair assumption because both planets may have been habitable at their surface at some point in solar-system history. However, the assumption is very probably over-generous, considering Venus and

Mars today. Can we use our understanding of heliophysical processes to determine which planets are likely to have climates conducive to surface habitability? Can we apply models tuned to Venus, Earth, and Mars, and validated against spacecraft observations of atmospheric-loss processes, to narrow the list? It will certainly be exciting to try.

8

Upper atmospheres of the giant planets

LUKE MOORE, TOM STALLARD, AND MARINA GALAND

All celestial bodies are surrounded by gaseous envelopes, at least to some degree. When the gas is gravitationally bound to a parent body's nucleus it is called an atmosphere, whereas if the gas is not confined by gravity, such as at a comet, it is called a coma (Strobel, 2002). At one atmospheric extreme, such as Mercury or the Moon, the extremely tenuous atmosphere originating from the surface is referred to as a surface-bound exosphere, as the atmospheric atoms and molecules are much more likely to escape to space or to collide with the surface rather than collide with each other. At the other extreme, such as at the gas giants (Jupiter, Saturn, Uranus, Neptune), the rocky core about which the atmosphere is gravitationally bound is on the order of 0.1 planetary radii and gas constitutes the majority of the planet. A dense atmosphere is typically divided into two broad categories: the lower and upper atmospheres. The study of the lower regions (troposphere and stratosphere) forms the discipline of meteorology, while the study of the upper regions (mesosphere, thermosphere, exosphere) and their ionized component (ionosphere) forms the discipline of aeronomy.

Atmospheres play vital roles in planetary and satellite evolution, as they help to insulate the surface of a body from external influences. In particular, the upper atmosphere represents a key transition region between a dense atmosphere below and a tenuous space environment above. An array of complex coupling processes from below, such as waves, and from above, such as forcing by solar extreme ultraviolet (EUV) photons and energetic particles, means that aeronomy deals with the highly coupled system of neutrals, plasmas, and electromagnetic processes that link planets, moons, and comets from their surfaces to their magnetospheres, to the solar wind, and ultimately to the Sun itself (Mendillo *et al.*, 2002).

Evidence of these coupling processes include various upper-atmospheric emissions, such as dayglow and nightglow, resulting from the absorption of solar photons, and aurorae, which are produced by the energy deposition of energetic particles from the space environment. Such emissions can be detected remotely, and have consequently allowed detailed study of the planets in the solar system. In addition to a host of ground-based observations, a number of spacecraft have also been used to study the giant planets. Spacecraft encounters with the outer planets include Pioneer 10 and 11 in the 1970s, Voyager 1 and 2 in the 1970s and 1980s, and New Horizons at Jupiter in 2007. More in depth studies have also been enabled by orbiting spacecraft: Galileo at Jupiter (1995–2003) and Cassini at Saturn (starting in 2004). The coupled atmosphere–magnetosphere systems at the giant planets are smaller scale representations of electromagnetic interaction regions that occur elsewhere in the universe. Furthermore, many hot Jupiters and hot Neptunes have been discovered so far (see Ch. 5; also, e.g., Fogg and Nelson, 2007; McNeil and Nelson, 2010), and by studying the giant planets in our own neighborhood we can improve our understanding of the rapidly accumulating zoo of exoplanets.

Much of early outer planet science was guided by our knowledge of the terrestrial system. Now, following nearly 40 years of spacecraft exploration and continually improving Earth-based capabilities, we understand enough about the giant-planet atmosphere–magnetosphere systems to categorize them based on the different processes that dominate each. Such a comparative approach has proven to be beneficial for study of all of the solar system planets, and will serve as a useful platform for initializing study of exoplanets and furthering our understanding of planetary formation and evolution. Subtle differences in atmospheric composition, magnetic field, internal magnetospheric plasma sources, and external forcing have led to significant differences in atmospheric and auroral morphology and dynamics at Jupiter, Saturn, Uranus, and Neptune. Similar exciting differences can be expected at other Jupiter like exoplanets, in addition to the differences in stellar forcing. The future of comparative aeronomy is likely to be an exciting and enriching one.

In this chapter we give an overview of the current state of knowledge of giant-planet upper atmospheres (an overview of terrestrial upper atmospheres follows in Ch. 9). We focus first on the thermosphere and ionosphere, next on the processes coupling planetary atmospheres and magnetospheres, and finally on the auroral emissions resulting from those coupling processes. In addition to the references cited herein, further basic concepts are explored in more detail in related review chapters in the Heliophysics series (cf., Table 1.2), such as in Vol. I, Ch. 12, in Vol. III, Ch. 13, and in Ch. 9 in this volume for upper atmospheres, and in Vol. I, Chs. 10, 11, and 13, and Vol. II, Ch. 10 for magnetospheres.

8.1 Thermospheres of the giant planets

The atmospheres of the solar system giant planets are predominantly molecular hydrogen and (mostly inert) helium. Consequently the resulting photochemistry differs significantly from other classes of atmospheres, such as the N_2-dominated atmospheres of Earth, Pluto, and Titan, or the CO_2-dominated atmospheres of Venus and Mars. Trace amounts of heavier cosmically abundant elements are also present in the deep atmospheres of the giant planets, though primarily in the form of hydrides due to the profusion of hydrogen (e.g., CH_4, NH_3, H_2O, H_2S, etc.). While turbulent (or eddy) diffusion acts to mix atmospheric constituents in the lower atmosphere (referred to as the homosphere) and thereby maintain constant vertical mixing ratios, there is a transition region – called the turbopause or homopause – above which less frequent collisions allow molecular diffusion to dominate and atmospheric constituents begin to separate according to their masses (a region referred to as the heterosphere). At the giant planets, the lower atmosphere includes a minimum in temperature with a negative temperature gradient in the troposphere and a positive gradient in the stratosphere. The upper atmosphere, primarily within the heterosphere, is characterized by a positive temperature gradient due to the absorption of EUV solar radiation and energy deposition from above and below. This region can further be separated into two coincident fluid components – the charged ionosphere and the neutral thermosphere – beneath a mixed kinetic component, the exosphere. While atmospheric species are still largely gravitationally bound within the exosphere, collisions are too infrequent to lead to a collective fluid behavior. It is a region where escape occurs and it is associated with a roughly isothermal temperature referred to as the exospheric temperature. The upper atmosphere, therefore, represents the boundary between a dense atmosphere below and a magnetosphere above, and mediates the exchange of particles, momentum, and energy between these two regions.

Dominant atmospheric constituents at the homopause of giant planets are molecular hydrogen, helium, and methane (CH_4). Above the homopause methane and other hydrocarbons are quickly separated out due to their relatively high masses and confined to the lower portions of the upper atmosphere; the atomic hydrogen fraction consequently continually increases with altitude. Therefore, while there is a complex array of hydrocarbon chemistry at work in the giant-planet atmospheres, it is primarily important for the formation of clouds, hazes, and aerosols in the lower atmosphere, leaving hydrogen photochemistry to dominate over the majority of the upper atmosphere. Despite the apparent simplicity of having one atmospheric constituent dominate the chemistry of a region there are a number of observations of giant-planet upper atmospheres that have yet to be explained by theory, including global thermal structure and ionospheric variability.

Measurements of giant-planet thermospheric properties are typically made remotely using occultations in the ultraviolet (UV), infrared, and radio spectral regions (visible occultation observations are hampered by the bright background of reflected solar light; radio occultation is described in Ch. 13). Jupiter is unique

Table 8.1 *Planetary and atmospheric properties for the gas giants*

	Jupiter	Saturn	Uranus	Neptune
Mass ($\times 10^{27}$ kg)	1.9	0.57	0.087	0.10
R_{eq} ($\times 10^3$ km)	71.5	60.3	25.6	24.8
R_{pole} ($\times 10^3$ km)	66.9	54.4	25.0	24.3
Day (Earth hours)	9.925[a]	10.656[b]	17.24	16.11
Year (Earth years)	11.86	29.24	83.75	164.7
Semi-major axis (AU)	5.20	9.58	19.2	30.0
Obliquity (°)	3.13	26.73	97.77	28.32
Dipole moment ($\times 10^{17}$ T m^3)[c]	1600	47	3.8	2.8
Dipole tilt (°)	9.6	<0.1	58.6	46.9
Dipole offset (R_{planet})	0.13	0.04	0.30	0.55
T at 1 bar (K)	165	134	76	72
T at homopause (K)	176	160	~200	~250
T at exobase (K)	900	420	800	750
Atmospheric composition by volume				
H_2	89.8%	96.3%	82.5%	80.0%
He	10.2%	3.25%	15.2%	18.5%
CH_4	3000 ppm	4500 ppm	2.3%	1.5%
Energy input (GW) from EUV photons and from precipitating particles[d]				
Photons: solar EUV	800–1200	150–270	8	3
Precipitating particles: auroral	10^5	10^4	100	1

[a] System III (1965) spin period of 9h 55m 29.711s (e.g., Bagenal *et al.*, 2014).
[b] Saturn's rotational period is not as well defined (e.g., Hubbard *et al.*, 2009; Read *et al.*, 2009).
[c] Schunk and Nagy (2009).
[d] Strobel (2002); Cowley *et al.* (2004); Müller-Wodarg *et al.* (2006).
Most tabulated values are from NASA Planetary Fact Sheets: http://nssdc.gsfc.nasa.gov/planetary/planetfact.html.
Occultation references: Jupiter: Eshleman *et al.*, 1979; Atreya *et al.*, 1981; Broadfoot *et al.*, 1981; Festou *et al.*, 1981; Hubbard *et al.*, 1995; Yelle *et al.*, 1996; Hinson *et al.*, 1997, 1998; Saturn: Festou and Atreya, 1982; Smith *et al.*, 1983; Lindal *et al.*, 1985; Hubbard *et al.*, 1997; Koskinen *et al.*, 2013; Uranus: Herbert *et al.*, 1987; Lindal *et al.*, 1987; Neptune: Broadfoot *et al.*, 1989; Lindal, 1992.

among the giant planets in that the Galileo Probe also made in situ atmospheric measurements (Seiff *et al.*, 1998). Stellar and solar UV occultations use the variation in the transmission of starlight/sunlight – as seen by an observer passing behind the planet – to derive the altitude profile of horizontal column density, which can be converted to an H_2 pressure–temperature profile by applying the ideal gas law ($p = \rho kT/m = nk_BT$) and hydrostatic equilibrium ($dp/dz = -\rho g$) assumptions (for a review see Smith and Hunten, 1990). In the preceding equations, p, ρ,

n, m, and T are the pressure, mass density, number density, mean molecular mass, and temperature of H_2, respectively; and z, g, R, and k_B represent altitude, gravity, the universal gas constant, and the Boltzmann constant, respectively. Longward of 110 nm, hydrocarbon species are the primary absorbers in giant-planet upper atmospheres, and so CH_4 mixing ratios and the homopause location can also be derived from UV occultations. Infrared ground-based stellar occultations can probe the lower thermospheres of giant planets by measuring the attenuation (via defocusing) of starlight caused by atmospheric refractivity gradients (e.g., Hubbard *et al.*, 1995). Similarly, radio occultations track the diminution of a signal emitted by a spacecraft and measured by a radio telescope at Earth. The refractive defocusing of radio signals can be caused by the neutral atmosphere and by free electrons, and so radio occultations probe both lower atmospheric properties and ionospheric electron densities. A list of key references of giant-planet occultation observations is given in Table 8.1, along with basic planetary and upper-atmospheric properties.

Upper-atmospheric mixing ratios (Fig. 8.1), number densities, and temperatures (Fig. 8.2) have been derived from the Galileo Probe measurements at Jupiter; they are qualitatively representative of other giant-planet mixing ratios and thermal profiles. The Jovian homopause is clearly identifiable in both figures as the region where mixing ratios suddenly diverge from their constant lower atmospheric values ($\sim$350 km above the 1 bar pressure level). Figure 8.2 serves as a useful guide for basic thermospheric structure across all of the giant planets: e.g., the presence of methane and low temperatures near the homopause, and – in the upper atmosphere – the dominance of light species (H_2, H, and He) and a positive temperature gradient transitioning into an isothermal domain.

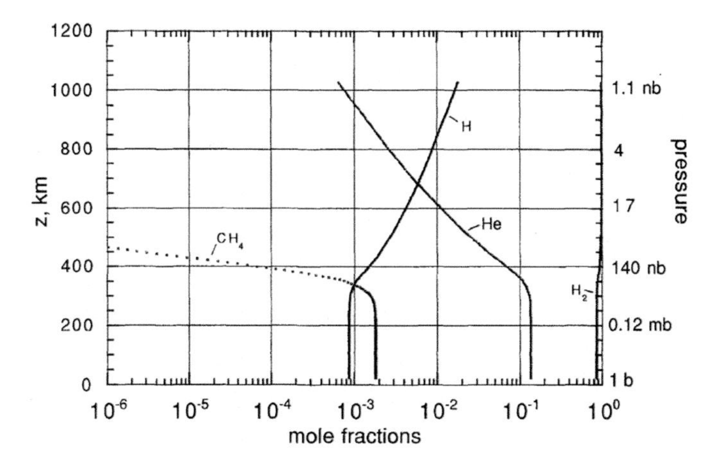

Fig. 8.1 *Galileo Probe* results showing Jupiter upper-atmospheric mixing ratios. Altitudes refer to radial distance above the 1 bar pressure level. (From Seiff *et al.*, 1998).

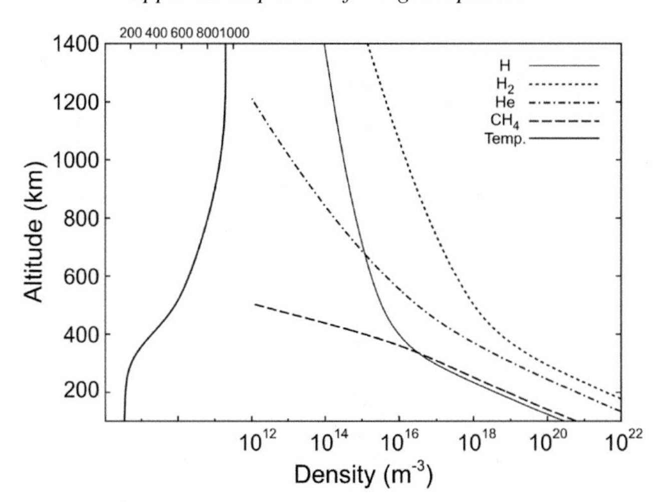

Fig. 8.2 Jupiter thermospheric parameters, based on *Galileo Probe* measurements. Altitudes refer to radial distance above the 1 bar pressure level. (From Barrow and Matcheva, 2011.)

Unlike at Earth, where the main energy source at non-auroral latitudes is usually solar radiation, the energy balance required to maintain the observed giant-planet thermospheric thermal structures at low- and mid-latitudes remains a puzzle. Possible sources of heating in giant-planet upper atmospheres include absorption of solar energy, precipitation of charged particles from the magnetosphere, and dissipation of kinetic energy in winds and waves. In general, absorption of solar EUV photons at thermospheric altitudes leads to downward heat conduction, generating a positive temperature gradient above the homopause. However, as illustrated in Fig. 8.3, calculations based solely on solar-energy inputs fall significantly short of reproducing the observed upper-atmospheric temperatures. Global solar EUV energy inputs at giant planets (e.g., ~1 TW and ~0.2 TW at Jupiter and Saturn, respectively), are dwarfed by magnetospheric energy inputs, estimated to be of order 10 TW or more (Miller *et al.*, 2005; Müller-Wodarg *et al.*, 2006). Consequently, the observed high temperatures at low latitudes may be related to a redistribution of energy inputs at auroral latitudes, though there remain significant problems with overcoming the powerful zonal winds generated by Coriolis forces on the rapidly rotating giant planets (e.g., Smith *et al.*, 2007; Müller-Wodarg *et al.*, 2012; see also Majeed *et al.*, 2005). Precipitation of charged particles can lead to heating primarily via the dissipation of energy resulting from currents in a resistive ionosphere. Commonly referred to as Joule heating, this process represents two components, the thermal heating of the atmosphere by electrical currents and the change in kinetic energy of the atmospheric gases which results from momentum change due to ion drag effects (Vasyliūnas and Song, 2005). The quantitative impact on giant-planet

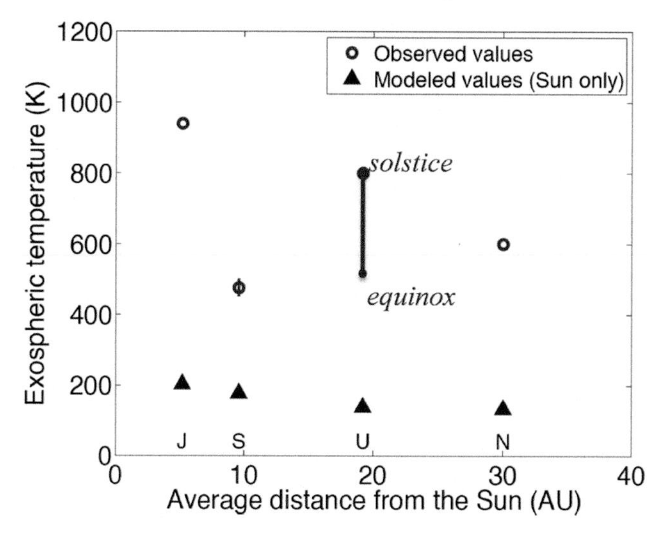

Fig. 8.3 Upper-atmospheric temperature as a function of heliocentric distance for the giant planets, comparing observations with model values using only solar irradiance as energy input (after Yelle and Miller, 2004; Melin *et al.*, 2011a, 2013). Note that these values represent a combination of measurements from a range of latitudes with different seasonal and solar conditions. In addition, the Uranus values include both neutral exobase temperatures as well as H_3^+ temperatures near the ionospheric peak altitude.

upper-atmospheric thermal structure of the third category of energy inputs – the dissipation of waves – remains unclear. Though Galileo Probe temperature measurements are consistent with upwardly propagating gravity waves (Young *et al.*, 1997), there is no consensus regarding their effects on thermospheric energetics (e.g., Matcheva and Strobel, 1999; Hickey *et al.*, 2000).

8.2 Ionospheres of the giant planets

Ionizing radiation at the giant planets comes primarily in two forms: solar EUV and soft X-ray photons, and energetic particle precipitation from the planetary magnetosphere, which is mostly concentrated at high magnetic latitudes. Roughly 90% of the ionizing radiation is absorbed directly by H_2, leading to the production of H_2^+ ions and electrons, usually suprathermal. These electrons, referred to as photoelectrons in the case of photoionization and secondary electrons in the case of particle impact ionization, possess enough energy to excite, dissociate, and further ionize the neutral atmosphere as well as to heat the ambient plasma. Therefore, in order to accurately model an ionosphere, it is necessary to track the evolution of both photons and electrons throughout the atmosphere. Photoionization production rates as a function of altitude and wavelength follow from application of the Lambert–Beer

Law assuming neutral atmospheric densities, incident solar fluxes, and photoabsorption and photoionization cross sections are known (see also Ch. 12 in Vol. I). The Lambert–Beer Law, also known as Beer's Law, the Beer–Lambert Law, or the Beer–Lambert–Bouguer Law, describes the attenuation of light through a medium (Houghton, 2002). In order to track the transport, energy degradation, and angular redistribution of suprathermal electrons – including photoelectrons as well as secondary electrons – a kinetic approach is typically applied by solving the Boltzmann equation (e.g., Perry *et al.*, 1999; Moore *et al.*, 2008; Galand *et al.*, 2009). The ion-production rates and thermal-heating rates generated by energetic electrons are dependent upon the ambient atmospheric parameters, which themselves are altered by the electron-energy deposition, and so iteration between atmospheric fluid and kinetic codes is required to close the solution.

The dominant giant-planet ionospheric ions are H^+ and H_3^+, though their relative importance is mostly unconstrained at present and is likely to vary with a range of parameters, including season, solar flux, latitude, and local time. While the rapid production of H_2^+ is balanced by an equally rapid loss via charge exchange with H_2, producing H_3^+, the slow production of H^+ is offset by a very slow radiative recombination loss process and a short day. In fact, early giant-planet ionospheric modelers predicted H^+ would be completely dominant (e.g., McElroy, 1973; Capone *et al.*, 1977), a notion that was challenged only when measured electron densities from radio occultations were found to be an order of magnitude smaller than predicted and to exhibit strong dawn/dusk asymmetries. This model–data discrepancy required the introduction of new photochemical loss chemistry that would reduce electron densities by converting long-lived atomic H^+ ions into short-lived molecular ions. Representative giant-planet ionospheric electron-density profiles derived from spacecraft radio occultation experiments are presented in Fig. 8.4. Ionization fractions at the giant-planet ionospheres are roughly of order 10^{-6}, smaller on average than ionization fractions in much of Earth's ionosphere (cf., Sect. 12.8 in Vol. I). In general, the EUV-driven peak electron densities near 1000 km decrease with heliocentric distance. Profiles shown in Fig. 8.4 are, however, associated with different solar conditions (solar activity, solar zenith angle) and magnetospheric conditions (e.g., particle precipitation).

The two additional most commonly suggested pathways for chemical loss of protons in giant-planet ionospheres are charge exchange with molecular hydrogen (McElroy, 1973) and charge exchange with water group and/or ice particles (Connerney and Waite, 1984). While the former – $H^+ + H_2(\nu \geq 4)$ – is exothermic only when H_2 is excited to the fourth or higher vibrational level, the latter – e.g., $H^+ + H_2O$ – depends on an external influx of water group particles such as H_2O or OH. Neither process is well constrained at present. There have been a few first-principles calculations of the vibrational levels of H_2 at Jupiter and Saturn (e.g.,

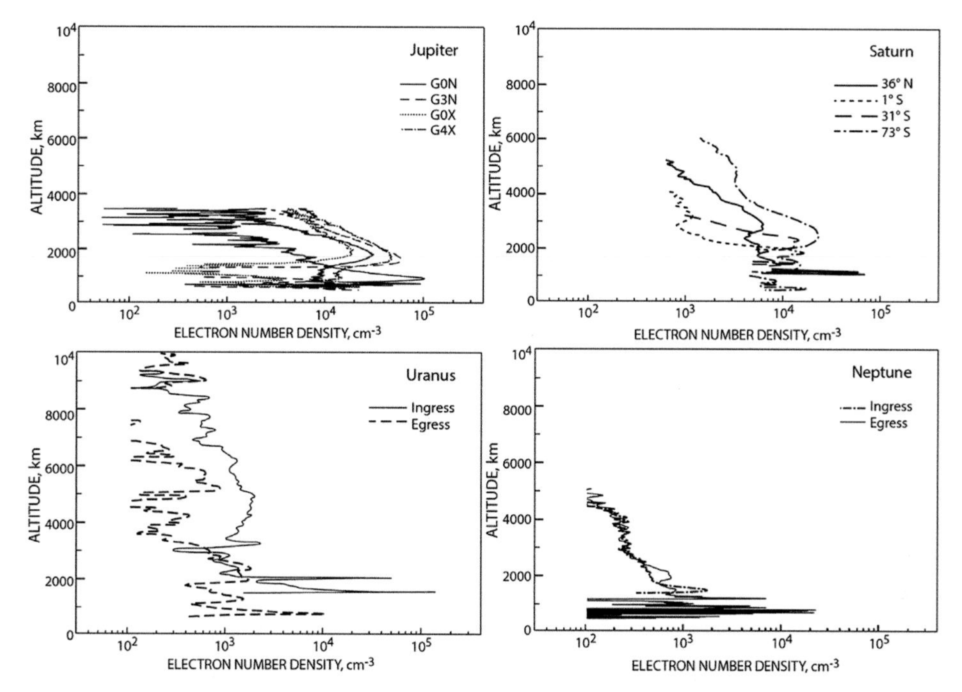

Fig. 8.4 Ionospheric electron-density profiles derived from spacecraft radio occultation experiments at (top left) Jupiter, (top right) Saturn, (bottom left) Uranus, and (bottom right) Neptune. Altitudes refer to radial distance above the 1 bar pressure level. Profiles are from the Galileo spacecraft for Jupiter (Yelle and Miller, 2004), and from the Voyager spacecraft for Saturn (Lindal *et al.*, 1985), Uranus (Lindal *et al.*, 1987), and Neptune (Lindal, 1992).

Cravens, 1987; Majeed *et al.*, 1991), but those estimates did not lead to consistent reproductions of observed electron densities. Consequently, contemporary models typically use some parametrization of those calculations in order to specify $H_2(\nu \geq 4)$ populations. Estimates of external particle influxes at the giant planets, on the other hand, are based primarily on Infrared Space Observatory observations of stratospheric carbon and oxygen bearing compounds (Feuchtgruber *et al.*, 1997; Moses *et al.*, 2000) and on electron density model–data discrepancies (e.g., Connerney and Waite, 1984; Moore *et al.*, 2010). Possible external sources of oxygen at all of the giant planets include direct atmospheric ablation of interplanetary dust particles, deposition of material following cometary impacts, and an influx of materials from rings or satellites. In particular, at Saturn, the water vapor plumes of Enceladus (Porco *et al.*, 2006) and the water ice rings imply ample sources of external water group particles are available (e.g., Tseng *et al.*, 2010; Fleshman *et al.*, 2012), and infrared observations of thermal H_3^+ emissions revealed a direct ring–atmosphere connection (Connerney, 2013; O'Donoghue *et al.*, 2013).

In addition to the major ionospheric ions, H^+ and H_3^+, giant-planet ionospheres are expected to exhibit an additional ledge of ionization above the homopause and below the ionospheric electron-density peak, initialized by charge-exchange reactions between the major ions and methane and also by direct photoionization of methane. This low-altitude ionospheric region is predicted to be dominated by hydrocarbon ions – wherein a complex array of hundreds of photochemical reactions culminates with ions such as $C_3H_5^+$ and CH_5^+ – and possibly also to play host to metallic ions derived from meteoroid ablation, such as Mg^+ (e.g., Kim and Fox, 1994; Moses and Bass, 2000; Y. Kim *et al.*, 2001, 2014). Narrow layers in electron density are observed frequently in and just above this region (Fig. 8.4), possibly caused by vertical shears in neutral winds (Lyons, 1995; Moses and Bass, 2000), such as might result from atmospheric gravity waves (Barrow and Matcheva, 2011). Similar electron density layers are also observed at other planets, such as sporadic-E layers at Earth (e.g., see Vol. III, Ch. 13). Vertical plasma drifts may also be behind the drastic variations in the observed altitudes of peak electron density (e.g., McConnell *et al.*, 1982). Sample model vertical ionospheric profiles that include various combinations of the above processes are shown in Fig. 8.5. Note that these model profiles are meant to be representative of the expected overall structure, and do not necessarily correspond to the specific conditions associated with the measured profiles shown in Fig. 8.4. However, by carefully specifying combinations of unconstrained parameters – such as oxygen influx, vibrational H_2 populations, and vertical plasma drifts – it is possible to reproduce much of the observed electron density structure.

Giant-planet low-altitude ionospheres (e.g., below ~600 km at Jupiter and below ~1000 km at Saturn), where hydrocarbon and H_3^+ ions dominate, are generally in photochemical equilibrium. The fast dissociative recombination rates of these molecular ions – tens of seconds to a few minutes – means that they are neutralized before any transport effects can have an impact, and consequently their densities can be derived directly from a series of ion continuity equations which balance ionization and chemical production with chemical loss. Assuming charge neutrality in a single ion component ionosphere (with $n_i = n_e$), photochemical equilibrium is represented schematically by $q = \alpha_{\text{eff}} n_e^2$, where q is the ion production rate, α_{eff} is the dissociative ion–electron recombination-rate coefficient (of order $\sim 10^7\,\text{cm}^3\,\text{s}^{-1}$ for H_3^+ in giant-planet ionospheres), and n_e is the electron density. At higher altitudes – where dissociative recombination rates are slower due to reduced electron densities, and where H^+ charge-exchange loss processes are slow due to reduced neutral densities – ion transport plays an important role in determining ionospheric densities. Ion–neutral collision frequencies are much smaller than ion gyrofrequencies at high altitudes; consequently ions are primarily constrained to move along magnetic field lines, and so ion transport is modified

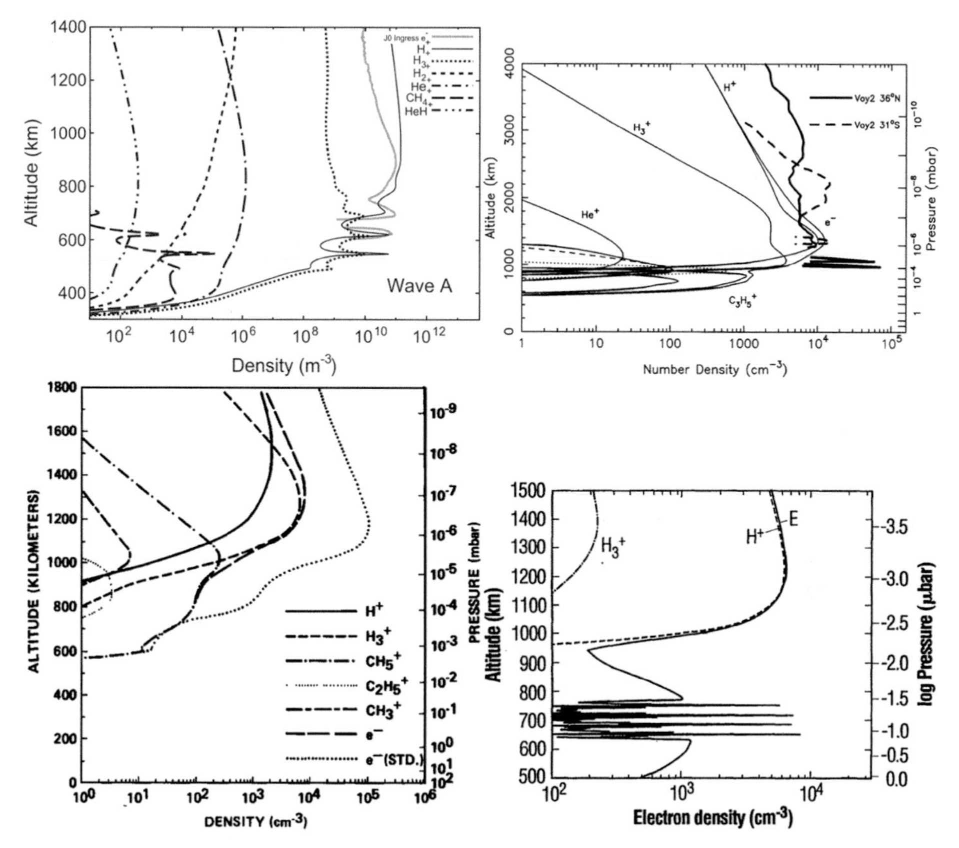

Fig. 8.5 Ionospheric model calculations for (top left) Jupiter, (top right) Saturn, (bottom left) Uranus, and (bottom right) Neptune. Altitudes refer to radial distance above the 1 bar pressure level. Note that electron density profiles are labeled as e^- in the Jupiter, Saturn, and Uranus panels, and as E in the Neptune panel. (Sources: Jupiter: Barrow and Matcheva, 2011; Saturn: Moses and Bass, 2000; Uranus: Chandler and Waite, 1986; Neptune: Lyons, 1995.)

by the planetary magnetic field. Ion-transport processes include drifts driven by neutral winds and ambipolar diffusion, wherein the electrical interaction between (relatively) heavy ions and nearly mass-less electrons leads to a coupled diffusion process and an ion scale height roughly twice that of the corresponding neutral.

Ionospheric structure at the giant planets is relatively unconstrained by observation. Radio occultations are sparse – with only nine published for Jupiter, 65 for Saturn, two for Uranus, and two for Neptune – and furthermore are all limited to measurements near the terminator, due to the geometry required by radio occultation observations of superior planets. Dawn and dusk are periods of rapid change in an ionosphere, as photoionization is turned on or off by the rising or setting of the Sun, and so are not ideal times to sample unknown ionospheric structure. At

Saturn, where the Cassini spacecraft (Jaffe and Herrell, 1997) has been in orbit since 1 July 2004, the increased number of radio occultation measurements has allowed identification of two main global ionospheric structures. First, there is a dawn/dusk asymmetry, with peak electron densities larger at dusk and the altitudes of the electron-density peak higher at dawn (Nagy *et al.*, 2006, 2009). Second, there is a counter-intuitive latitudinal behavior in electron density: peak electron densities are smallest at the equator – where solar ionization rates are largest – and increase with latitude (Kliore *et al.*, 2009). Both of these behaviors can be explained by an external oxygen influx from Saturn's rings and icy satellites, such as H_2O or O_2^+, if the influx maximizes at low latitude (e.g., Moore *et al.*, 2015).

Voyager and Cassini radio-wave measurements have also allowed a derivation of the diurnal variation in peak electron density at Saturn. Broadband radio signatures of powerful lower-atmospheric lightning storms, which for presently unknown reasons occur at only a select few latitudes, are refracted and attenuated by Saturn's ionosphere before being detected by a spacecraft. As the storm system rotates with the planet, the minimum measured Saturn Electrostatic Discharge (SED; note that this is a very different process from terrestrial storm enhanced densities for which the same acronym is used) frequency yields the peak electron density of the intervening plasma as a function of local time (Kaiser *et al.*, 1984; Fischer *et al.*, 2011). The strong diurnal variation in peak electron density derived from SED measurements has yet to be explained by models, however, as the ion production and loss rates implied by the observations are much larger than current best estimates (Majeed and McConnell, 1996; Moore *et al.*, 2012). No such similar emissions have been detected at the other giant planets, possibly due to a lack of active lightning discharge during spacecraft flybys, or to attenuation of the radio waves by the planetary ionospheres (Zarka, 1985), or to the slow nature of the lightning discharge itself (Farrell *et al.*, 1999).

One additional remote diagnostic of giant-planet ionospheres that has proven to be remarkably fruitful is the measurement of thermal H_3^+ emission in the infrared. First detected at Jupiter (Drossart *et al.*, 1989), it has since been observed regularly at Jupiter, Saturn, and Uranus, but not – so far – at Neptune. There are a number of strong H_3^+ rotational–vibrational emission lines available in the mid-IR, particularly in the L-band (3–4 micron) atmospheric window, and those emissions are strongly temperature dependent. Conveniently, this spectral region (e.g., near 3.4 micron) also corresponds to a deep methane absorption band, such that light from giant-planet interiors cannot escape, and H_3^+ emission therefore appears as bright emission against a dark background. Because H_3^+ is expected to be thermalized to the surrounding neutral atmosphere, it can be used to track both thermospheric temperatures and ionospheric H_3^+ column content. These factors combine to make H_3^+ an excellent probe of giant-planet upper atmospheres, and the fact that important emission lines fall within atmospheric transmission windows allows cost effective observations from ground-based telescopes (see Miller *et al.*, 2010, and Stallard *et al.*, 2012b, for reviews).

Emission from giant-planet H_3^+ ions is strongest in the auroral regions, where particle precipitation enhances the local ionization far above solar-produced levels and where temperatures are largest. For this reason, a majority of H_3^+-related science has focused on understanding auroral structure and behavior. Giant-planet auroral UV emission is predominantly caused by inelastic collisions of energetic electrons with atmospheric molecular hydrogen, and represents a prompt response to changes in magnetospheric inputs. In contrast, auroral IR H_3^+ emissions represent the temporally integrated response of the upper atmosphere to those inputs above the homopause. Therefore, UV and IR emissions are highly complementary in studies of giant-planet auroral ionospheres (e.g., Melin *et al.*, 2011b; Radioti *et al.*, 2013). At Jupiter the hydrogen excitation aurora and the H_3^+ thermal aurora appear to be well separated in altitude ($\sim$250 km for the visible (Vasavada *et al.*, 1999) and $\sim$1000 km for the IR (Lystrup *et al.*, 2008)) whereas at Saturn they peak at a similar altitude near 1150 km (in the UV; Gérard *et al.*, 2009; Stallard *et al.*, 2012a). No observational constraints are available at present regarding the altitude distribution of H_3^+ at Uranus or Neptune.

Though weak compared to auroral emissions, H_3^+ has also been detected across the dayside disk of Jupiter (Lam *et al.*, 1997), and the measured latitudinal variations indicate additional sources of non-solar mid-latitude ionization may be required (Rego *et al.*, 2000), though no magnetospheric source for this ionization has been suggested. Similarly, long-term H_3^+ observations have been made at Uranus – likely representing a range of combinations of auroral and non-auroral emission – revealing an unexplained cooling trend of the upper atmosphere, persisting past equinox (Melin *et al.*, 2013). At Saturn, the cooler thermospheric temperatures and the lower ionospheric densities meant that the prospect for observing non-auroral H_3^+ emission was slim. A low-latitude detection of H_3^+ made using the Keck telescope (O'Donoghue *et al.*, 2013), however, has reignited hopes of probing upper-atmospheric properties across the visible disk of Saturn. Even more intriguing, the O'Donoghue *et al.* measurements revealed significant latitudinal structure in H_3^+, with local extrema in one hemisphere being mirrored at magnetically conjugate latitudes, and mapping along magnetic field lines to regions of increased or decreased density in Saturn's rings, implying a direct ring–atmosphere connection (Connerney, 2013).

8.3 Ionosphere–thermosphere–magnetosphere and solar wind coupling

Particles, energy, and momentum are exchanged between planetary upper atmospheres and magnetospheres via currents that flow through the high magnetic latitude ionosphere. Birkeland currents, or currents which flow along planetary magnetic field lines, supply angular momentum to the magnetosphere – by allowing closure of magnetospheric currents in the ionosphere – and energy – e.g., in the form of particle precipitation and associated Joule heating – to the atmosphere. Charged particles are accelerated into the atmosphere as a result of the varying charged particle density at different positions along planetary magnetic field lines. Both ions and electrons are concentrated in two locations, the ionosphere of the planet and close to the equatorial plane of the magnetosphere. There is, as a consequence, a lack of current-carrying plasma part-way along the field lines, particularly far from the magnetic equator, on the field line close to the planet above the exosphere. A circuit that closes through the ionosphere requires field-aligned potentials to develop in order to augment the electron distribution in this low-density region, allowing increased field-aligned currents to flow between the ionosphere and magnetosphere. The resultant current–voltage relation is nonlinear, depending upon the density and temperature of the electron population (Knight, 1973; Ray *et al.*, 2009). This acceleration of magnetospheric electrons not only increases the field-aligned current density, but also the energy and energy flux of the electrons precipitating into the atmosphere.

Ionospheric currents allow closure to the magnetospheric current system; they depend on local conditions, which are in turn strongly affected by the enhanced ionization brought by Birkeland currents. Electrical, ionospheric conductivities are associated with particle mobility in the direction perpendicular to the planetary magnetic field and parallel (Pedersen) or perpendicular (Hall) to the ionospheric electric field. Pedersen conductivities, associated primarily with a current carried by ions, peak in the lower ionosphere near the homopause, where the ion gyrofrequency is approximately equal to the ion–neutral collision frequency, and where molecular hydrocarbon ions begin to dominate. Hall conductivities, associated primarily with a current carried by electrons, peak at lower altitudes below the homopause, within a region of complex hydrocarbon chemistry. Pedersen and Hall conductances – or height-integrated conductivities – therefore depend upon local sources of ionization, ionospheric chemistry, and planetary magnetic field strength.

There are no direct observational constraints on ionospheric electrical conductances at the giant planets for a number of reasons. First, only roughly half of the relatively few electron-density altitude profiles retrieved from radio-occultation measurements extend down to the ionospheric conducting layers. Furthermore, there are no obvious auroral ionization signatures in the handful of high-latitude outer-planet radio occultations, therefore limiting their application to magnetosphere–ionosphere coupling studies. Second, the other main remote diagnostic of giant planet ionospheres – emission from H_3^+ – is a column integrated measurement, and therefore typically possesses little altitude information. To date H_3^+ altitude information has been derived only twice, once at Jupiter (Lystrup *et al.*, 2008) and once at Saturn (Stallard *et al.*, 2012a). Finally, even a

Table 8.2 *Pedersen conductances Σ_P calculated using energy deposition and ionospheric models and presented as a function of the ionization source (Sun, auroral electrons) over the main auroral oval. The characteristics of the auroral electrons are given in terms of the initial mean energy E_{prec} and energy flux Q_{prec}.*

Energy source [E_{prec} (keV), Q_{prec} (mW m^{-2})]	Pedersen conductance Σ_P (mho)	Reference [atmospheric model]
	Jupiter	
Electrons [10, 1]	0.04	
Electrons [10, 10]	0.12	Millward *et al.*, 2002
Electrons [10, 100]	0.62	[3D GCM]
Electrons [60, 10]	1.75	
Electrons [22, 100] and [3,10]	9 (NH)[a]	Bougher *et al.*, 2005
and [0.1,0.5]	12.5 (SH)[a]	[3D GCM]
Electrons [1, 1]	0.008	Hiraki and Tao, 2008
Electrons [10, 10]	0.5	[1D ionospheric model]
	Saturn	
Solar only (Main oval: noon,	0.7	
78°, equinox, solar minimum)		Galand *et al.*, 2011
Solar + Electrons [10, 1]	11.5	[1D ionospheric model
Solar + Electrons [10, 0.2]	5	using 3D neutral output]
Solar + Electrons [2, 0.2]	10	

[a] NH and SH stand for northern hemisphere and southern hemisphere, respectively.

complete H_3^+ altitude profile only provides a lower limit to the ionospheric conductance, as the dominant contribution from the hydrocarbon ion layer would still be missing. Therefore, in practice, ionospheric electrical conductances are commonly estimated from models – either magnetospheric models that require certain conductances to explain observations of magnetospheric phenomena (e.g., Cowley *et al.*, 2008) or ionospheric models which rely on assumed precipitation sources to calculate conductances from the resulting ionization (e.g., Millward *et al.*, 2002; Galand *et al.*, 2011).

Based on model calculations, ionospheric conductances are expected to be largest at low- and mid-latitudes on the dayside due to solar-induced ionization, and in the auroral regions at all local times due to particle precipitation. Additionally, outside of the auroral precipitation regions, where solar-induced ionization sources dominate, there are likely strong seasonal and local time variations corresponding to changes in the solar zenith angle. Table 8.2 summarizes some of the different estimates of ionospheric Pedersen conductances in the literature (which have only been made for Jupiter and Saturn thus far).

The variations between the Pedersen conductances derived for Jupiter and Saturn (Table 8.2) are driven primarily by differences in planetary magnetic field strength, affecting ion gyrofrequency, and the particle precipitation energies and fluxes. On the one hand, more-energetic particles deposit their energies at lower altitude within an atmosphere, and so there is a range of energies that will enhance ionization in the conductance layer most efficiently, with higher- and lower-energy electrons ionizing below or above the conductance layer, respectively. Enhancements in the precipitating particle fluxes, on the other hand, amplify the degree of ionization in the altitude regime where particles of a specified energy are deposited. In fact, the electron density in auroral precipitation regions is proportional to the square root of the energy flux, and consequently the conductance is also approximately proportional to the square root of the energy flux (e.g., Millward *et al.*, 2002; Müller-Wodarg *et al.*, 2012). Therefore, any temporal variations in the precipitating energy flux will be closely tracked by corresponding variations in the ionospheric electrical conductance, though with a delay that depends on the chemical timescales within the conductance layers (Galand *et al.*, 2011).

8.3.1 Solar-wind interactions

Auroral emission on the Earth is driven by the interaction between the Earth's magnetic field and the solar wind, which, in turn, is dominated by a process known as reconnection. Reconnection occurs when two plasmas with non-parallel frozen-in magnetic fields are pushed together. This is visualized as if magnetic field lines from each plasma "break" to "reconnect" with each other, magnetically linking the plasma regions. On Earth, magnetic field lines within the magnetosphere reconnect with field lines carried within the solar wind along the noon edge of the magnetosphere, opening field lines in Earth's polar region to the solar wind (cf., Ch. 10 of Vol. I). This open flux is carried across Earth's polar cap with the solar wind, to the nightside. On the nightside of the polar cap, open field lines are closed and removed from the equatorial plane of the tail by reconnection, at times associated with substorm formation. As these field lines are closed, the polar cap contracts and flows redistribute flux and plasma within the polar regions. These processes drive a twin-cell ionospheric convection pattern, with antisunward flow across the poles, and sunward flows around the equatorward edge of the polar cap (Dungey, 1961; see Fig. 10.5 in Vol. I).

This process is complicated at the giant planets (cf., Ch. 13 in Vol. I), as the scale of the magnetospheres of these planets is significantly larger (as a result of both stronger magnetic fields and weaker solar wind) and the planets rotate more quickly, so that the solar wind takes many planetary rotations to cross the magnetosphere. For example, at Saturn this results in a rotationally dominated

Dungey cycle, where rotating components are stronger than the twin-cell flows, such that a single-cell flow occurs across the polar region, with a return flow only occurring on the dawnside of the planet (see discussion surrounding Fig. 9.5 of Gombosi *et al.*, 2009). Ionospheric flows produce an ionospheric current system that drives current along magnetic field lines, with downward currents on open field lines just poleward of the main aurora, and associated diffuse upward currents near the pole. Equatorward of this, a narrow ring of upward current exists on closed field lines which, in turn, close at lower latitudes. This upward current is directly related to the downward flow of accelerated electrons, driving the main auroral emission at Saturn. As the rotation of the planet drives the ionospheric flows onto the dawnside of the polar region, this results in a dawn enhancement in the auroral emission.

The extent of solar-wind interaction with Jupiter's auroral region remains a matter of significant scientific debate. Ionospheric flows within the dawnside of Jupiter's polar region are clearly held in the solar-wind reference frame, resulting in a strong ionospheric flow relative to the neutral atmosphere (Stallard, 2003). Some have suggested this flow results from a modified Dungey cycle flow similar to that seen at Saturn (Cowley *et al.*, 2003). Although this hypothesis matches with many of the observed conditions within this region, one of the major problems with this interpretation is that Jupiter's dawn polar region sees significant variable auroral emission in a region that might be expected to be free of plasma (Grodent *et al.*, 2003). An alternative explanation for how such "Swirl emission" can exist is that the ionospheric flows observed are driven by solar-wind-driven magnetospheric flows caused by viscous processes at the magnetopause boundary (Delamere and Bagenal, 2010). This hypothesis allows closed field lines, filled with plasma that can produce "Swirl" aurora, to drive flows within the ionosphere that are held within the solar-wind reference frame. Which of these hypotheses actually dominates solar-wind interaction at Jupiter is hotly contested.

Observations of UV auroral emissions from Uranus (Lamy *et al.*, 2012) have shown that these weak auroral emissions are associated with changes in the solar-wind density. The lack of a strong response to solar-wind compressions may reflect the lack of a well-developed tail structure, which would be expected for Uranus' current equinox configuration (e.g., Tóth *et al.*, 2004). This contrasts with Uranus at solstice, when open flux could be produced continuously and flow in the slowly rotating nightside tail (cf., Vol. I, Ch. 13, Fig. 13.9; Cowley, 2013).

8.3.2 Internal current systems

The magnetospheres of both Jupiter and Saturn differ significantly from that of the Earth in the distribution and quantity of plasma contained within them. The major source of internal plasma at Earth is its ionosphere. As a result, plasma is concentrated very close to the planet, and is mostly found on magnetic field lines close to the planet. Both Jupiter and Saturn have significant sources of plasma away from the planet. At Jupiter, the volcanic moon Io contributes ~1000 kg/s of mass to a torus of equatorially bound plasma that orbits Jupiter close to Io's orbit, at 6 R_J (Dessler, 1980). This torus is by far the dominant source of all plasma within the magnetosphere, with ionized material forming a significant plasmasheet that extends from Io outwards. Saturn has a wider distribution of plasma sources, including the rings close to the planet, the cryovolcanic water plume from Enceladus at 3.95 R_S, as well as scattering from the surfaces of various moons. This results in the concentration of significant plasma in the equatorial regions of both planetary magnetospheres. Because this plasma is generated from neutral material orbiting with Keplerian velocities, it has a nonzero velocity relative to the magnetic field of the planet, which rotates with the planet's interior. In a collisionless MHD approximation of this process, the plasma is frozen into the magnetic field, and the deviation from the planet's rotation rate results in the magnetic field lines being azimuthally bent back near the equatorial plane, driving a radial current outwards through the plasmasheet. This produces a current that closes by producing an equatorward current within the ionosphere as well as field-aligned currents into and out of the ionosphere. In the rotation-dominated magnetospheres of Jupiter and Saturn, the equatorial plasmasheet rotates with the planet on a time scale of ~10 h while small-scale, diffusive interchange of magnetic flux tubes leads to net radial transport of mass outwards and return of magnetic flux inwards on time scales of tens of days (Vol. I, Ch. 13). However, the frozen-field approximation breaks down in the ionosphere of the planet, as charged particles within the ionosphere are accelerated by the surrounding neutral atmosphere. This results in currents that transfer angular momentum from the neutral atmosphere out to plasma within the magnetosphere, producing a steady-state coupling between the atmosphere and magnetosphere that drives magnetospheric plasma into co-rotation with the planet.

There are two limiting factors to this current system.

(i) The Keplerian orbital velocity decreases with distance from the planet, so that maintaining co-rotation requires larger forces. At the same time, the magnetic field strength also decreases with distance, requiring a greater current. As a result, the currents required to maintain co-rotation within the magnetosphere increase with distance from the planet.
(ii) The ionospheric Pedersen conductance represents particle mobility within the ionosphere, parallel to the electric field and perpendicular to the magnetic

field. Because the Pedersen conductivity is finite, the magnitudes of the currents that flow through the ionosphere depend on the conductivity for a given ionospheric electric field. When the current required by the above process grows too large, the force produced begins to make the ionosphere sub-rotate relative the surrounding atmosphere, which in turn allows plasma within the magnetosphere to no longer co-rotate with the planet.

Therefore, as magnetospheric plasma moves radially outwards (a radial flow that occurs as the direct result of charged particles being accelerated away from their Keplerian orbital velocity), the ionospheric Pedersen conductivity required to maintain that plasma's co-rotation in the magnetosphere continually increases. Ultimately, the drifting plasma reaches a tipping point at which ionospheric currents are no longer able to maintain co-rotation, and the plasma begins to rotate significantly slower than the magnetic field of the planet. This nonzero azimuthal plasma velocity results in currents that flow radially outwards through the plasmasheet within the equatorial magnetosphere. This current again closes along magnetic field lines into the ionosphere, producing broader downward currents and ion precipitation in the poleward mapping of this breakdown in co-rotation, and a narrow region of upward currents and electron precipitation just equatorward of the boundary of sub-rotating magnetospheric plasma.

At Jupiter, this breakdown in co-rotation drives a continuous bright main auroral emission (Cowley and Bunce, 2001; cf., Fig. 13.7 in Vol. I) and results in a significant sub-rotational ion wind flow, $\sim$1.5 km/s, in the region of the main emission, caused by Hall drift from the Pedersen currents that drive the aurora (Rego *et al.*, 2000; Stallard *et al.*, 2001). At Saturn, the magnetosphere is significantly more mass-loaded than Jupiter, due to the weaker magnetic field (Vasyliūnas, 2008), and so the breakdown in co-rotation occurs at $\sim 3R_S$, inside the orbit of Enceladus. The resultant aurora produced by this process at Saturn has been shown to be far weaker (Cowley and Bunce, 2003) and has thus far not been observed in UV emission, but may have been seen in the IR (Stallard *et al.*, 2008, 2010).

This transfer of energy from within the magnetosphere down into the upper atmosphere is universal to any planet with a significant source of plasma within the magnetosphere. This process has been suggested as the potential future source of detectable auroral emission from exoplanets, with strong radio emission resulting from a higher planetary rotation rate (and a presumably stronger magnetic field), and a higher stellar XUV luminosity. Similar current systems have also been suggested as the possible source of already observed radio emission from brown-dwarf stars (Schrijver, 2009; Nichols *et al.*, 2012).

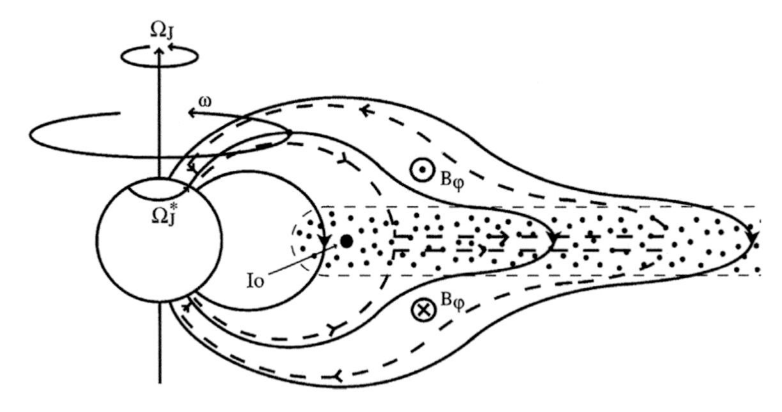

Fig. 8.6 Sketch of a meridian cross section through the Jovian magnetosphere, showing the principal features of the inner and middle magnetosphere regions. The arrowed solid lines indicate magnetic field lines, which are distended outwards in the middle magnetosphere region by azimuthal currents in the plasmasheet. The plasmasheet plasma originates mainly at Io, which orbits in the inner magnetosphere at 6 R_J, liberating $\sim 10^3$ kg s^{-1} of sulphur and oxygen plasma. This plasma is shown by the dotted region, which rotates rapidly with the planetary field due to magnetosphere–ionosphere coupling, while more slowly diffusing outwards. Three separate angular velocities associated with this coupling are indicated. These are the angular velocity of the planet Ω_J, the angular velocity of a particular shell of field lines ω, and the angular velocity of the neutral upper atmosphere in the Pedersen layer of the ionosphere, Ω_L^*. The latter is expected to lie between ω and Ω_J because of the frictional torque on the atmosphere due to ion–neutral collisions. The oppositely directed frictional torque on the magnetospheric flux tubes is communicated by the current system indicated by the arrowed dashed lines, shown here for the case of sub-co-rotation of the plasma (i.e., $\omega \leq \Omega_J$). This current system bends the field lines out of meridian planes, associated with azimuthal field components B_ϕ as shown. (From Cowley and Bunce, 2001.)

A sample sketch of the Jovian magnetospheric current systems is given in Fig. 8.6.

8.3.3 Vasyliūnas cycle

Plasma is continually added to the magnetospheres of both Jupiter and Saturn. Plasma confined to the equatorial region is transported radially outwards (via centrifugally driven flux tube interchange), it is ultimately lost from the magnetosphere through the tail, resulting in a downtail outflow of magnetic field lines that are stretched out and eventually pinched off. This forms a plasmoid, containing trapped closed field lines and released downtail, as well as closed, relatively empty field lines that propagate back onto the dayside of their magnetosphere, a process described as the Vasyliūnas cycle (Vasyliūnas, 1983). The relatively few particles on the closed flux tubes are accelerated as the flux tube springs back to a more dipole configuration on releasing the plasmoid. This process has been invoked to explain the region of bright auroral emission in the duskside of Jupiter's polar region, often leading to bright polar flare emission (Grodent *et al.*, 2003).

8.3.4 Moon–magnetosphere–atmosphere interaction

Giant planets also have aurorae directly related to the magnetic interaction between the magnetosphere and moons. These auroral emission fall into two types, a spot of emission located within the planetary ionosphere at the magnetically mapped position of the moon itself, and a tail of emission that extends away from the moon, mapping to the moon's orbital path (e.g., Kivelson *et al.*, 2004). The localized moon spots are created by the Alfvénic disturbance imposed by the moon upon the magnetic flux tube sweeping past the moon and accelerating electrons in both directions along the field lines (Bonfond *et al.*, 2008). Such spot emission was first identified for Io at Jupiter (Connerney *et al.*, 1993), but similar spot features have now been identified for Europa, Callisto, and Ganymede at Jupiter (Clarke *et al.*, 2002, 2011), and for Enceladus at Saturn (Pryor *et al.*, 2011). The trailing emission, seen in the wake of both Io (Clarke *et al.*, 2002) and Europa (Grodent *et al.*, 2006), represents a steady state current system which accelerates localized plasma from the moon into co-rotation with the surrounding magnetic field. At Saturn, a weak H_3^+ emission aurora is observed at latitudes that map magnetically to between 3–4 R_S, extending around the entire planet. This aurora might be considered a proxy for Jupiter's breakdown in co-rotation aurora, but because mass loading means plasma from Enceladus is never fully accelerated into co-rotation, it might equally be described as a satellite wake emission that extends to all longitudes in the ionosphere (Stallard *et al.*, 2008, 2010).

8.3.5 Additional magnetosphere–atmosphere interaction

Charged particles that are not part of a current system within a magnetosphere tend to be confined near the equatorial plane, through magnetic mirroring, because particles propagating up the field line towards the planet receive a repulsive force due to the increased magnetic field strength along the field line, leading to a bounce motion between the magnetic poles. However, if the particles are energetic enough, or if their pitch angles small enough, the magnetic mirror points of the particles can extend into the atmosphere, precipitating the particles into the atmosphere. While this process produces aurorae, both the particle energies and the total flux of the precipitating particles are typically smaller than those corresponding to discrete aurora, resulting in weaker and more diffuse emission. This precipitation process will also vary if the magnetic field strength varies with longitude, resulting in longitudinal enhancements in precipitation.

At Jupiter, this form of aurora has been suggested as one source of the mid-to-low latitude emission (Miller *et al.*, 1997), and the radiation belt has been modeled as a possible source for both H_3^+ and X-ray emission (Abel and Thorne, 2003). At Saturn, hot plasma located between 8–15 R_S has been suggested as a source of an auroral arc observed in both IR and UV on the nightside of the planet, equatorward of the main oval (Grodent *et al.*, 2010). In addition, observations of the equatorial region of the planet have shown variations in the H_3^+ emission at latitudes that magnetically map to Saturn's rings (O'Donoghue *et al.*, 2013). It is likely that ionized water group particles are entering Saturn's ionosphere along magnetic field lines, leading to reductions in the local electron density, and in turn resulting in an increase in the corresponding H_3^+ density (Moore *et al.*, 2015).

8.4 Auroral emissions

Within the auroral region, field-aligned currents drive particle precipitation into the atmosphere, ultimately driving a variety of auroral emissions. The sources for this auroral power come from two major types of precipitation: discrete aurorae are formed from precipitating charged particles that have been accelerated into the atmosphere along magnetic field lines, while diffuse aurora is formed from energetic precipitating particles resulting from plasma interaction in the magnetosphere. Aurorae are produced through energy released by precipitating energetic particles in their interaction with an atmosphere. At the giant planets, auroral emissions are produced in four different ways: emission can be released directly from the precipitating particle itself; the precipitation process can transfer energy into the atmosphere, resulting in either the energetic excitation or the ionization of the atoms or molecules within the atmosphere; and the precipitation process heats the atmosphere both directly and indirectly, leading to a thermalized auroral emission from the atmosphere. Auroral emissions from the giant planets are discussed in detail within Bhardwaj and Gladstone (2000), Kurth *et al.* (2009), and Clarke *et al.* (2004). In addition, an overview of Jupiter's magnetosphere and auroral processes is given by Bagenal *et al.* (2014). Figure 8.7 shows a representative comparison between the UV and IR aurora at Jupiter, with major features indicated.

8.4.1 Emission from precipitating particles

Radio and X-ray emissions are produced directly from the precipitating particle itself. Radio emission is generated by precipitating electrons as they are accelerated into the atmosphere along the magnetic field lines. They are thought to be produced

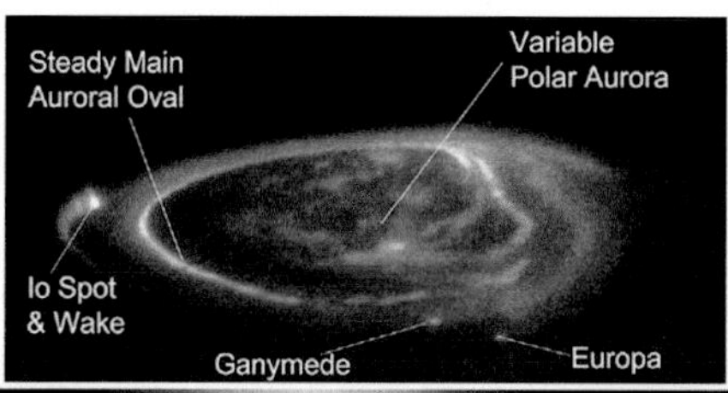

Fig. 8.7 UV (top) and IR (bottom) auroral images of Jupiter, with major features indicated. (Top image from Ch. 13 in Vol. I (Clarke *et al.*, 2004); bottom image from T. Stallard, personal communication, 2014.)

by cyclotron maser instabilities, which rely on the motion of energetic electrons around the magnetic field and produce a resonant radio emission. Because such instabilities require very-high-energy electrons, radio aurorae appear to originate in the low-density region above the planet, where potential differentials along field lines significantly increase the mean energy of electrons. This is the cause of the significant auroral radio emission observed at all the giant planets (Zarka, 1998; Lamy *et al.*, 2009).

At Earth, auroral X-ray emission is produced most commonly by bremsstrahlung resulting from high-energy precipitating electrons scattered by the atmosphere. At Jupiter, X-ray bremsstrahlung is present and overlaps spatially with the bright auroral oval seen in the UV, also induced by energetic electrons (e.g., Branduardi-Raymont *et al.*, 2007). However, the majority of the observed X-rays are too energetic to have been produced by such a process (Metzger *et al.*, 1983). Instead, they result from precipitating energetic heavy ions, which have become highly charged through electron stripping by interaction with atmospheric neutrals and can subsequently undergo charge-exchange (i.e., electron capture) through further collisions, leading to emission of an X-ray photon associated with K-shell lines. The production of Jovian X-rays by precipitating ions is a process identified by spectroscopy (Branduardi-Raymont *et al.*, 2007) and by the location of a hot spot of X-ray emission poleward of the main auroral emission (Branduardi-Raymont *et al.*, 2008), at a location that is thought to map to downward currents (Gladstone *et al.*, 2002).

8.4.2 Atmospheric excitation

Particle precipitation leads to significant excitation of the underlying atmosphere, in turn producing prompt emission from excited molecules and atoms. It is this process that dominates the auroral emission seen on Earth, and though the atmospheric composition is significantly different, similar aurorae are observed on the giant planets. These auroral emissions are directly controlled by the precipitation process and, as such, provide an instantaneous view of the particle precipitation process. The composition of the upper atmospheres of giant planets is dominated by hydrogen, and so the observed prompt emission is largely due to hydrogen excitation. The brightest prompt emission from giant planets is associated with the excitation of hydrogen atoms (already present in the atmosphere or produced from the dissociation of H_2), resulting in the strong UV Lyman-α emission (at 121.6 nm) and, to a lesser extent, visible light Balmer series emission (including lines at 410.2, 434.1, 486.1, and 656.3 nm; Dyudina *et al.*, 2011). There are also significant emissions from molecular H_2, produced by electronically excited hydrogen molecules in the Lyman and Werner bands (dominating over ~90–170 nm), and a weak continuum emission from the H_2 a-b dissociation transitions (200–250 nm). Hydrogen may also be electronically excited by the Sun, with some evidence of solar fluorescence (Shemansky *et al.*, 1985), though the observed transitions may also be caused by scattering of sunlight (Yelle *et al.*, 1987). Vibrational excitation of hydrogen also produces infrared H_2 quadrupole emission, but under most ionospheric conditions these molecules are thermalized, and so evidence of such excitation is lost. However, excitation within the collisionless exosphere, at the top the atmosphere, is not removed by thermalization, and so IR excitation aurora caused by cold electrons may occur in this rarefied region (Hallett *et al.*, 2005).

8.4.3 Thermal auroral emission

Molecules produce thermalized auroral emission wherever the atmosphere is heated through the interaction between the atmosphere and the magnetosphere, as long as the spontaneous emission time scale is significantly longer than the time scale for collisions (otherwise local thermal equilibrium is lost). Molecular hydrogen, hydrocarbons, and hydrogen ions all emit infrared light when thermalized within the surrounding neutral atmosphere, with this emission representing one of the major energy sinks for the upper atmosphere (Drossart *et al.*, 1993). Observations of this energy loss process have concentrated upon cooling from the H_3^+ molecule, as this is the most easily observed of the thermalized auroral emitters in giant-planet atmospheres. These observations have shown that auroral H_3^+ is hotter than H_3^+ in non-auroral regions (Lam *et al.*, 1997; Stallard *et al.*, 2002), though exact measurements of the local temperature can be difficult, as changes in observed temperature are at least partially driven by changes in the source altitude, combined with the heat gradient; sudden changes in temperature may be due to changes in the H_3^+ peak altitude, rather than the actual temperature of the neutral atmosphere (Lystrup *et al.*, 2008).

Although observations have concentrated on H_3^+, it is the underlying hydrocarbon molecules that produce the majority of cooling within the upper atmospheres. However, observations of the heat produced within these molecules has not progressed significantly since early measurements on Jupiter, which showed an auroral hot spot at ~60° N and 180° W (Caldwell *et al.*, 1980; Livengood *et al.*, 1990), with temperatures of ~250–320 K at 2–34 microbar (Kostiuk *et al.*, 1993). Quadrupole H_2 auroral emission contributes relatively little to the overall cooling of the atmospheres of giant planets, but provides a measure of temperatures in a layer somewhere between that of high-altitude H_3^+ and low-altitude hydrocarbons. On Jupiter, they emit at a few microbar and have temperatures not inconsistent with those found for H_3^+ (530–1220 K; Kim *et al.*, 1990).

8.4.4 Ionization aurora

Particle precipitation is the dominant source of ionization within the auroral regions of the gas giants. Because there is significant thermal inertia within the upper atmospheres of giant planets (Müller-Wodarg *et al.*, 2012), any resultant thermalized aurora will vary slowly, both temporally and spatially. However, ionization occurs on much smaller scales, and is only limited by the lifetime of the resultant ions. As a result, aurorae in infrared lines are dominated in structure by the ionization process, while overall brightness across the entire auroral region is more strongly controlled by temperature.

The dominant ionic products caused by particle precipitation depend on where in the atmosphere the peak ionization is occurring. At Jupiter, the peak ionization occurs beneath the homopause, resulting in a significant amount of both hydrogen and hydrocarbon ions being formed. However, because H_3^+ is easily destroyed by neutral hydrocarbons, the H_3^+ density peaks at a higher altitude, above the homopause. At Saturn, the peak ionization occurs above the homopause, so that there are few hydrocarbon ions and H_3^+ becomes a dominant product within the auroral ionosphere.

Infrared emission from these molecules is dominated by temperature changes. However, because this temperature varies over long temporal and spatial time scales, the localized H_3^+ auroral morphology is controlled by the density of H_3^+, which in turn is controlled by the particle precipitation process (Stallard *et al.*, 2001). An H_3^+ aurora thus closely follows the morphology seen within prompt UV emission, with particle precipitation driving ionization in the same location as hydrogen excitation. The main auroral differences come from precipitation energy, where UV aurorae are formed beneath the homopause at Jupiter; short-time-scale effects, where the ~10–15 min recombination rate of H_3^+ smooths out short-term changes in precipitation; and localized heating, where strong thermal gradients can actually influence localized intensity variations (Clarke *et al.*, 2004; Radioti *et al.*, 2013).

9

Aeronomy of terrestrial upper atmospheres

DAVID E. SISKIND AND STEPHEN W. BOUGHER

As one moves upward in altitude in a planetary atmosphere, several important changes in composition and structure are apparent. Most notably, as a consequence of hydrostatic equilibrium, the gas density decreases, i.e. the air becomes "thinner". The decrease in density is exponential and governed by a scale height which typically varies in the range of about 5–50 km. Concomitant with this density decrease, the atmosphere becomes increasingly transparent to shorter wavelengths in the solar (or stellar, for exoplanets) spectrum. These shorter wavelengths, typically in the mid, far, and eventually, extreme ultraviolet (MUV, FUV, and EUV respectively), can first dissociate and then at higher altitudes, ionize, various gases in the atmosphere and this alters the composition of the atmosphere. Furthermore, with decreasing density, the frequency of collisions between atmospheric molecules decreases to the point where bulk motions such as turbulence are no longer able to mix the atmosphere. Instead, molecular diffusion becomes the more rapid process and this also leads to a composition change whereby the lighter constituents, typically atomic species such as atomic oxygen, diffuse upwards more rapidly than their heavier counterparts such as O_2, N_2, or CO_2. The region where the atmosphere is well mixed is known as the homosphere; the region where diffusive separation dominates is known as the heterosphere. Although this transition takes place over a range of altitudes, it is common to define some reference boundary altitude known as the homopause to divide the two regimes.

A second transition occurs in the thermal structure. The increased exposure of the atmosphere to energetic UV radiation and the greater dominance of atomic species which are typically inefficient infrared radiators means that the temperature increases markedly with increasing altitude. The altitude regime where the temperature exhibits a large positive temperature gradient is known as the thermosphere. Because that portion of the solar UV spectrum which forms the thermosphere is more variable than the longer wavelengths which heat lower altitudes, thermospheres respond much more strongly to solar variability than atmospheres at lower altitudes. While the thermosphere and heterosphere are closely related and generally overlap in altitude, the physical processes which govern their variability are not precisely identical. In this chapter we will discuss both "spheres", while lumping the two together under the more general label of "upper atmosphere". The lower atmosphere (or often further divided into the lower and middle atmosphere) consists of the troposphere, stratosphere, and mesosphere. A classic reference which summarizes this is Chamberlain and Hunten (1987). The middle atmosphere, and its possible variability with changing climate, is discussed in Ch. 16 of Vol. III.

The study of the upper atmosphere is important for two reasons. First, it is where satellites orbit. Our ability to track and forecast satellite trajectories and quantify satellite lifetimes depends critically on an understanding of how the density of the thermosphere varies (Emmert and Picone, 2010; Pilinski *et al.*, 2013). Further, and of relevance for this chapter, planetary upper atmospheres co-exist with a weakly ionized plasma known as the ionosphere. As described in Ch. 13 of Vol. III, the ionization fraction varies from one part in a million at the homopause to about 1% at the base of the exosphere (defined loosely as where the molecular mean free path exceeds a scale height, about 600 km altitude; cf. Eq. (7.4)). That chapter also discusses the nomenclature for the various ionospheric regions (e.g., D, E, and F) and makes the point that among terrestrial planets, the Earth's ionosphere is unique. This is because at its peak, it is dominated by an atomic ion (O^+) which recombines fairly slowly (an F_2 ionosphere). By contrast, the Martian and Venusian ionospheres are dominated by molecular ions which recombine rapidly and thus are of the chemically controlled F_1 type. This means that the terrestrial ionosphere will present dynamical variability that is not seen on the other planets.

Our emphasis is on the inner terrestrial planets, with specific emphasis on Earth and Mars. Chapter 8 of this volume provides an overview of thermospheres and ionospheres of the giant planets. Both Earth and Mars are being studied by NASA remote sensing missions. In the case of Earth, this includes the ongoing data from the NASA Thermosphere Ionosphere Mesosphere Energetics and Dynamics (TIMED) mission and upcoming data from the ICON (Ionospheric Connection) and GOLD (Global Observations of Limb and Disk) missions. For Mars, the MAVEN (Mars Atmosphere Volatiles Experiment) mission is focusing on the aeronomy of the Martian upper atmosphere and ionosphere. In the sections below, we present an overview of the basic structure of planetary upper atmospheres and then focus on ways that variations in these neutral atmospheres can influence planetary ionospheres.

9.1 Global mean upper-atmospheric structure

9.1.1 Composition

As noted above, and as discussed in standard textbooks (e.g., Chamberlain and Hunten, 1987; Banks and Kockarts, 1973) the two distinguishing characteristics of planetary upper atmospheres are the transition to diffusive equilibrium and a positive temperature gradient. While the altitudes marking these transitions are physically near each other, they should not be confused as being identical since, to first order, they are governed by different physical processes. First, we consider the homopause. At altitudes above the homopause, the composition varies as a function of altitude due to molecular diffusion. Since molecular diffusion coefficients (D) vary inversely as molecular mass, the molecular diffusion velocities are greater for the lighter constituents and smaller for heavier constituents. Furthermore, they vary inversely as the total density (i.e. diffusion of a gas is more rapid if collisions are less frequent), thus D increases with altitude.

By contrast, in the lower atmosphere, collisions are sufficiently rapid that bulk atmospheric motions dominate. These bulk motions have historically been characterized by an eddy diffusion coefficient known as K (Colegrove *et al.*, 1965, 1966) and the homopause historically has been defined as that altitude where D equals K. The term "turbopause" is often used synonymously for this layer and expresses the concept that turbulence, with associated constituent mixing, effectively ceases at this altitude (Hall *et al.*, 2008). Physically, this turbulence has been linked to the breaking of small-scale gravity waves (Garcia and Solomon, 1985) which are typically unresolved in global models. Since the resultant diffusion is primarily in the vertical, K is typically expressed as either K_z, $K(z)$ or K_{zz} (the latter because middle atmosphere waves are often parametrized as a horizontal eddy diffusion, K_{yy}; Garcia (1991). As we will discuss in this chapter, for upper-atmospheric applications, this view of atmospheric mixing as being solely due to turbulent diffusion is too simplistic. As theoretical models have become more sophisticated in resolving atmospheric wave motions, it has become recognized that there is a hierarchy of atmospheric chemical and dynamical phenomena that can produce mixing. For example, Fuller-Rowell (1998) showed that the large-scale height–latitude circulation of the thermosphere at solstice effectively increased the constituent mixing by bring up molecule-rich air from lower altitudes. A model which did not capture this circulation-induced mixing (for example, a purely one-dimensional globally averaged model) would necessarily require a larger value of K_{zz} to properly simulate the resultant vertical constituent profiles. Nonetheless, the concept of eddy diffusion and a turbopause remains quite useful as a tool to understand the variability of planetary upper-atmospheric structure.

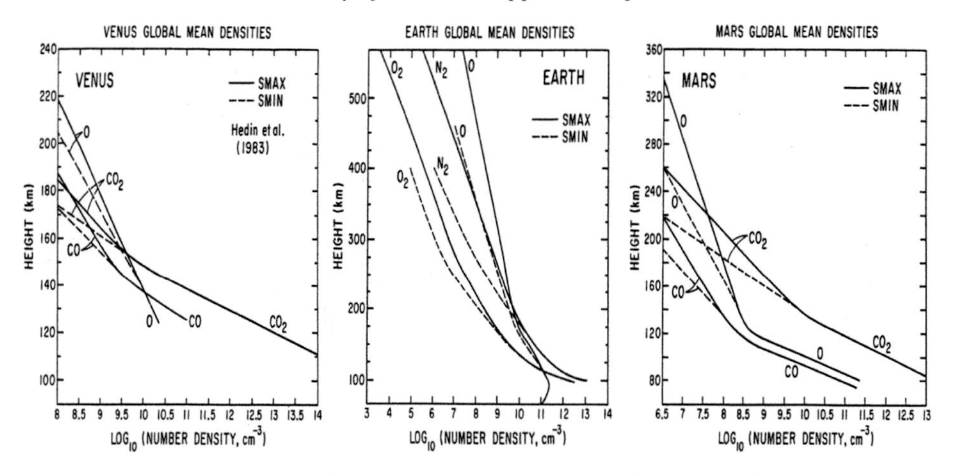

Fig. 9.1 Compositions of the upper atmospheres of Venus, Earth, and Mars. (From Bougher and Roble, 1991.)

Above the homopause or turbopause, the relative abundance of the lighter atomic constituents, such as atomic oxygen, increase at the expense of molecular constituents. Altitude profiles of upper-atmospheric constituents for the Earth, Venus, and Mars are given in Ch. 13 of Vol. III; a similar figure is given here as Fig. 9.1. In all three planets, the relative abundance of atomic oxygen increases with increasing altitude; however, this transition occurs much more rapidly for the Earth. General turbopause altitudes for Venus are in the range 130–136 km (von Zahn *et al.*, 1980), for Mars, 120–130 km (Nier and McElroy, 1977), and for the Earth about 100–110 km (Colegrove *et al.*, 1965, 1966). The difference between the Earth and the other two terrestrial planets is probably linked to more vigorous mixing from the lower atmospheres of Mars and Venus, hence higher K values. However, in the case of Venus, it was hypothesized by von Zahn *et al.* (1980) that the global circulation might provide a contribution to vertical mixing that could mimic high K much as we discussed above in the terrestrial case described by Fuller-Rowell (1998).

It should also be noted that strictly speaking the designation of a single altitude (or narrow range of altitudes) as the homopause is an oversimplification for two reasons. First, since the transition to diffusive equilibrium depends upon the mean molecular mass of each constituent, we expect the $K = D$ criteria to be different for different constituents. Garcia *et al.* (2014) recently discussed the case of CO_2 on Earth. Since CO_2 is heavier than the background atmosphere, it has a smaller vertical scale height (defined for the ith constituent as $H = kT/m_i g$ where k is Boltzmann's constant, m_i is the mean molecular mass, and g is gravity) and also experiences a net downward diffusion velocity (by contrast, atomic species such as atomic hydrogen and helium, which are lighter than the background atmosphere experience a net upward diffusion velocity) (cf., Garcia *et al.*, 2014, their equation (5)). The net result is that the CO_2 density decreases more rapidly with altitude than does the background atmosphere and thus CO_2 departs from a well-mixed condition at a relatively low altitude ($\sim$ 0.01 hPa or 80 km) compared with the conventional turbopause at 100 km. Second, there is a large variability in K such that the $K = D$ criteria can vary significantly with latitude and season (Smith, 2012).

The sensitivity of the upper-atmospheric composition to variability in eddy diffusion was first considered by Colegrove *et al.* (1965, 1966). They showed the consequences of varying K_{zz} by a factor of 4 on the calculated O, O_2, and N_2 profiles of the atmosphere. They showed that for higher values of K_{zz}, more atomic oxygen is transported downwards, out of the thermosphere to where chemical recombination can more rapidly occur. Through diffusive equilibrium, this depletion gets transmitted up to higher altitudes so that the net effect of increasing K_{zz} is to decrease the atomic oxygen in favor of the molecular constituents. This is replicated in Fig. 9.2, which shows two K_{zz} profiles (top panel) and the calculated neutral constituents (middle panel). The model we use is the global averaged thermosphere–ionosphere model first introduced by Roble (1987). The case with the increased K_{zz} profile displays consistently lower atomic oxygen from 100–250 km. O_2 is seen to vary inversely.

These neutral atmospheric changes from varying K_{zz} can then be transmitted to the ionosphere through ion–neutral chemistry. The importance of varying recombination rates on the ionosphere is discussed in Ch. 13 of Vol. III. Briefly, because the radiative recombination of O^+ with free electrons is very slow, the rate limiting pathway for O^+ recombination is the reaction with either O_2 or N_2, followed by dissociative recombination. The dissociative recombination of NO^+ or O_2^+ is approximately 10^6 times faster than the radiative recombination of O^+. The bottom panel of Fig. 9.2 shows that for higher K_{zz}, the O_2^+ and NO^+ is increased and the O^+ is decreased. Thus Fig. 9.2 demonstrates why we expect the electron density near the peak ionosphere to correlate with the O/N_2 ratio and to anticorrelate with eddy diffusion at the base of the thermosphere.

The above demonstration is important for two reasons. First, K_{zz} continues to be a favorite free parameter of thermosphere–ionosphere modelers as a mechanism to parametrize uncertain dynamical effects emanating from the lower atmosphere on thermosphere neutral constituents. Qian *et al.* (2009) postulated a seasonal variation of a factor of 5 in K_{zz} to explain variations in the neutral density sampled by satellites at 400 km. Much as we have done here, they showed how the O/N_2 ratio varied inversely with their assumed K_{zz} and argued that the resultant variations agreed with observations from the TIMED Global UV Imager (GUVI) instrument. Subsequently, Qian *et al.* (2013) showed that variations in the neutral

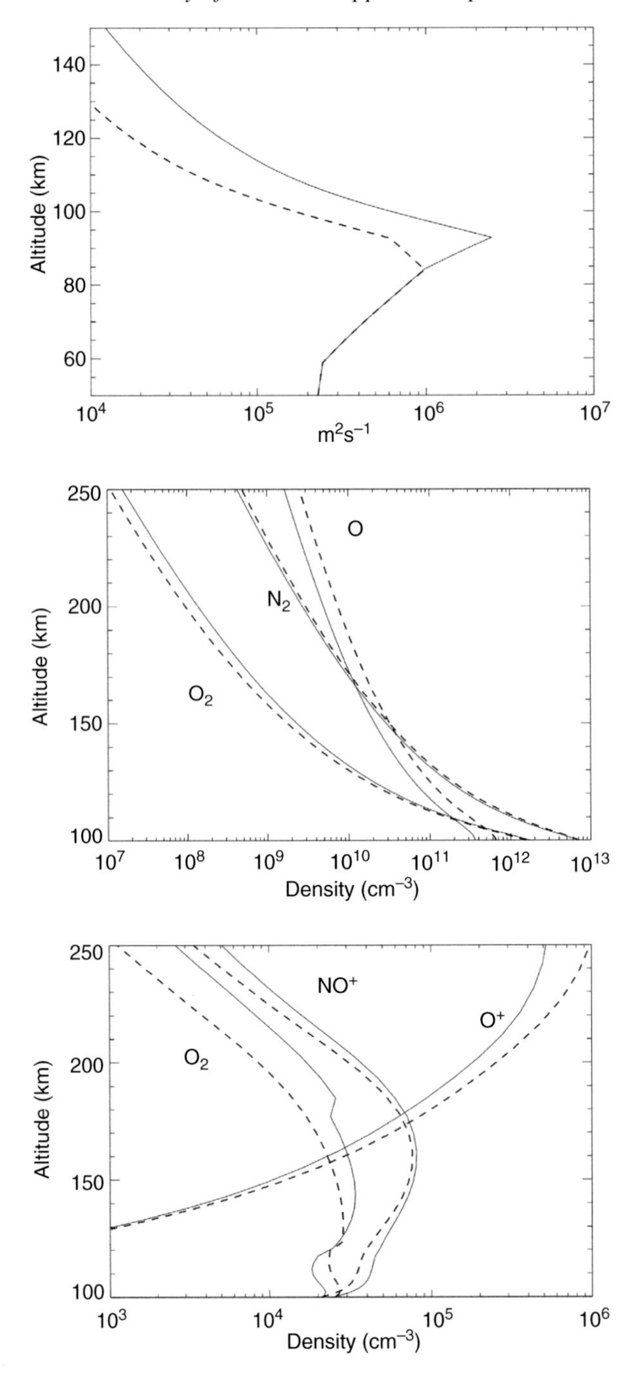

Fig. 9.2 Effect of varying K_{zz} on neutral thermospheric and ionospheric constituent profiles. The top panel shows the two K_{zz} profiles. The center panel shows the resultant neutral constituent profiles where the dashed curves are associated with the dashed K_{zz} profile. The bottom panel shows the resultant ionospheric profiles where using the dashed K_{zz} profiles leads to the largest value of O^+ at 250 km.

constituents led to variations in the ionospheric F_2 peak densities and ultimately to improved agreement with radio-occultation measurements of the ionospheric electron-density profile. Although they applied this K_{zz} change to a 3D model and we used a simple 1D globally averaged model here, the mechanism is the same. Ultimately however, K_{zz} is merely a proxy for complex dynamical effects. In the next section we give an example of one such process to illustrate how the inclusion of a more complete representation of dynamics in three dimensional general circulation models (GCMs) can obviate the need for large values of assumed K_{zz} in models.

Second, our example illustrates the important link between vertical motion of constituents and the ionosphere. This motion can be either upwards or downwards. While increasing K_{zz} decreases O by transporting it down to the lower atmosphere, increases in upward bulk motion (vertical wind) will have similar consequences because it will transport molecule-rich air up to higher (or down to lower) altitudes and locally decrease (or increase) the O/N_2 ratio. This phenomenon is well known in cases where there is added heat deposited in the lower thermosphere. For example, during strong geomagnetic activity, the increased heating from auroral particles and electric fields acts to cause upwelling of N_2-rich air that will change the composition of the ionosphere (Hays *et al.*, 1973; Burns *et al.*, 1989). More recent work has emphasized that both upwelling and downwelling can occur with concomitant changes in the O/N_2 ratio and F region electron density that vary together (Fuller-Rowell *et al.*, 1996; Immel *et al.*, 2001; Crowley *et al.*, 2006). Prolss (2012) reviewed the long history of research into the response of thermospheric and ionospheric composition to solar and auroral energy input.

For Mars, the analog to the terrestrial O/N_2 ratio is the O/CO_2 ratio. Like on Earth, ionospheric chemistry is strongly impacted by the O abundance. Specifically, the conversion of CO_2^+ (from CO_2 ionization) to O_2^+ (primary ion) via atomic O is a very fast reaction. The measured O_2^+/CO_2^+ ratio at the dayside F_1-ion peak can be used as a diagnostic of the local O/CO_2 ratio itself (see Hanson *et al.*, 1977). Also, like on Earth, the O/CO_2 ratio above the homopause is sensitive to the magnitude of eddy diffusion.

Unfortunately, Martian atomic oxygen abundances are not presently well measured. The limited data we have dates back to the Viking 1 and 2 Landers (e.g., Nier and McElroy, 1977). The two sets of density profiles, corresponding to solar minimum near aphelion conditions, reveal CO_2 to be the major species, followed by N_2 and CO. Atomic O could not even be measured; instead, ionospheric model calculations (to match the Viking Lander 1 ion composition measurements) have estimated that the O density exceeds the CO_2 density at $\sim$200 km (e.g., Hanson *et al.*, 1977). Likewise, the O/CO_2 mixing ratio at $\sim$130 km (the altitude of the dayside F_1-ion peak) was estimated to be 1.2%. There have also been some uses of Martian dayglow emission to retrieve neutral densities from both the Mariner and Mars Express missions (e.g., Strickland *et al.*, 1972, 1973; Stewart *et al.*, 1992; Chaufray *et al.*, 2009; Gronoff *et al.*, 2012). Extracted O/CO_2 ratios range from ~0.6% to 2.0% at a similar F_1-peak altitude, reflecting a mixture of solar cycle and local time variations. Atomic O should also serve as a tracer of the Martian global thermospheric circulation for which larger O/CO_2 ratios should appear on the nightside than the dayside. Finally, variations in the global distribution of the O/CO_2 ratio should accompany solar-flare events, similar to the O/N_2 variations seen in the Earth's upper atmosphere. Acquiring comprehensive O/CO_2 data from both airglow and mass spectrometry is a key focus of the MAVEN mission.

As summarized by Nair *et al.* (1994), the data that we do have on Mars' thermospheric composition described above have led to K_{zz} estimates which are quite high, well in excess of 10^7 cm^2 s^{-1}. As compared with Fig. 9.2, this is over a factor of 10 greater than typical global mean values assumed for the Earth and is linked to the correspondingly higher homopause on Mars relative to the Earth. This factor of 10 is consistent with recent work on gravity waves on Mars (Barnes, 1990; Fritts *et al.*, 2006) which suggests about a 10 $\times$ greater flux of gravity waves relative to Earth, ultimately driven by the more prominent topography on Mars.

9.1.2 Temperature

The second transition process which distinguishes planetary upper atmospheres is the steep increase in temperature with increasing altitude (and which leads to the term thermosphere). This arises because the peak EUV and FUV absorption (i.e. the heat sources) occurs at altitudes above the infrared cooling which balances this heating globally. Thus heat is transported downward via thermal conduction. The altitude at which the entire conducted heat is radiated away is defined as the mesopause, where $\frac{\delta T}{\delta z} = 0$. On all three of the terrestrial planets, the mechanism for IR cooling at the mesopause is radiation by CO_2 at 15 μm. This emission is enhanced by the collision of atomic oxygen with CO_2 which excites the $\nu_2 = 1$ bending mode of CO_2. This process proceeds with different efficiencies on the Earth, Venus, and Mars and this impacts the altitude where the $\frac{\delta T}{\delta z} = 0$, is satisfied. This is another area where the O/CO_2 ratio and its variability are believed to be of importance on planets such as Venus and Mars. The magnitude of this IR cooling at the base of the Martian thermosphere is throttled by this O abundance, and both should vary with the solar cycle (see Bougher *et al.*, 1999, 2000). Such variability can serve to limit the solar cycle variation in dayside thermospheric temperatures.

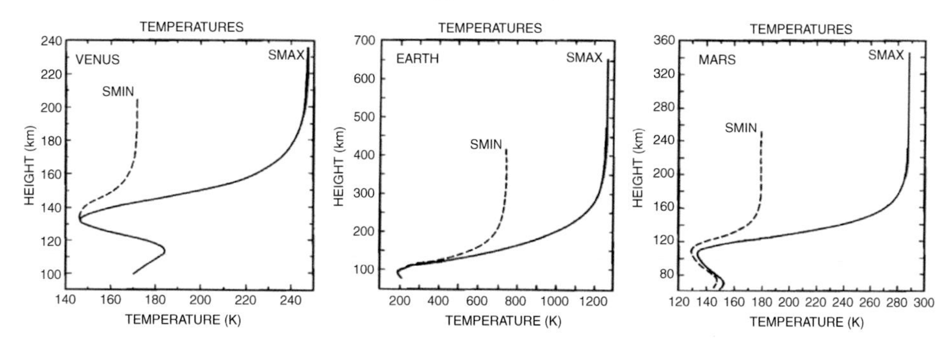

Fig. 9.3 Three planet global mean temperature profiles for solar minimum (SMIN) and maximum (SMAX) conditions. (From Bougher and Roble, 1991.)

Figure 9.3 illustrates the globally averaged temperature profiles from Venus, the Earth, and Mars. Some important differences between the three planets are apparent. First the mesopauses of the three planets are at different altitudes, lowest on the Earth, highest on Venus, and intermediate on Mars. This reflects the efficiency of the CO_2 cooling. On Earth, CO_2 is a minor constituent and the thermospheric heat is not fully radiated away until transported down below 100 km. Second, the relative paucity of CO_2 in the terrestrial thermosphere means that the Earth's thermosphere gets much hotter than those of either Venus or Mars. Finally, because of the inefficiency of cooling in the terrestrial thermosphere, it is more sensitive to variations in the input heat source from the Sun. Thus the solar-cycle variation in the terrestrial thermosphere is larger than on Venus and Mars (Bougher *et al.*, 1999, 2000, 2009) and this has been confirmed by densities derived from orbital drag measurements from the Mars Global Surveyor (MGS) spacecraft (Forbes *et al.*, 2008) as well as earlier such data from Venus spacecraft (Kasprzak *et al.*, 1997).

There is one complication in the terrestrial thermospheric energy budget which is not present on Mars and Venus, namely the existence of an additional major cooling term. This is the emission from vibrationally excited NO at 5.3 μm. Although, as noted in Ch. 13 of Vol. III, the triple bond of molecular nitrogen is strong, energetic electron impact from both photo and auroral electrons can dissociate N_2 and create atomic nitrogen (Siskind *et al.*, 1989a, b). This atomic nitrogen is rapidly oxidized in the lower thermosphere to make nitric oxide (NO). The key characteristic of NO that is relevant for the thermal budget is that there is a pronounced solar-cycle variation in its abundance (Barth *et al.*, 1988). Thus its role in the thermospheric heat budget maximizes at solar maximum. Roble (1987) quantified the relative roles of NO and CO_2 in the energy budget. Figure 8 of that paper is reproduced here as Fig. 9.4. Note that the CO_2 cooling curve changes little from solar maximum to solar minimum whereas at solar maximum, the NO profile is greatly

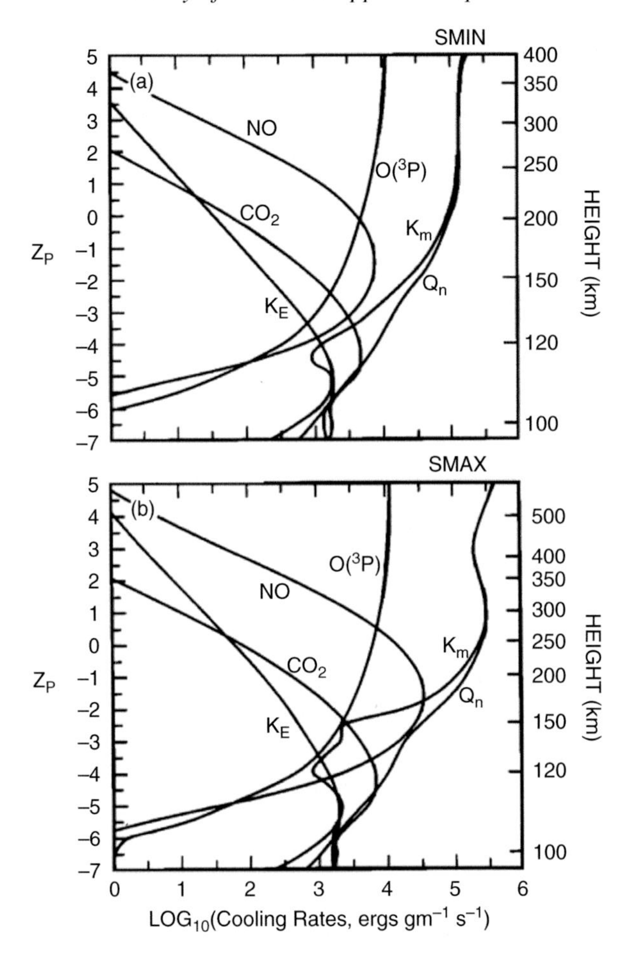

Fig. 9.4 Calculated $\log_{10}$ neutral gas heating and cooling rate profiles (ergs gm^{-1} s^{-1}) for (a) solar minimum and (b) solar maximum conditions. Q_n is the total neutral heating rate, K_m is the cooling rate by downward molecular thermal conduction, K_E is for eddy thermal conduction; NO is radiative cooling from the 5.3 μm emission from nitric oxide, CO_2 is radiative cooling from 15 μm emission from CO_2 and $O(^3P)$ is cooling from the fine structure of atomic oxygen. (From Roble, 1987.)

enhanced. Observations of the infrared emission from these two cooling terms by the Sounding of the Atmosphere by Broadband Emission Radiometry (SABER) instrument on TIMED (Mlynczak *et al.*, 2010) have confirmed this theory.

The varying significance of NO and CO_2 cooling in the terrestrial thermosphere has important consequences for one variation in the Earth's thermosphere that is without any analog on Mars or Venus, namely anthropogenic global change. Since the Roble and Dickinson (1989) paper, it has been recognized the upper atmosphere should cool and thus become less dense at a given altitude in response to increases in the CO_2 abundance. Evidence for density decreases has emerged from a study of 40 years of orbit data (Emmert *et al.*, 2008). It is also apparent that there is a solar-cycle dependence to this trend with the largest density decreases at solar minimum when the CO_2 cooling term is most dominant. It also has been proposed that this should affect the ionosphere, most evidently as lowering of the F_2 layer height (Qian *et al.*, 2009, and references therein). As with the density trend, this should be most apparent at solar minimum. Chapter 14 of Vol. III presents an overview of a wide range of scenarios for long term geospace climate change, including, among others, solar Maunder minimum conditions and geomagnetic field reversals. The conclusion is that the recent finding of neutral density decreases is more robust than the purported changes in the F_2 layer height due to uncertainties in the interpretation of ionosonde data.

9.2 How do neutral dynamics affect planetary ionospheres?

Having considered the global mean structure and variability of terrestrial thermospheres and ionospheres, we now review in greater detail the dynamics by which neutral thermospheres will affect planetary ionospheres. The dynamics of the upper atmosphere have been discussed in Ch. 15 of Vol. III (Walterscheid, 2010). Here we present recent observations and models of coupling processes between the neutral and ionized fractions of planetary upper atmosphere. Of particular interest is the coupling with the lower atmosphere. As we will discuss, many of the perturbations which are manifested in ionospheric variability can be tracked back to meteorological disturbances in the troposphere and middle atmosphere. Thus the differences in the lower and middle atmospheres of the terrestrial planets have ionospheric consequences.

Three differences, relative to the Earth, that we encounter in studying Mars are (1) the absence of a significant dipole magnetic field on Mars, (2) the large orbital eccentricity that provides a $\pm 22\%$ variation in the net solar flux received at the planet during the year, and (3) the unique role of dust. The electrodynamic consequences of the first factor are discussed in Sect. 9.2.1 below. Of the latter two effects, they are coupled in that dust heating is an important atmospheric forcing function and the variable heating throughout the year affects this forcing such that the height of a dayside reference pressure level rises/falls by 10–15 km over the course of the year. This seasonal expansion and contraction of the lower atmosphere impacts the upper atmosphere as well. For example, since the dayside F_1 (primary ion peak) occurs at approximately a constant pressure level, its altitude also varies by 10–15 km in step with the lower/middle-atmospheric forcing (see Stewart, 1987; Zhang and Busse, 1990). In general, the photochemical control of the Martian ionosphere (below 200 km altitude) means that perturbations are often manifested as changes in the height of the electron density peak and we will see other examples of this below.

Concerning dynamics, we consider both the background zonal mean wind and the waves which represent the variability about this mean state. The background mean wind can affect ionospheric plasma through direct perturbations to the O/N_2 (or O/CO_2) ratio, through ion–neutral coupling and associated electrodynamic effects, and by changing the background stability of the atmosphere. Of wave perturbations, following Ch. 15 of Vol. III, we consider three classes: rotational low-frequency planetary waves, divergent high-frequency gravity waves, and tides, which combine aspects of both planetary and gravity waves. Wave phenomena can impact the ionosphere in one of two ways. First, wave dissipation is a means for transferring angular momentum from one region of the atmosphere to another. This momentum deposition can change the circulation field and impact neutral constituent transport. As we will discuss below, what is considered as K_{zz} variability in a one-dimensional global mean framework above, can often be resolved as a specific wave mode in a three-dimensional framework. An example of the second wave effect, from non-dissipating waves, is those that result from oscillations of the ion–neutral dynamo. We consider each dynamical category in turn below.

Finally, we stress that the discussion below is merely to highlight some similarities and differences between the Earth and Mars. It is by no means fully comprehensive, particularly for the Earth. A comprehensive modern discussion of the Earth's ionosphere is given by Kelley (2009). For Mars, there are recent reviews of the Martian ionosphere by Withers (2009), Haider *et al.* (2011), and Bougher *et al.* (2014).

9.2.1 Mean winds and electrodynamics

As we have noted above, in any comparative analysis of terrestrial planetary upper atmospheres and the coupling between the ionic and neutral constituents, the fact that the Earth has a large intrinsic magnetic field, and Mars (and Venus) do not, is of primary importance. One interesting consequence is that, as observed from deep space, the Earth's ionosphere reveals structure that is directly the result of its magnetic field. Observations of the far-ultraviolet airglow from the Apollo 16 astronauts on the Moon (Carruthers and Page, 1976) show this nicely (Fig. 9.5). One sees two bands of airglow on the nightside that are roughly symmetric about the equator. This is the so-called Appleton anomaly, or the Equatorial Ionospheric Anomaly (EIA). It is an electrodynamic effect that is due ultimately to E–W (zonal) neutral winds in the E-region (110 km) which through collisions with the ions,

Fig. 9.5 FUV airglow of the Earth as observed with the Apollo 16 lunar camera. (From Carruthers and Page, 1976.)

create an E–W electric field. This is known as a dynamo electric field and it leads to a vertical ion drift according to

$$\mathbf{v} = \frac{\mathbf{E} \times \mathbf{B}}{B^2}. \tag{9.1}$$

As outlined by Fig. 2 of Immel *et al.* (2006), plasma gets lifted upwards over the equator and then, at F-region altitudes, diffuses poleward and downward along magnetic field lines. Thus the peak electron densities are typically found on either side of the equator, even though the photoionization source peaks at the equator. As we discuss in Sect. 9.2.2, the variability of the E-region dynamo caused by oscillations in neutral tidal winds has been of great recent interest.

Mars has no analog to the above ionospheric structure. However, it is untrue to say that there are no magnetic field effects on the Martian ionosphere. Rather, complex crustal fields, a remnant of the formerly strong Mars intrinsic magnetic field, were observed by the MGS (Acuña *et al.*, 1998). Field strengths at ionospheric altitudes can exceed 1000 nT (several percent of Earth's field strength) which is sufficient to counter the solar-wind pressure. These regions, where both ends of the crustal field lines that thread the ionosphere intersect with the surface, rather than being open to space have thus been termed "mini-magnetospheres" (Mitchell *et al.*, 2001). Electron density enhancements have been indicated both above these crustal field regions (Nielsen *et al.*, 2007; Duru *et al.*, 2008) and within the mini-magnetospheres (Krymskii *et al.*, 2009). An example of one such enhancement is

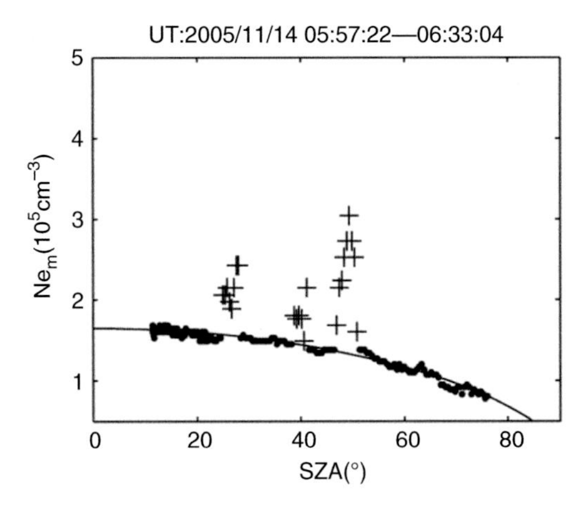

Fig. 9.6 Electron densities from the Mars Express as function of solar zenith angle (SZA). (From Nielsen *et al.*, 2007.)

given in Fig. 9.6. Explanations for these enhancements have focused on elevated electron temperatures which would reduce the rate of dissociative recombination (Schunk and Nagy, 2009). In turn, the causes of these electron temperature enhancements have been attributed to a two-stream plasma instability above the fields and to trapping of hot photo-electrons within the mini-magnetospheres.

9.2.2 Tides

Tides are the response to periodic thermal or mechanical forcing of the atmosphere, typically on diurnal time scales. Tides grow in amplitude as they propagate upwards until by thermospheric altitudes, they are the dominant wave mode in the neutral wind. They are labeled by their direction of propagation (eastward or westward), their period (i.e. diurnal, semi-diurnal), and their wavenumber. There are two categories of tides: migrating or non-migrating. Migrating tides are forced by atmospheric heating that follows the diurnal rising and setting of the Sun and propagate westward. Mathematically, tides are characterized by their zonal wavenumber (e.g., $s = \ldots - 2, -1, 0, 1, 2 \ldots$) and temporal harmonic ($n = 0, 1, 2$), where diurnal tides are $n = 1$, semi-diurnal are $n = 2$, etc. Westward-propagating tides have $s > 0$, eastward-propagating tides, $s < 0$, and zonally symmetric tides have $s = 0$. Migrating tides are therefore defined as $s = n$ and are independent of longitude while non-migrating tides have $s \neq n$ and capture the longitudinal dependence (Bougher *et al.*, 2004). As discussed by Forbes *et al.* (2002), when observed by a satellite in Sun-synchronous orbit, non-migrating tides appear as stationary waves varying in longitude with a zonal wavenumber of $|s - n|$.

Table 9.1 *Summary of primary tidal forcings for Earth and Mars.*

Tidal component	Earth	Mars
Migrating diurnal	NIR abs. of H_2O	VIS and IR abs. of CO_2
Migrating semi-diurnal	Ozone heating	Dust heating
Non-migrating	Latent heat release from tropical convection	Topographic modulation of near surface heating

As first discussed by Lindzen (1970) and noted by Moudden and Forbes (2010), the similar rotation rates of the Earth and Mars implies that the mathematical characteristics of the tidal modes for the Earth and Mars are similar. However, the excitation of these tides differs greatly between both planets and results from unique characteristics of each planet's atmosphere and surface. For both planets, absorption by IR active gases drives the migrating tides; on the Earth, this is H_2O and stratospheric ozone; on Mars, it is CO_2. Additionally, absorption by dust in the Martian atmosphere contributes significantly to the excitation of the semi-diurnal tide. An additional source for the diurnal tide in the thermospheres of both planets arises from absorption of solar EUV at altitudes above about 120 km. Regarding non-migrating tides, on the Earth, they are excited by the latent heat release from tropical convection; the general concentration of this convection in three rain forests across the tropics (Amazon, African, SE Asian) leads to a wavenumber 3 pattern. On Mars, non-migrating tides are forced by the modulation of the daily cycle of solar surface heating by the very large surface pressure variations forced from the prominent variations in Martian topography (Zurek, 1976). Table 9.1 summarizes these differences. In general, due to these large topographic variations, the manifestation of these tides in the Martian upper atmosphere is greater on Mars than the Earth.

On Mars, tidal variability has shown up as an obvious feature in accelerometer data from the Mars Global Surveyor orbiter (Keating *et al.*, 1998; Withers *et al.*, 2003; Haider *et al.*, 2011 (their Fig. 7)) and are attributed to the topographic modulation of solar thermal tides excited near the surface. Figure 9.7 shows an example of these waves. The amplitudes of these waves appear to maximize at about 130 km (just above the mean homopause) and decay with increasing altitude up to approximately 160 km (presumably due to dissipation). These altitudes are near the peak of the Martian ionosphere and the neutral thermospheric waves have a corresponding impact on the ionospheric structure: e.g., the height of the main ionospheric peak varies in step with these wave features (Bougher *et al.*, 2001, 2004; Withers, 2009). Figure 9.8 shows an example of these longitude-fixed features in the MGS data analyzed by Bougher *et al.* (2004). A spectral fit to these data showed the presence of a wave 3 variation, consistent with a semi-diurnal wave frequency

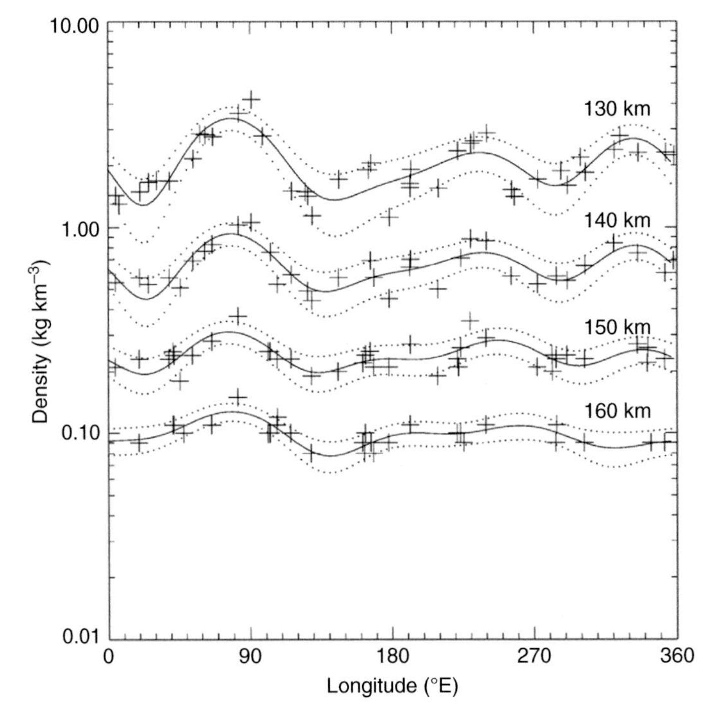

Fig. 9.7 Mass density measurements at 130, 140, 150, and 160 km altitude obtained between 10°–20° N during MGS aerobraking. Model fits to data (waves 1–3) from each altitude are plotted as solid lines and $1-\sigma$ uncertainties about each fit as dotted lines. All data are taken from a local time of 15 h. Measurements at each altitude were taken over about 1 week. All statistically significant peaks and troughs appear fixed in longitude. (From Withers *et al.*, 2003.)

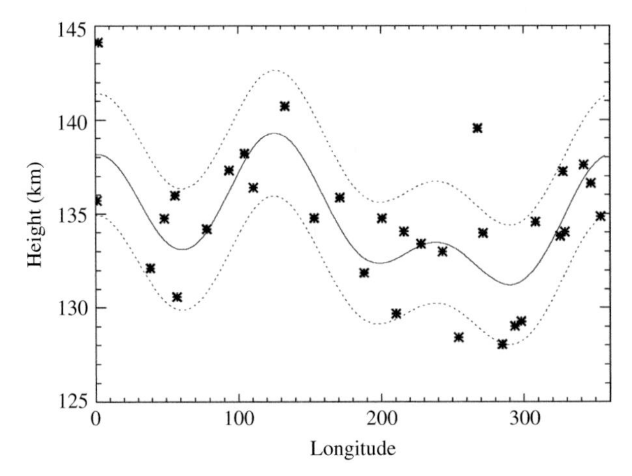

Fig. 9.8 Heights of primary electron-density peak showing a wave pattern. (From Bougher *et al.*, 2004.)

indicating non-migrating (i.e. $s \neq n$) tidal forcing. These altitude changes were considered notable because such variability was unexpected given that the peak of the Martian ionosphere is in photochemical equilibrium and the background solar flux, latitude, and local time had all remained constant.

On the Earth, motivated by the observations by Immel *et al.* (2006) of longitudinal variations in ionospheric airglow, there has been an explosion of interest in interactions between tides and the ionosphere. Indeed, this interest was a key motivating factor behind the selection of the Ionospheric Connections (ICON) Explorer by NASA for a 2017 launch. Immel *et al.* showed that that there was a distinctive wave 4 pattern in the brightness pattern of the F-region airglow and identified the non-migrating diurnal eastward-propagating wave 3 (DE3) tide as the cause (note, as discussed above, a diurnal tide ($n = 1$) with $s = -3$ (eastward-propagating), will be observed as a wave 4 pattern in a fixed local time satellite measurement). Their key result is shown as Fig. 9.9. They proposed that tidal oscillations in the neutral wind at E-region altitudes will modulate the zonal electric field which in turn, modifies the $E \times B$ vertical ion drift given above in Eq. (9.1). Because this redistributed plasma produces airglow through O^+ chemistry, any variations in the source of this plasma will be reflected as airglow variability. These results have sparked many studies on the coupling between tropospheric tides and the ionosphere. Hagan *et al.* (2009) looked at the effects of tidal dissipation on the E-region winds. Other studies that identified other waves which could contribute the patterns found by Immel *et al.* include Pedatella *et al.* (2012) and Liu and Richmond (2013).

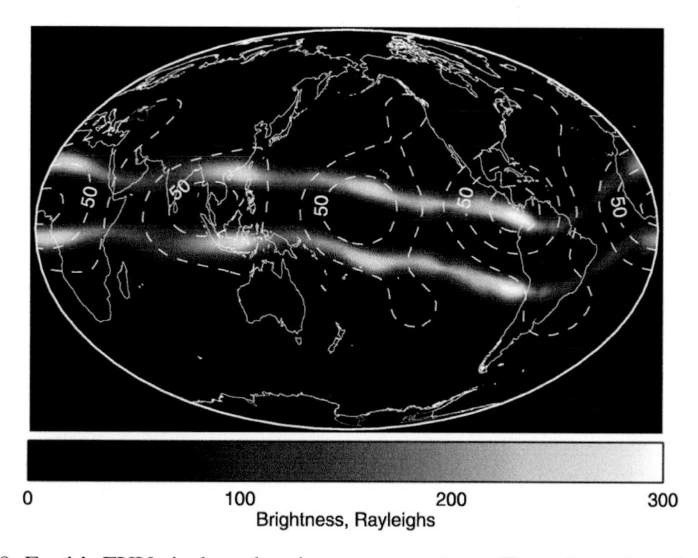

Fig. 9.9 Earth's FUV airglow showing a wave pattern. (From Immel *et al.*, 2006.)

Global whole-atmosphere models have had partial success in capturing this feature. For example, Jin *et al.* (2011) demonstrated the quantitative link between the wavenumber spectrum of tropical rainfall and that of key ionospheric parameters such as peak F_2 density and vertical $E \times B$ drifts.

Recent work has highlighted a second mechanism by which tides can influence the ionosphere, through direct modulation of the vertical transport of atomic oxygen. Siskind *et al.* (2014) used the NCAR Thermosphere Ionosphere Electrodynamics GCM (TIEGCM) and compared calculated electron densities for a case where the TIEGCM was forced at the boundary by an idealized representation of migrating tides versus a case where it was forced by realistic meteorological variability (winds, temperatures, and geopotential heights) from the NOGAPS-ALPHA high-altitude weather model (Navy Operational Global Atmospheric Prediction System – Advanced Level Physics High Altitude; Eckermann *et al.*, 2009). Figure 9.10 shows spectra of the calculated equatorial vertical wind from the TIEGCM near 115 km for the case with the NOGAPS-ALPHA boundary forcing and compares with the standard NCAR formulation that solely uses forcing from migrating tides. The figure shows that the NOGAPS/TIEGCM combination displays a rich spectrum of diurnal, semi-diurnal, and terdiurnal components which

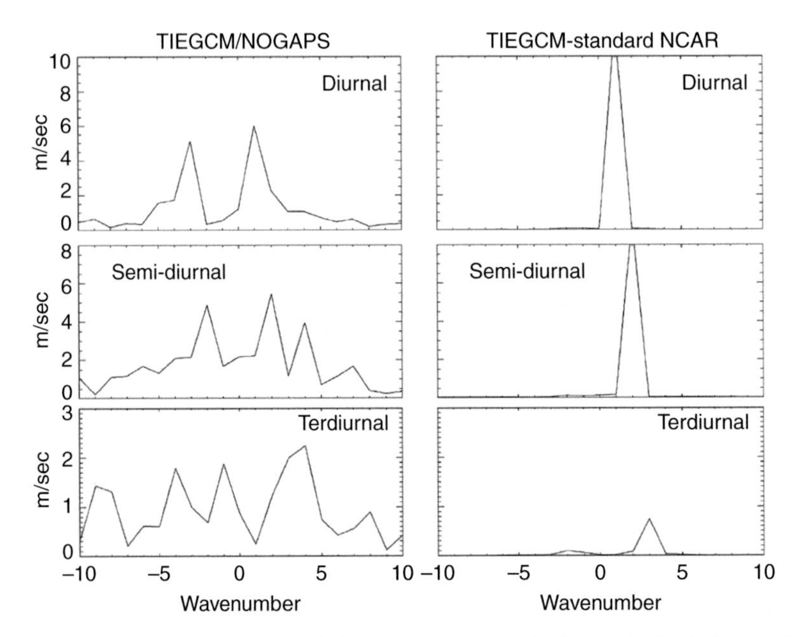

Fig. 9.10 Calculated spectrum of the lower-thermospheric vertical wind, monthly averaged (March) at the equator near 115 km. The calculation on the left uses a model forced by realistic meteorological conditions from the NOGAPS-ALPHA assimilation system. The calculation on the right assumes only forcing from migrating diurnal and semi-diurnal tides.

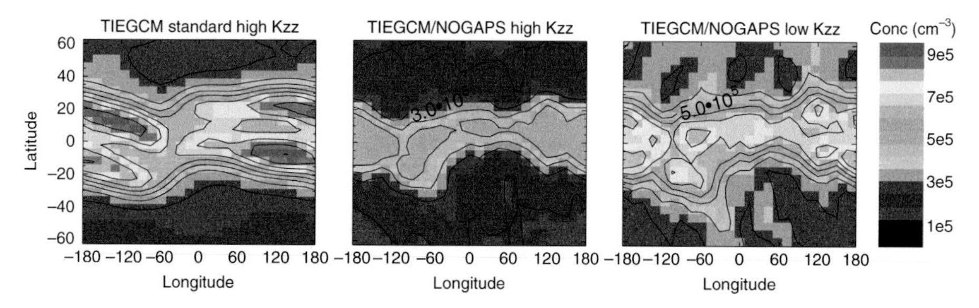

Fig. 9.11 Calculated averaged peak F_2 electron density for March, 1300 local for three TIEGCM simulations. Left column uses standard K_{zz} = 125 m^2 s^{-1} with vertical winds from the right column of Fig. 9.7. Middle uses NOGAPS-ALPHA vertical winds. Rightmost field is with NOGAPS-ALPHA vertical winds and K_{zz} divided by 5. A black and white version of this figure will appear in some formats. For the color version, please refer to the plate section.

when taken together drive a more vigorous upwelling than that obtained with solely migrating tides. Siskind *et al.* (2014) found that the calculated ionosphere using the NOGAPS-ALPHA forcing showed surprisingly low electron densities unless the assumed K_{zz} was divided by 5 from the standard NCAR input. Figure 9.11 shows a sample of their results, comparing the monthly averaged model output for the standard NCAR case (migrating tides + high K_{zz}), the NOGAPS-ALPHA run with high K_{zz}, and the NOGAPS-ALPHA run with lower K_{zz}. Much as we demonstrated for the one-dimensional case in Fig. 9.2, Siskind *et al.* (2014) traced the effect of lowering K_{zz} on atomic oxygen and showed that the case with the lower K_{zz} gave atomic oxygen mixing ratios near the homopause which were in better agreement with SABER observations. This illustrates two points. First, that models which incorporate a greater representation of lower atmospheric dynamics should use a lower value of K_{zz} than models which parametrize a portion of these dynamics and, second, that an accurate representation of the neutral atmosphere at the homopause is important for an accurate calculation of the ionosphere.

9.2.3 Planetary waves

Planetary waves can be classified as either traveling or stationary. Traveling waves are normal modes of the atmosphere. Typical periods for the Earth are centered near 2, 5, 10, and 16 days. Stationary waves are forced in the troposphere by topographic variations either from land–sea contrasts or orography (mountain ranges). In both cases, when these waves dissipate they can transfer momentum to the background flow. As noted in Ch. 15 of Vol. III and by Liu *et al.* (2010), while the middle atmosphere supports a rich variety of planetary waves, it is typically believed that the stationary waves either dissipate or are absorbed by a critical level (where the background wind equals the phase speed of the wave) before reaching thermospheric altitudes. Cahoy *et al.* (2006) came to an analogous conclusion for Mars, i.e. thermal tides dominate the variability above 75 km altitude.

Given the limitations on stationary-wave propagation into the thermosphere, it was therefore somewhat of a surprise that pronounced changes in tropical ionospheric vertical drifts were observed in response to the sudden stratospheric warmings (SSWs) (which are essentially triggered by stationary planetary waves in the stratosphere) in January 2008 (Chau *et al.*, 2009) and 2009 (Goncharenko *et al.*, 2010). Figure 9.12 shows the salient result: the normal diurnal variation of the upward ion drift is significantly perturbed during SSW events. Since it is unclear how stationary waves in the polar middle atmosphere could affect the tropical ionosphere, attention has focused on interactions between these waves and tides, which can then propagate into the thermosphere. Through the dynamo mechanism outlined above, tidal wind variability could then couple to the ionosphere. There have been several model simulations of the effects of SSWs on the thermosphere and ionosphere; not all of them are consistent. Pedatella *et al.* (2014a,b) analyzed the results of four different so-called whole-atmosphere model simulations of the dramatic 2009 SSW. They concluded that whereas the diurnal tide variability was similar in all four models, the semi-diurnal tide differed amongst the four. Ultimately, they tracked back many of the differences to differing approaches towards gravity-wave drag parametrizations. Gravity-wave drag indirectly affects the upper atmosphere by controlling the mesospheric wind field through which upward-propagating planetary waves and tides propagate.

Concerning traveling planetary waves, an example of traveling planetary wave effects on the ionosphere that has recently received new attention was given by Chen (1992). Reproduced as Fig. 9.13, he documented the occurrence of a pronounced two-day oscillation in the peak electron density at the crest of the equatorial anomaly region. Also note from the figure that during the period of the two-day oscillations (March 6–16), that there is an overall net decrease in the electron density (given by foF2, the maximum ordinary-mode radio frequency which can be reflected by the F_2 layer decreases from about 14 MHz prior to the event to an average of about 11–12 MHz during the event). The westward-traveling two-day wave (more properly, quasi-two-day wave since the period can vary, hereinafter Q2DW) has long been of interest to middle-atmospheric scientists and is considered to be an atmospheric normal mode that is excited by baroclinic instabilities in the mesospheric summer jets. Recent interest has been stimulated by the suggestions that it is implicated in facilitating the rapid transport of space shuttle exhaust (Nicejewski *et al.*, 2011) and that it contributes to the variability of Polar Mesospheric Clouds (Siskind and McCormack, 2014). Coordinated observations of the Q2DW in the neutral lower-thermospheric temperatures and in total electron

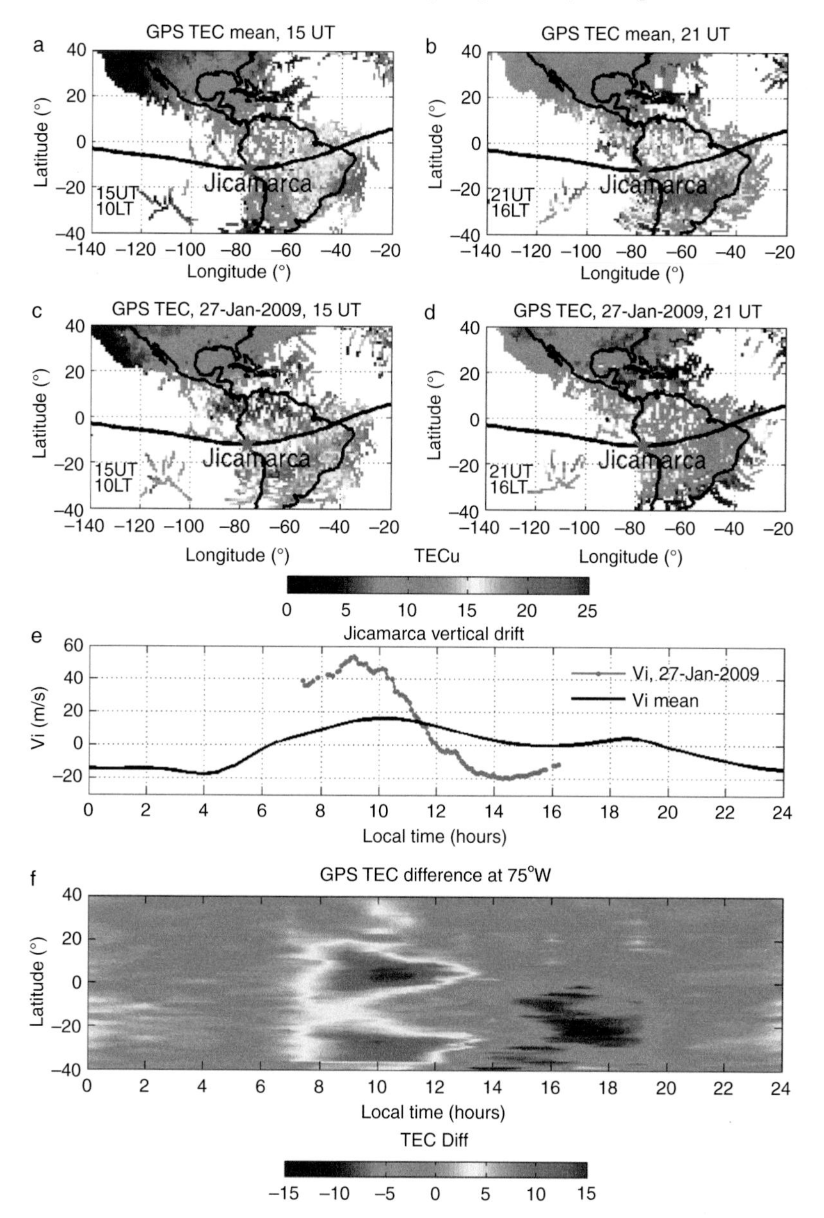

Fig. 9.12 Perturbations to the ionosphere, both total electron content (TEC) and vertical ion drift from the sudden stratospheric warming (SSW) of January 2009. The top row shows typical morning (15 UT = 10 local time at 75° W) and afternoon (21 UT) TEC fields over South America. The second row shows these fields after the SSW with a notable enhancement of TEC in the morning. The third panel shows the difference in the vertical ion drift as measured from Jicamarca Peru. The bottom panel shows difference fields between the SSW perturbation and the mean case as a function of local time, emphasizing the morning TEC enhancement and the afternoon depletion. (From Goncharenko *et al.*, 2010.) A black and white version of this figure will appear in some formats. For the color version, please refer to the plate section.

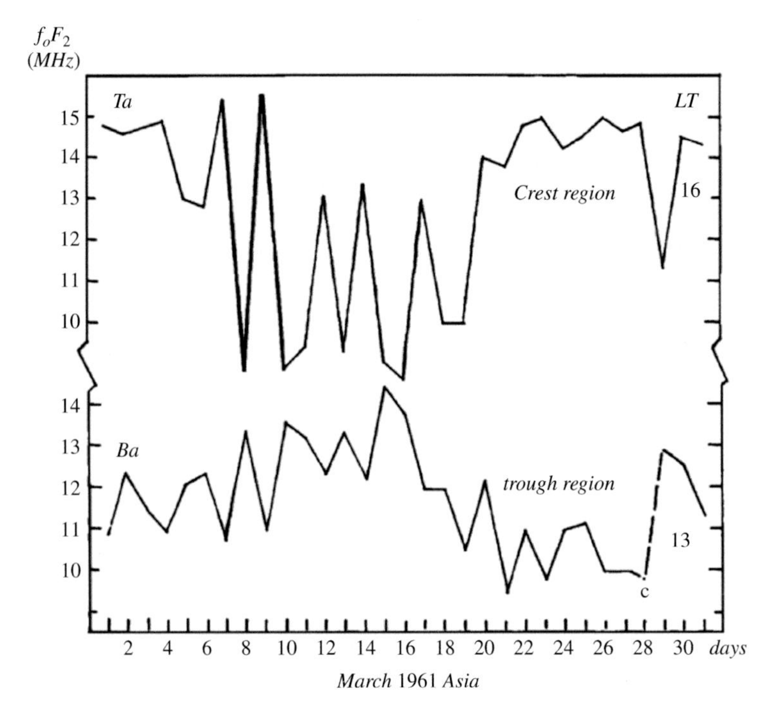

Fig. 9.13 Ionosonde data illustrating the oscillation of the peak F_2 electron density variations with a quasi-two-day wave. (From Chen, 1992.)

content (TEC) were presented by Chang *et al.* (2011). Yue *et al.* (2012) modeled the coupling between the neutral Q2DW and the ionosphere through modulation of the E-region dynamo. In this manner, the Q2DW effects on the ionosphere act similarly to the tidal modulations described above. Most recently Yue and Wang (2014) have proposed that the dissipation of the Q2DW in the lower thermosphere acts as a momentum source and drives a secondary circulation which changes the ratio of atomic oxygen to molecular nitrogen which, as we have seen, is a primary way the terrestrial neutral atmosphere modulates the terrestrial ionosphere.

9.2.4 Gravity waves and instabilities

Because of their small scale, gravity-wave effects on the thermosphere and ionosphere of both the Earth and Mars are the least well understood of the dynamical phenomena discussed here. Fundamental questions about their sources and consequences remain. Owing to the larger topographic variability coupled with faster middle-atmospheric wind speed, gravity-wave amplitudes and effects (such as turbulent mixing) appear to be much larger on Mars than on Earth (Fritts *et al.*, 2006). Accelerometers on MGS and Mars Odyssey revealed significant small-scale

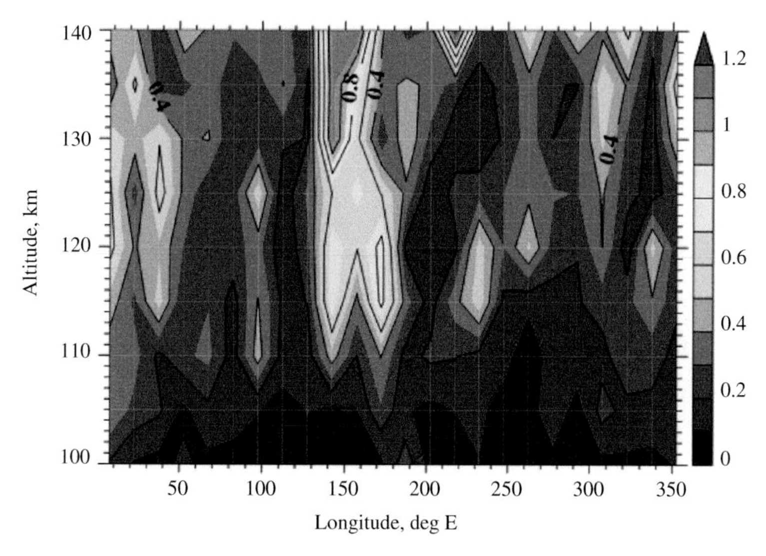

Fig. 9.14 Density variance data from the Mars Odyssey accelerometer in 15° and 5-km bins as function of longitude over about 127 orbits. (From Fritts *et al.*, 2006.) A black and white version of this figure will appear in some formats. For the color version, please refer to the plate section.

fluctuations that can be attributed to gravity waves. A sample of these data is shown as Fig. 9.14, taken from Fritts *et al.* (2006, their Fig. 10). Fritts *et al.* note the localization in longitude and also that the altitudinal growth in variance is less than ρ^{-1}, which suggests dissipation, mixing, and momentum deposition. The altitudes shown correspond to the peak of the Martian F_1 ionosphere; unfortunately, there are not the data to link these data to ionospheric variability. Clearly this is an area for future research.

For Earth, there have been numerous studies linking thermospheric gravity waves both to thermospheric circulation and ionospheric instabilities. For example, gravity waves have been linked to the phenomenon of Equatorial Spread F (ESF) (also termed Convective Equatorial Ionospheric Storms; Kelley (2009)). ESF is understood as a Rayleigh Taylor instability whereby bubbles of depleted plasma can develop after sunset and rise to high altitudes. Kelly *et al.* (1981) suggested that gravity waves might be the triggering mechanism although this is unconfirmed (Woodman, 2009). ESF is almost undoubtedly sensitive to the background neutral wind. This was first shown in the two-dimensional study of Zalesk *et al.* (1982) and more recently in the three-dimensional study of Huba *et al.* (2009). Traveling Ionospheric Disturbances (TIDs) are seen as fluctuations in the F-layer electron density which are also likely linked to gravity waves. Here, however, since these disturbances propagate away from the auroral zones, it is typically believed that auroral heating is a primary source rather than the lower atmosphere, although disturbances

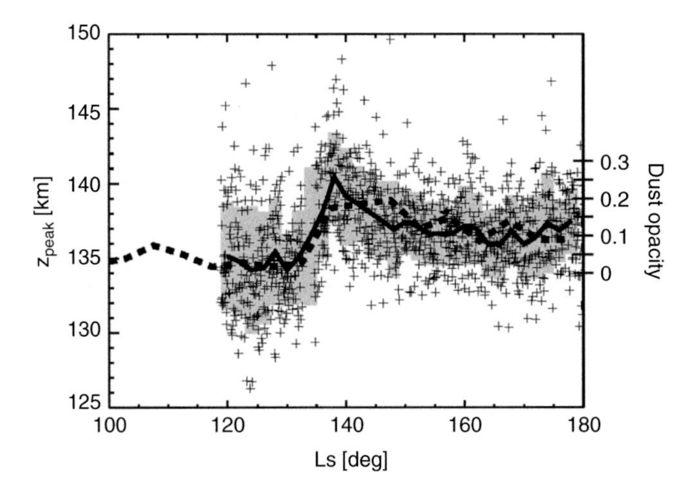

Fig. 9.15 Martian ionospheric peak altitudes measured at latitudes 62°–80° N by the MGS radio occultation instrument during a dust storm. Data are shown as crosses, the solid line is a 2° average, and the gray shading represents the 1σ variation. The dashed curve is the average dust opacity referenced to the right-hand axis. (From Withers and Pratt, 2013.)

recorded during geomagnetically quiet times may have a source from tropospheric convection (Vadas and Liu, 2009) or even ocean waves (Djuth *et al.*, 2010).

9.2.5 Dust storms on Mars

Dust storms provide an episodic forcing on the Martian upper atmosphere and ionosphere that has no clear terrestrial analog. Dust storms can be regional or global and are generally unpredictable, although there is a "dust storm season" which covers the second half of the Martian year (Zurek *et al.*, 1992; Kass *et al.*, 2013). During the first MGS aerobraking phase in 1997–1998 there was a notable regional dust storm which was associated with a factor of three enhancement in thermospheric densities (Keating *et al.*, 1998; Bougher *et al.*, 1999). This corresponds to a 10 km change in the height of a reference upper-atmosphere pressure surface (see Keating *et al.*, 1998). For subsequent dust-storm events (e.g., in July 2005) there was no upper-atmospheric sampling with accelerometers; however there were radio occultation data that observed changes in the F_1 ion peak height (Withers and Pratt, 2013). As we have discussed, for the photochemically controlled Mars F_1 ionosphere, perturbations in the neutral atmosphere are registered as changes in the peak height. Figure 9.15 shows that for the 2005 event, the dayside heights rose by 5 km.

9.3 Summary and outlook

As we have noted above, our understanding of the relative roles of neutral atmospheric effects on the ionospheres of both Earth and Mars are expected to significantly improve in the 2015–2020 time frame. On Earth, the ICON mission will make simultaneous observations of both E- and F-region neutral wind variability and the F-region electric fields and particles to quantify the link between lower-atmospheric wave forcing and equatorial ionospheric variability. GOLD hopes to be able to image the variability of thermospheric and ionospheric weather from geostationary orbit. Both ICON and GOLD are slated for launch in 2017. On Mars, the MAVEN mission's overall goal is to understand the processes that have shaped the evolution of the Martian atmosphere, in particular its escape. As part of fulfilling that overall goal, we will have comprehensive global data on the composition of the neutral thermosphere and the ionosphere for the first time. We will finally have sufficient data on atomic oxygen from which to study the global variability of the O/CO_2 ratio. Thus we expect to observe much of the same dynamical variability that we see on Earth.

10

Moons, asteroids, and comets interacting with their surroundings

MARGARET G. KIVELSON

A body such as a planet, a moon, an asteroid, or a comet, typically enveloped in a tenuous neutral gas, perturbs its surroundings in a flowing, magnetized plasma. The structure of field and plasma resulting from the interaction depends on properties of the plasma and of the body onto which the plasma flows. This chapter addresses the interaction of solar-system plasmas with a number of small bodies of the solar system. Size is taken as criterion, but we do not include small planets even though the biggest moons are similar in size to the smallest planet, Mercury. For reviews of planetary magnetospheres, see Chs. 10 and 13 in Vol. I.

Given that the Moon has no atmosphere, one might suppose that our discussion focuses on bodies that lack atmospheres, but that is not the case. Some moons, including small ones, are enveloped in neutral clouds that are not gravitationally bound but are, in many ways, similar to atmospheres. Other important properties differ from one body to another. Some moons have icy outer layers while others have outer layers of silicates and other non-icy materials; some have conducting regions in their interiors whereas others do not; and some are magnetized but others are not. Like the planets, the bodies that we discuss are spread throughout the solar system. Correspondingly, the plasma conditions vary from one to another. The focus on moons and comets is arbitrary, but the examples that we shall discuss are diverse and illustrate a large range of physical processes that can occur in the solar system. The interaction regions surrounding the small bodies of interest vary in global geometric configuration, in spatial extent relative to the size of the central body, and in the nature of the plasma disturbances. The objective of the discussion is to understand the physical processes that account for the observed features of the interaction regions.

10.1 Physics of large-scale processes in space plasmas

It is useful to start by discussing the physical principles that govern the behavior of the flowing, magnetized plasmas in which moons and comets are embedded.

The plasma of the space environment is a partially ionized gas, usually dominated by the ionized component. The physical principles that govern the behavior of the flowing medium differ from those that describe a neutral gas. In a neutral gas, collisions among the constituent atoms and/or molecules play a major role in the dynamics. In a magnetized space plasma, collisions are generally very infrequent and electromagnetic interactions control the behavior of the system. Symmetry is broken from the start by the presence of a magnetic field. Strong Coulomb forces maintain net charge neutrality within spatial volumes large enough to contain many ions and electrons, but ion and electron motions may differ enough to generate electric current. In the presence of a magnetic field, currents exert forces, adding to the complexity of the physical environment.

On large spatial and temporal scales, the properties of a magnetized plasma are governed by a combination of fluid equations and Maxwell's equations of electromagnetism, i.e., the equations of magnetohydrodynamics (MHD). Here, we examine the basic equations, not with the intention of mathematical manipulation, but with a desire to make clear how forces and responses are linked. The MHD equations are provided in Ch. 3 of Vol. I (see Table 1.2 for a chapter listing in all volumes), but here we are interested in somewhat different conditions. We neglect gravitational forces but include the sources and losses linked to ionization of neutrals, charge exchange, and associated processes. The equations that express the conservation of mass and momentum can be written as

$$\frac{\partial \rho}{\partial t} + \nabla \cdot \rho \mathbf{u} = S - L \qquad \text{(mass)} \tag{10.1}$$

$$\left(\frac{\partial \rho \mathbf{u}}{\partial t} + \nabla \cdot \rho \mathbf{u}\mathbf{u}\right) = \rho\left(\frac{\partial \mathbf{u}}{\partial t} + \mathbf{u} \cdot \nabla \mathbf{u}\right) + \mathbf{u}(S - L)$$
$$= \nabla p + \mathbf{j} \times \mathbf{B} + (\mathbf{S}_p - \mathbf{L}_p) \quad \text{(momentum)} \tag{10.2}$$

where the first equality in Eq. (10.2) makes use of Eq. (10.1). To these equations, we add two of Maxwell's equations,

$$\frac{\partial \mathbf{B}}{\partial t} = -\nabla \times \mathbf{E} \qquad \text{(Faraday's law)} \tag{10.3}$$

$$\nabla \times \mathbf{B} = \mu_0 \mathbf{j} \qquad \text{(Ampere's law).} \tag{10.4}$$

In Eq. (10.4), we have assumed that the processes of interest occur on time scales that are sufficiently long to justify omission of the displacement current. In this, as yet incomplete, set of equations, ρ is mass density, $\mathbf{u}$ is bulk flow velocity, S and L are the source and loss rates of ion mass per unit volume, respectively, p is thermal pressure, $\mathbf{j}$ is current density, $\mathbf{B}$ is magnetic field, $\mathbf{S}_p$ and $\mathbf{L}_p$ are source and loss rates of momentum density per unit volume, $\mathbf{E}$ is the electric field, and μ_0 is the permeability of vacuum. Sources and losses are of exceptional importance for comets, which emit gases and dust when their surfaces heat up as they approach the sun. Some moons have atmospheres or geysers, and for them, too, sources and losses may be extremely important.

One can write Eq. (10.2) that governs the momentum density in the following useful form

$$\frac{\partial \rho \mathbf{u}}{\partial t} = -\nabla \left(p + \frac{B^2}{2\mu_0} + \rho u^2 \right) + \mathbf{B} \cdot \nabla \mathbf{B}/\mu_0 + (\mathbf{S}_p - \mathbf{L}_p). \tag{10.5}$$

Here, the quantities $B^2/2\mu_0$ and ρu^2 enter as additions to thermal pressure. These terms are referred to as the magnetic pressure and the dynamic pressure, respectively. The term $\mathbf{B} \cdot \nabla \mathbf{B}/\mu_0$ is referred to as the curvature force (actually force density) and acts like tension in a string to accelerate plasma towards the center of field line curvature.

The electric field can be related to $\mathbf{u}$ and $\mathbf{j}$ through a form of Ohm's law: $\mathbf{E} = -\mathbf{u} \times \mathbf{B} + \eta \mathbf{j}$, where η is the resistivity. When the resistivity becomes sufficiently small, as is the case in most parts of the systems that we consider, the last term in this equation can be omitted and the electric field is given by

$$\mathbf{E} = -\mathbf{u} \times \mathbf{B}. \tag{10.6}$$

Equation (10.6) can be used to rewrite Eq. (10.3) as

$$\frac{\partial \mathbf{B}}{\partial t} = \nabla \times (\mathbf{u} \times \mathbf{B}). \tag{10.7}$$

The resulting system of equations is referred to as ideal MHD, which is a description that we will find particularly useful.

Equations (10.1), (10.2), (10.4), and (10.7) are still incomplete as one can confirm by counting the number of unknowns (11) and the number of equations (10), recognizing that a vector or a vector equation has three components. In Ch. 3 of Vol. I, an approximation to energy conservation (their Eq. (3.3)) was used to complete the set of equations, but here we complete the description using the adiabatic assumption that relates pressure to density:

$$p\rho^{-5/3} = \text{constant} \quad \text{(adiabatic condition)}. \tag{10.8}$$

Now we have 11 equations in 11 unknowns.

One may wonder why we have omitted two of Maxwell's equations, i.e., Poisson's equation that relates $\mathbf{E}$ to the charge density and the requirement that $\mathbf{B}$ be divergence free ($\nabla \cdot \mathbf{B} = 0$). Actually the latter can be incorporated in our analysis as an initial condition. The divergence of Eqs. (10.3) or (10.7) requires that

$\partial(\nabla \cdot \mathbf{B})/\partial t = 0$ so if the divergence of $\mathbf{B}$ is set to zero at any time, it will remain zero at all times. Poisson's equation is not used because we require the charge density, and correspondingly $\nabla \cdot \mathbf{E}$, to be negligibly small, i.e. of order the terms we have dropped in the approximations used to write the equations that we retain.

From the ideal MHD equations it becomes evident that currents affect flows (Eq. (10.2)), but also that flows change the magnetic field (Eq. (10.7)) and changes of the magnetic field modify the current (Eq. (10.4)). The complex coupling leads to interesting physics. When a plasma flows onto a solid body, or even a dense atmosphere, it is usually slowed and diverted, just as the flow of a stream of water is slowed and diverted by an obstacle in its path. As the flow changes, the local structure of the magnetic field changes and generates current that modifies the flow. In space, multiple properties are inextricably linked.

Although the equations provided in this section treat plasmas as fluids, some concepts are better presented in terms of the behavior of ions and electrons. For example, the Lorentz force law, $\mathbf{F} = q(\mathbf{v} \times \mathbf{B})$, with $\mathbf{F}$ the force on a particle of charge q moving at instantaneous velocity $\mathbf{v}$ in a magnetic field, tells us that magnetic forces do not change the component of velocity along a field but impose force across the field. (Note the distinction between the velocity of an individual charged particle, $\mathbf{v}$, and the bulk flow velocity, $\mathbf{u}$, which is the mean velocity of a collection of particles.) Charged particles move in helical paths along the field, right-handed for electrons and left-handed for ions. The effect is to tie the plasma to the magnetic field. The tight link between a flux tube and its plasma content is described as the frozen-in field condition and is discussed at length in Ch. 4 of Vol. I. The theorem, valid for ideal MHD, states that if two fluid elements lie on a common field line at one time, then they lie on a common field line at all times. The concept can be exploited to account for changes of the flux tube structure imposed by the properties of the plasma flow. More important, the theorem makes clear that interpenetration of distinct magnetized plasmas requires a breakdown of ideal MHD.

10.2 Characterizing the plasma that interacts with solar-system bodies

As obstacles to the flow, moons and comets create local disturbances in the plasmas in which they are embedded. The specific structure that envelopes the body depends on properties of the ambient plasma and these properties vary greatly depending on whether the external plasma is the rapidly flowing and temporally variable solar wind or the slowly flowing plasma of a planetary magnetosphere. The typical plasma density, temperature, and magnetic field intensity also vary from one case to another. Consequently, it may seem unlikely that general rules can describe the interaction regions.

We are rescued from the need to treat each case as totally distinct by recognizing that physical theories often incorporate a small set of dimensionless parameters that control important aspects of a system, even if such properties as spatial scale, temperature, and flow velocity vary by many orders of magnitude. For a flowing plasma incident on an obstacle, the form of the interaction depends critically on how the flow speed is related to the speed of waves that transmit information about changes of plasma properties from one part of the system to another. An analogy to waves in neutral gases helps to clarify the concept. In the frame of an airplane in flight, the atmosphere flows onto the plane at some velocity, call it $\mathbf{u}$. As the gas encounters the plane, pressure perturbations develop. Pressure perturbations launch sound waves that travel at the sound speed, c_s. If such waves can move away from the plane, they can divert the atmosphere upstream of the plane. But the waves are swept back toward the plane at the flow speed of the plasma. Only if $u < c_s$ is it possible for the waves to begin to divert the atmosphere well upstream of the plane. If $u > c_s$, as for a supersonic jet, the waves pile up in front of the plane, causing a shock to develop upstream. Only downstream of the shock is the flow diverted. Assuming that the plane is large compared with distances characteristic of atmospheric properties, the parameter that determines whether or not a shock will form is the (dimensionless) sonic Mach number of the surrounding atmosphere, u/c_s.

In a plasma, much as in a neutral gas, compressional perturbations develop when there is an obstacle in the flow. The sound speed is relevant to understanding how such perturbations propagate through the system, but in a plasma, there are waves that differ from sound waves and change both the field and the plasma. Because the form of the disturbances in the vicinity of a moon or other small body depends so strongly on the nature of the waves that carry perturbations through the plasma, we digress to describe the properties of some characteristic wave modes.

10.2.1 Magnetohydrodynamic waves

The waves that carry information through a magnetized plasma differ from the sound waves of a neutral gas, partly because of the anisotropy imposed on the fluid by a magnetic field and partly because the waves must be capable of carrying currents that modify the properties of both matter and magnetic field. The properties of such waves can be derived from the MHD Eqs. (10.1)–(10.8) by analyzing the evolution of small perturbations.

Consider a uniform plasma with constant pressure and density (p and ρ) whose center of mass is at rest ($\mathbf{u} = 0$). Assume that a constant background field ($\mathbf{B}$) is present and that neither sources nor losses need be considered. Small departures from this background state are taken to vary with space ($\mathbf{x}$) and time (t) as $e^{i(\mathbf{k}\cdot\mathbf{x}-\omega t)}$. Here, $\mathbf{k}$ is the wave vector and ω is the angular frequency of the wave. Perturbations occur in density $d\rho$, velocity $\mathbf{du}$, pressure dp, current $\mathbf{j}$, and field $\mathbf{b}$. Terms linear in small quantities in Eqs. (10.1) and (10.5) satisfy

$$-\omega d\rho + \rho \mathbf{k} \cdot \mathbf{du} = 0 \tag{10.9}$$

$$-\omega \rho \mathbf{du} = -\mathbf{k} dp + \mathbf{b}(\mathbf{k} \cdot \mathbf{B})/\mu_0 - \mathbf{k}(\mathbf{b} \cdot \mathbf{B})/\mu_0. \tag{10.10}$$

Equation (10.8) gives the pressure perturbation in terms of the density perturbation as

$$dp/p = \gamma d\rho/\rho \tag{10.11}$$

and Eq. (10.7) implies

$$\omega \mathbf{b} = \mathbf{du}(\mathbf{k} \cdot \mathbf{B}) - \mathbf{B}(\mathbf{k} \cdot \mathbf{du}). \tag{10.12}$$

The solutions to Eqs. (10.9) to (10.12) are the roots of the equation

$$(\omega^2 - v_A^2 k^2 cos^2\theta)[\omega^4 - \omega^2 k^2 (c_s^2 + v_A^2) + k^4 v_A^2 c_s^2 cos^2\theta] = 0, \tag{10.13}$$

where θ is the angle between $\mathbf{k}$ and $\mathbf{B}$, and the Alfvén speed (v_A) and the sound speed (c_s) have been introduced. These quantities characterize the speed of propagation of waves in a magnetized plasma and are defined by

$$v_A^2 = B^2/2\mu_0\rho \tag{10.14}$$

$$c_s^2 = \gamma p/\rho. \tag{10.15}$$

The sound speed has the form familiar for a neutral gas. The Alfvén speed is a second natural wave speed characteristic of a magnetized plasma. Just as we introduced the (dimensionless) sonic Mach number as the ratio of the flow speed to the sound speed, it is useful to define a dimensionless Mach number, the Alfvénic Mach number ($M_A = u/v_A$), related to the Alfvén speed.

As mentioned previously, the quantity $B^2/2\mu_0$ is the pressure exerted by the magnetic field, so both of the basic wave speeds are proportional to the square root of a pressure divided by a density. The ratio of the thermal pressure to the magnetic pressure is another useful dimensionless quantity, for which we use the symbol, β, defined as $\beta = p/(B^2/2\mu_0)$. When $\beta < 1$, magnetic effects dominate the effects of the thermal plasma, but in a high-β plasma, the plasma effects dominate.

Equation (10.13) is of sixth order in ω/k with three pairs of roots. One pair results from setting the first factor in Eq. (10.13) to zero; the resulting dispersion relation is

$$(\omega^2 - v_A^2 k^2 cos^2\theta) = 0. \tag{10.16}$$

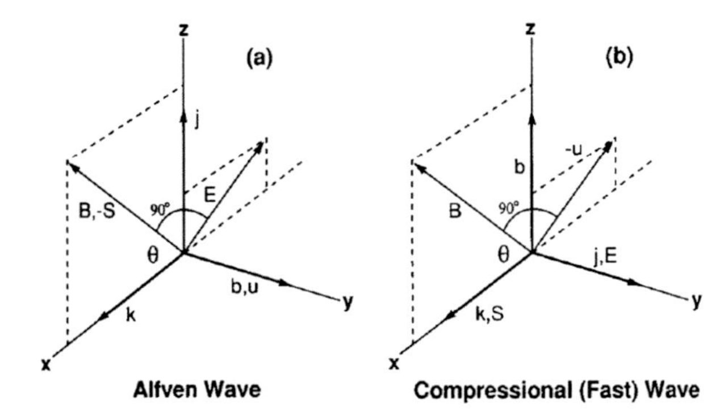

Fig. 10.1 Polarization of MHD waves in a uniform medium. The parameters other than **S** are defined in the text and are plotted with the wave vector along the x-axis and the background field (**B**) in the $x-z$ plane. The incompressible Alfvén wave polarization is plotted on the left and the compressional fast wave polarization is plotted on the right. **S** is the Poynting vector oriented in the direction in which energy density is transported by the wave: strictly field-aligned for the Alfvén wave and oblique for the compressional wave. (From Kivelson, 1995.)

This solution describes waves referred to as Alfvén waves. For this dispersion relation to apply, the magnetic perturbation must be perpendicular to both **B** and **k** (see Fig. 10.1). This orientation implies that to first order in small quantities, the Alfvén wave does not change the field magnitude $[(\mathbf{B}+\mathbf{b})^2 = B^2 + 2(\mathbf{B}\cdot\mathbf{b})^2 + b^2 \approx B^2]$. The wave phase speed is $v_{ph} = \omega/k$ and $v_{ph} = \pm v_A cos\theta$. Waves carry information at the group velocity, $\mathbf{v}_g = \nabla_k \omega$, where the subscript on the gradient indicates that the derivatives are taken in **k** space; the solution is $v_g = \pm\hat{\mathbf{B}} v_A$ where $\hat{\mathbf{B}}$ is a unit vector along the background field. The remarkable property of these waves is that they carry information only along the background field, and they bend the field without changing its magnitude. These properties are of considerable importance in interpreting the interaction of a flowing plasma with the solid bodies of the solar system.

Equation (10.13) has two more pairs of roots, the zeros of the fourth order polynomial in square brackets in Eq. (10.13), i.e., the solutions

$$v_{ph}^2 = \omega^2/k^2 = \frac{1}{2}\left(c_s^2 + v_A^2 \pm [(c_s^2 + v_A^2)^2 - 4v_A^2 c_s^2 cos^2\theta]^{1/2}\right). \quad (10.17)$$

The solutions (two pairs, one positive and one negative, of roots) correspond to what are unimaginatively referred to as fast (or magnetosonic) and slow mode waves. The wave perturbations of both modes may have magnetic perturbations along and across **B** (see Fig. 10.1b). Perturbations along **B** change the field magnitude and the thermal pressure. The fast-mode changes of thermal and magnetic pressure are in phase with each other; this implies that the total pressure fluctuates.

The slow-mode changes of thermal and magnetic pressure are in antiphase, and the total pressure fluctuations are very small. For waves propagating along the background field ($cos\theta = \pm 1$), the solutions to Eq. (10.17) are c_s^2 and v_A^2, with the larger of the two applying to the fast mode. For waves propagating at right angles to the background field ($cos\theta = 0$), the wave speeds are $c_s^2 + v_A^2$ and 0, indicating that only fast mode waves propagate across the field.

10.2.2 Selected properties of the upstream flow

Having identified some of the waves that carry information through a magnetized plasma, we are now able to introduce the dimensionless parameters that help us understand aspects of flow and field perturbations. The magnetosonic Mach number (M_{ms}) is the ratio of the flow speed to the fast-mode speed, taken as $(c_s^2 + v_A^2)^{1/2}$. M_{ms} reveals whether or not a shock is likely to form upstream in the flow. When $M_{ms} < 1$, compressional waves can travel upstream from the obstacle faster than the flowing plasma can sweep them back. These waves, moving upstream, can divert the incident flow around the obstacle, much as the bow wave of a ship diverts water to the sides, and no shock develops. However, as in the situation discussed in the context of supersonic flight, if $M_{ms} > 1$, compressional waves are unable to propagate upstream faster than they are swept back by the flow. They pile up to form a shock. Most bodies in the super-magnetosonic solar wind (exceptions are discussed in Sect. 10.8) create shocks standing somewhat upstream on their sunward sides. Downstream of the shock, plasma is heated, compressed, and diverted around the obstacle.

The Alfvén Mach number (M_A) is the ratio of the speed with which the ambient plasma flows towards an obstacle divided by the Alfvén speed. We will see that this quantity controls the shape of the interaction region in planes containing the unperturbed plasma flow and the background magnetic field. The plasma beta (β) is the ratio of the thermal pressure to the magnetic pressure. This quantity enables us to understand how significantly the magnetic field structure can be modified by changes of the plasma pressure.

The plasma environment differs greatly among the small bodies of the solar system. Some of the bodies are embedded in the solar wind, others in the plasma of a planetary magnetosphere, and some (such as Earth's Moon) move from one environment to another.[1] Table 10.1 lists some plasma properties relevant to the environment of selected bodies. In Sect. 10.4, we discuss further how M_A and β control aspects of the interaction. However, we shall first consider how various properties of the obstacle affect the interaction.

[1] The Moon orbits the Earth at a distance of 60 Earth radii. It spends part of each lunar month in Earth's magnetotail and the rest of the month in the solar wind.

Table 10.1 *Properties of the plasmas upstream of selected small bodies of the solar system.*

Obstacle	Ambient plasma	M_A	M_{ms}	β
Io, Europa, Ganymede	Jovian magnetosph.	< 1	< 1	< 3
Asteroids	Solar wind	> 1	> 1	~ 1
Comets	Solar wind	> 1	> 1	~ 1
Moon	Earth's magnetosphere or solar wind	either > 1 or < 1	either > 1 or < 1	~ 1 or < 1

10.3 Effects of the electrical properties of an obstacle in the flow

Currents flow readily in an ionized gas. If the plasma flows onto a solid body, one anticipates that the electrical properties of the obstacle will affect the interaction, and they do. Bodies like the Moon have no atmosphere and are very poor conductors but many solid bodies of the solar system are surrounded by an ionized region or a cloud of neutrals that can be ionized and can conduct electricity. Currents can also flow in plasmas trapped in the magnetic field of a magnetized body. The presence or absence of a conducting path through a body or its surroundings greatly affects the form of interaction with a flowing plasma.

10.3.1 Flow patterns at non-conducting or conducting bodies

A plasma flowing at velocity $\mathbf{u}$ generates an electric field, the $-\mathbf{u} \times \mathbf{B}$ electric field of Eq. (10.6). Correspondingly, in the unperturbed flow, an electric potential drop is present across distances separated perpendicular to the background field, such as the distance marked $2R_M$ in Fig. 10.2a. The form of the interaction of the plasma with a body immersed in the flow is significantly affected by the electrical conductivity of the body through responses that we consider here.

Earth's Moon is an excellent example of a very poor conductor. A poor conductor (like a capacitor) can develop surface charge consistent with the electric field across it. Despite the development of surface charge, no current will flow because of the low conductivity. The electric potential of the surface charges will cancel the potential drop in the solar wind. In this situation, no force is imposed on the plasma, which flows without obstruction onto the surface (see Fig. 10.2a). The magnetic field of the incident plasma penetrates the body and is present throughout the interaction region. The principal effect of the interaction is to deplete plasma in the region downstream in the wake. In this region the magnetic pressure increases to produce pressure balance with the plasma surrounding the wake. Below we discuss how the wake refills.

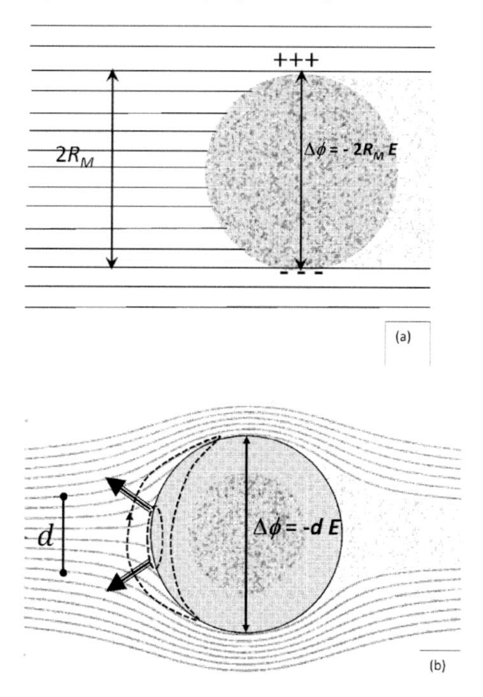

Fig. 10.2 Schematics of plasma flow (shown by lines of flow) at velocity **u** from the left onto (a) a non-conducting body and (b) a conducting body. In the plasma, **B** is into the paper, **E** is $-\mathbf{u} \times \mathbf{B}$ in both cases. Diagram (a) shows that a non-conducting body builds up surface charge that imposes a potential drop $\Delta\phi = -2R_M E$ across the diameter, producing an electric field that opposes the solar wind electric field. Diagram (b) shows the response of a conducting body that does not build up surface charge. Conducting paths allow current (shown schematically as a dashed line) to flow through the body and close in the incident flow. Heavy-banded arrows identify the orientation of the resultant $\mathbf{j} \times \mathbf{B}$ force that diverts part of the incident flow. Because much of the incident flow has been diverted, the potential drop across the body is only $\Delta\phi = dE$, where $d < R_M$ is the distance in the incident flow between the flow lines that just graze the body. The electric field that penetrates the body is a fraction of the upstream field determined by the fraction of the upstream flow that impacts the surface. In the wake region, gray in both diagrams, the plasma pressure is reduced and the magnetic pressure is increased relative to the upstream values.

If the body can conduct current, the response is quite different. For a sub-magnetosonic upstream flow, there is no bow shock. The schematic of Fig. 10.2b shows the plasma incident on a body with a conducting outer shell. In the absence of a local response, the upstream plasma within a cylinder of radius R_M about a line through the center would flow onto the moon's surface. However, in response to the electric field of the flowing plasma, current begins to flow either through a solid outer layer or through the ionized gas layers that envelope many solar-system bodies. Where the current emerges from the conducting layer, it comes in contact with the incident plasma and closes through it. In equilibrium, the current flowing through the plasma exerts a $\mathbf{j} \times \mathbf{B}$ force (see Eq. (10.2)) that diverts some of the incident flow as illustrated in Fig. 10.2b. Only a portion of the plasma that flows through the cylindrical volume that maps from upstream along unperturbed flow lines to the moon's cross section actually contacts the surface because some plasma is diverted around the sides. The maximum potential drop across the body then corresponds to the potential drop across the portion of the unperturbed flow (marked d in the figure) that actually reaches the surface; this potential drop is smaller than for a non-conducting body. In this case, the interaction modifies the flow well upstream of the moon, and the wake that develops is narrower than in the non-conducting case. If the conductivity of the body is extremely high, almost all of the plasma is diverted to the side, implying that d approaches 0 and correspondingly there is no potential drop across the body.

10.3.2 Pickup ions, neutrals, and associated currents

At comets and in the vicinity of moons, such as Io and Enceladus, that are significant sources of neutral gas, various processes that convert neutral atoms or molecules into ions are important to consider. Neutrals can be ionized by photons (photoionization) or by collisions with other particles, typically electrons (impact ionization). An additional process that affects the interaction region is charge exchange. In this process, a neutral gives up a charge to an ion. The original ion, now neutral, carries off its incident momentum while the original neutral becomes an ion at rest in the frame of the neutral gas.

The ions introduced into the plasma by ionization of neutrals modify the bulk properties of the plasma. Consider a situation in which the neutrals are at rest relative to the obstacle, towards which the plasma flows at (bulk) velocity $\mathbf{u}$. Photoionization and impact ionization add mass to the plasma whereas charge exchange between the ionized or neutral form of the same element does not change the mass density. All three processes slow the bulk flow because the new ions must be accelerated so that their average motion matches that of the bulk plasma and the process extracts momentum from the incident plasma. These processes also change the thermal energy of the plasma (e.g., Linker *et al.*, 1998) and may modify the plasma composition. The complex effects associated with pickup can significantly modify the interaction region surrounding a moon or a comet.

The relation between pickup and currents is shown schematically in the left-hand part of Fig. 10.3. The newly ionized ion senses the electric field of the flowing plasma and begins to move in the direction of this electric field. The electron that has separated from the ion is initially accelerated in the opposite direction. After

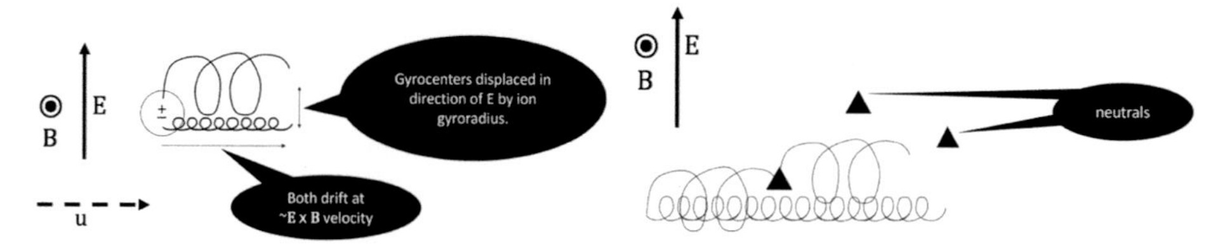

Fig. 10.3 Schematic of interactions with plasma with neutrals. Left: initial motion of pickup ions and electrons. The gray circle represents a neutral composed of a positively charged ion and a negatively charged electron. The directions of plasma flow velocity, **u**, of the magnetic field, **B**, and of the electric field, **E**, are indicated. In the image, following dissociation, the ion path starts upward and the electron path starts downward. Although initial motion is along **E** for the ion, the Lorentz force causes the path to twist, resulting in motion around **B** at the ion cyclotron period, leading to a net drift at a velocity of $\mathbf{E} \times \mathbf{B}/B^2$. The electron initially moves in the $-\mathbf{E}$ direction. Its motion also rotates around B, but at the electron cyclotron frequency. The net effect is a transient current in the direction of **E**. Right: schematic of the effect of collisions with neutrals for a case with the collision frequency of order the ion cyclotron frequency. The triangles represent neutrals. The effect of collisions is to slow the motion in the $\mathbf{E} \times \mathbf{B}$ direction of the ions but not of the electrons and to displace the ions in the direction of **E**. A net current arises, with one component along $-\mathbf{E} \times \mathbf{B}$ (a Hall current) and one component along **E**, a Pedersen current.

one gyroperiod, the average separation of the gyrocenters of the two charges is close to one ion gyroradius

$$\rho_g = m_{ion} v_{ion}/qB, \tag{10.18}$$

where m_{ion} is the ion mass, v_{ion} is its thermal velocity, and q is its charge. The result of the separation of charges is to produce a transient current density in the direction of the electric field. If the pickup is occurring at a rate $\dot{n}$, where $\dot{n}$ is the number of ionizations per unit volume and time, then the pickup current is

$$j_{pickup} = q\dot{n}\rho_g. \tag{10.19}$$

Because pickup current flows across the background field, a cloud of pickup ions acts much like a solid conducting obstacle in the flow and imposes the same types of perturbations, i.e., it slows and diverts the incident flow and produces bendback of the field in Alfvén wings in the manner described in Sect. 10.4.

Neutral gas density is often enhanced around a moon because material is continually sputtered off the surface, or because there are geysers (Enceladus) or volcanoes (Io), or because it has enough gravity to retain an atmosphere (Titan). In that case, current can arise in the plasma through collisions in a manner illustrated schematically on the left side of the right-hand diagram in Fig. 10.3. Current driven along **E** is referred to as Pedersen current. Current driven in the $-\mathbf{E} \times \mathbf{B}$ direction and carried by electrons is Hall current.

10.4 The interaction region and the role of MHD waves

In the discussion of Sect. 10.3.1, the slowing and diversion of the upstream flow was described in terms of currents, but it is more instructive to think of changes in the incident plasma being imposed by waves. Section 10.2 introduced the properties of the MHD waves that carry perturbations through a magnetized plasma. The fastest waves propagate farthest upstream in the flow. In this discussion, we assume that there are conducting paths either through the moon or other body or in its surroundings. Figure 10.1b shows that fast magnetosonic waves carry current density, **j**, perpendicular to the background field. The fast waves set up pressure gradients in the flow upstream of the obstacle, and Eq. (10.5) shows that, in the absence of sources and losses, a gradient of total pressure (the sum of thermal and magnetic pressure) can decelerate the flow. (The curvature term is insignificant in most of the situations that we consider.) Indeed, fast magnetosonic waves propagating upstream slow and divert the plasma upstream of a conducting obstacle. Because these waves can propagate along or across the field, the process is not greatly affected by the field orientation. As mentioned previously, if the upstream flow is super-magnetosonic, a shock forms upstream of a conducting body, and the shock, a steepened fast magnetosonic wave, imposes the slowing and the diversion of flow.

Once the flow has been modified by the effects of the fast mode waves, perturbations linked to the other two wave modes affect the interaction. Again, we start by describing the interaction in a sub-magnetosonic flow. Figure 10.4 presents two views of the interaction region, in (a) a cut through the center of the body in the plane containing the upstream flow and the unperturbed magnetic field ($x-z$ plane) and in (b) a cut through the center of the body and transverse to the upstream flow ($y-z$ plane). (Figure 10.2b completes the schematic illustration in a plane perpendicular to the two illustrated in Fig. 10.4.) In Fig. 10.4a, the slight bends of the field just upstream of the moon illustrate the compression imposed by the fast mode wave. This compression diverts the flow away from the central bulge in a direction perpendicular to the background field and the flow. Equation (10.12) shows that the flow diversion implies a magnetic perturbation transverse to the plane of the illustration in Fig. 10.4a. The perturbations propagate up and down along the background field as Alfvén waves. However, the propagation of these waves along the background field must be viewed in the rest frame of the plasma. (Recall that, in deriving the properties of Alfvén waves, we set the background flow velocity to 0, thus implying that the analysis was carried out in the plasma rest frame, where we

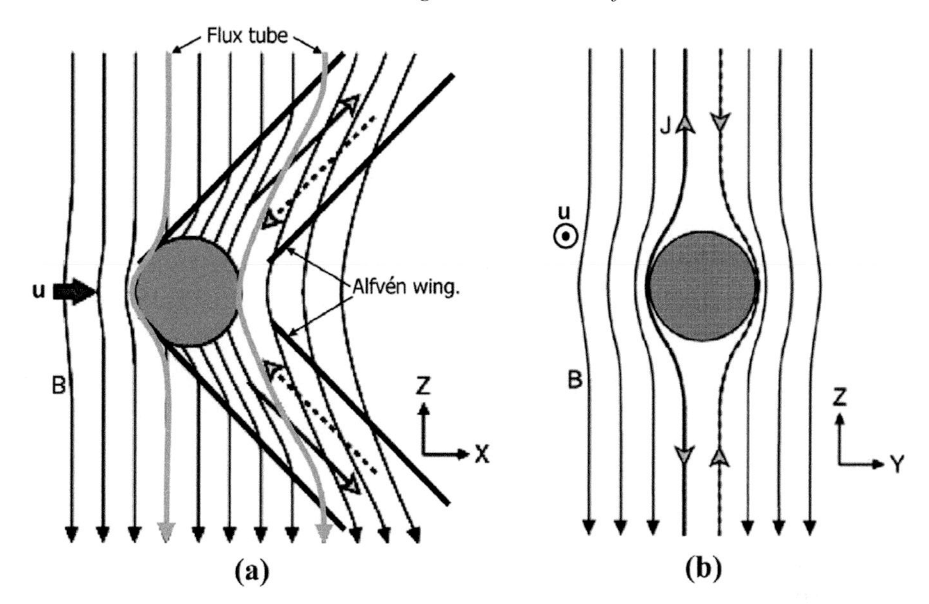

Fig. 10.4 Schematic representations of the interaction of a conducting body in a sub-magnetosonic plasma flow. (a) A cut through the center of the moon in the plane of the unperturbed field and the incident flow; (b) a cut through the center of the moon in the plane perpendicular to the incident flow. Field lines are thin black lines with black arrowheads but are shown in gray on the edges of the flux tube containing field lines that penetrate the obstacle, the Alfvén wing boundaries are thick black lines, the upstream flow is shown as a heavy black arrow labeled **u**, currents are black lines with open arrowheads, solid where the currents are directed away from the moon and dashed where they are directed towards the moon. (From Jia *et al.*, 2010.)

found that these waves carry information only in the direction $\pm\mathbf{B}$.) In Fig. 10.4, the plasma is moving at velocity **u** with respect to the body, taken to be at rest. Because the signal moves along the field in the moving rest frame, the signal can be present only downstream of the left-hand pair of heavy black lines in Fig. 10.4a. The region in which the strong field perturbations develop is called an Alfvén wing.

Figure 10.4a shows projections of bent field lines, although the actual bends are out of the plane of the image. The Alfvén wave does not change the field magnitude which is approximately constant across the boundaries in the illustration. Field-aligned currents flow on the boundaries of the region where Alfvénic perturbations are important and become weak in the region downstream of the right-hand pair of heavy lines on the right in Fig. 10.4a. Arrows in the figure indicate the direction in which current flows near the boundaries behind (dashed arrow) and in front (solid arrow) of the plane of the image. These currents are also shown in the plane perpendicular to the upstream flow in Fig. 10.4b. Downstream of the Alfvén wing, the slow mode begins to play a role in a region where the plasma pressure increases while the magnetic pressure decreases.

The situation illustrated in Fig. 10.4 represents interaction for M_A slightly less than 1. Near 1 AU (the radius of Earth's orbit around the Sun) in the solar wind, M_{ms} is typically of order 6. After passing through the bow shock, the flow speed diminishes, but still remains substantially larger than v_A. This means that downstream of Earth's bow shock the field must bend through an angle far greater than that shown in the figure. That is why Earth's magnetospheric boundary bends back so far that it forms the bullet-shaped cavity with which we are familiar.

The attentive reader will have noticed that Fig. 10.4 shows currents converging on the moon on one side and diverging from the moon on the opposite side. Currents are divergenceless, so there must be closure paths across the magnetic field direction that link the converging currents on the right of Fig. 10.4b to the diverging currents on the left, i.e., current must flow across the magnetic field in the near vicinity of the moon. If the moon itself is a good conductor, the current can close through it. Otherwise, it must close through the plasma in its immediate surroundings. The mechanisms that enable such closure currents to flow were discussed in Sect. 10.3.2.

10.5 Moons with magnetic fields permanent or inductive

We have discussed the interaction with small bodies assuming that the body is not magnetized and the plasma magnetic field is constant in time. Of course, these assumptions may not be valid. In particular, Jupiter's moon Ganymede, has a permanent internal magnetic moment and, for some moons, the periodic variation of the external magnetic field may drive an inductive response, thereby generating a varying internal magnetic field. Here, we consider these situations.

10.5.1 Ganymede, a magnetized obstacle

Prior to the Galileo Orbiter's first close pass by Ganymede, it seemed unlikely that so small a body (radius 2634 km) would have a magnetic field. Skepticism was justified because a planetary dynamo (see Ch. 6) requires an internal region of molten metal and it was thought that small bodies in the outer solar system would have cooled to a temperature at which even the deep interior would be solid. Thus it was a surprise when Galileo measurements revealed that Ganymede has a rather large dipole moment oriented nearly parallel to Jupiter's equatorial field, with equatorial surface intensity of $B_{eq} = 720$ nT. Ganymede is embedded in Jupiter's magnetospheric plasma, which flows towards the moon at a speed of approximately 140 km/s. Because to a good approximation, Jupiter's magnetospheric field is frozen into the flowing plasma (see Sect. 10.1), it cannot penetrate into the region dominated by Ganymede's internal field, but instead flows around it. In this situation, the interaction region is justifiably referred to as a magnetosphere. By definition, Ganymede's magnetosphere is the region embedded within Jupiter's magnetosphere that is threaded by the field lines with at least one end at the moon. The magnetosphere is unique. It is the only one that lies within a planetary magnetosphere because no other moon is known to possess a permanent internal magnetic moment. The size of Ganymede's magnetosphere is established by pressure balance, with the external pressure being $p + B^2/2\mu_0 + \rho u^2$ (see Eq. (10.5)) and the internal pressure well approximated as the magnetic pressure at distance r from its center of a dipole field centered within Ganymede. In Jupiter's magnetosphere near the orbit of Ganymede, the magnetic field magnitude is roughly 100 nT and magnetic pressure (4 nPa) dominates the contributions of thermal and dynamic pressure. Along the meridian containing the upstream flow, **u**, pressure balance requires $[B_{eq}/(r/R_G)^3]^2 2/2\mu_0 = 4\,\text{nPa}$ or $r \sim 2R_G$.

Figure 10.5 shows the form of Ganymede's magnetosphere extracted from an MHD simulation by Jia *et al.* (2008). Evident in Fig. 10.5b is the fact that the

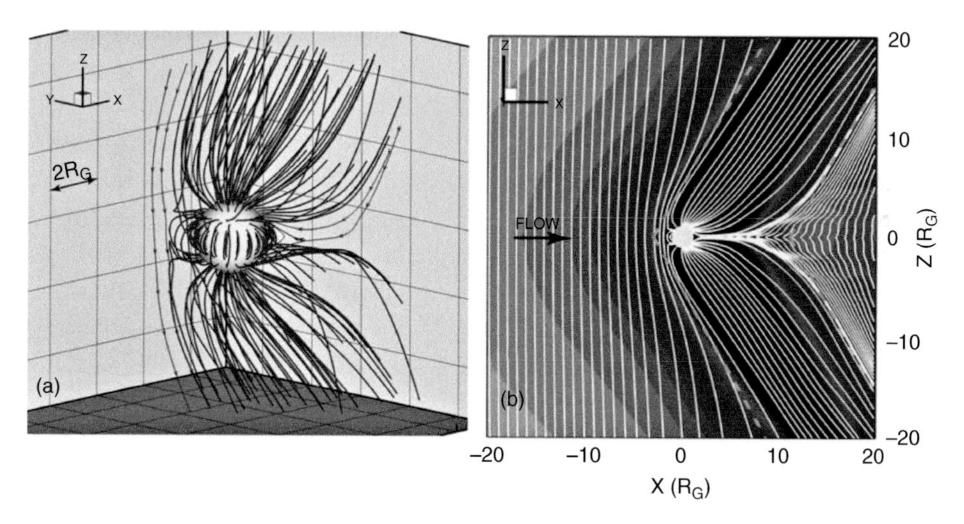

Fig. 10.5 (a) Selected magnetic field lines in Ganymede's magnetosphere from an MHD simulation. (b) Magnetic field lines projected onto the $x-z$ plane at $y = 0$. The x-component of the plasma flow velocity is shown in color. Orange dashed lines are tilted relative to the background field at the Alfvén angle and the flow is excluded from regions downstream of the left hand dashed lines, reappearing only in regions about 5 R_G further downstream. In the simulation, the sphere of radius 1.05 R_G is the inner boundary for plasma flow. (From Jia *et al.*, 2008.) A black and white version of this figure will appear in some formats. For the color version, please refer to the plate section.

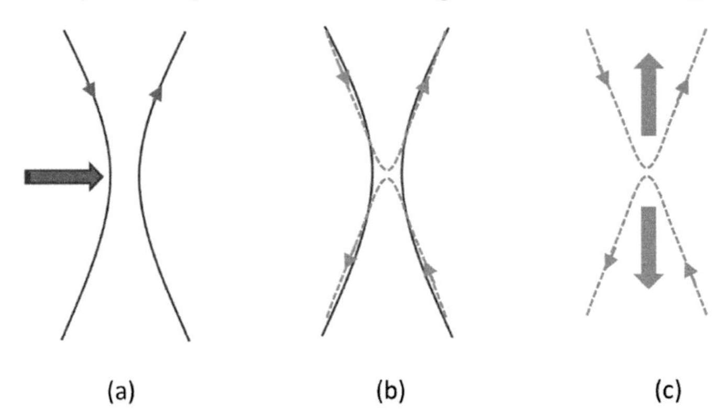

Fig. 10.6 Schematic of reconnection between oppositely oriented field lines. In (a) the southward-oriented field of Jupiter flows (heavy gray arrow) towards the northward-oriented field of Ganymede. In (b), the field lines reconnect so that the northern end of a Jovian field line continues into the northern end of a field line from Ganymede. In (c), the kinks in the newly connected field lines accelerate plasma to the north and the south, with gray arrows indicating the sense of flow.

upstream plasma does not flow into Ganymede's magnetosphere, but instead flows around it just as plasma would flow around a highly conducting body. A magnetic field acts as a barrier to plasma flow. The region in the equatorial plane from which the external plasma is excluded is substantially larger than the equatorial cross section of Ganymede, with the stand-off distance close to the $2R_G$ estimated. The shape of the magnetosphere differs greatly from the shape of Earth's magnetosphere because the symmetry of the system is dictated by the orientation of the external field, not by the direction of the plasma flow (see Table 10.1, noting that at Ganymede $M_A < 1$, whereas in the solar wind near Earth, $M_A > 1$).

At the orbit of Ganymede, Jupiter's field is oriented to the south, but Ganymede's equatorial field is oppositely oriented. This sets up the conditions for magnetic reconnection, a process in which oppositely directed field lines change the way they connect across a region in which their opposing orientation weakens the field. The process is shown schematically in Fig. 10.6. A fraction of the incident flow reconnects with the magnetic field of Ganymede, and produces the field lines seen in regions near the poles that are connected at one end to Ganymede and at the other end to Jupiter. (The field lines drawn to the upper and lower edges of the images in Fig. 10.5 should be regarded as continuing into Jupiter's ionosphere.)

Reconnection at Ganymede's magnetopause holds special interest for magnetospheric studies as an example in which the upstream plasma never rotates far from an orientation antiparallel to that of the internal field near the equator. This orientation implies that reconnection could occur in a relatively stable manner, as contrasted with the situation for planets in the solar wind exposed to fluctuating orientations of the component of the field aligned with their internal magnetic moments. However, studies of reconnection at Ganymede suggest that even in the case of Jupiter's stable magnetosphere, reconnection is not steady but intermittent with a typical recurrence time of tens of seconds. The short time between bursts of reconnection has a parallel in Mercury's magnetosphere, where reconnection has been found to occur intermittently and also to recur in tens of seconds. At other planets, reconnection can be intermittent but with much longer characteristic repetition intervals (~8 min at Earth).

10.5.2 Europa and response to a varying external magnetic field

The reader may wonder why the discussion of Ganymede in the previous section included a statement, carefully framed, that refers to a permanent internal magnetic moment. The reason is that some moons that lack a permanent magnetic moment may have a varying magnetic moment if they are embedded in a varying magnetic field. Such a situation can arise if a moon has a global-scale conducting layer, possibly below its surface, and is embedded in a varying magnetic field.

A striking example is Europa. Magnetic perturbations observed on the first flyby of Europa by the Galileo Orbiter were well modeled as those imposed by an internal dipole moment pointing along the radial direction to Jupiter and perpendicular to Europa's spin axis. Like the other Galilean moons (Io, Ganymede, and Callisto), Europa's orbit lies in Jupiter's spin equator. Because Jupiter's dipole moment is tilted by 10° relative to its spin axis, the orientation of the field embedded in the plasma flowing around all of these moons varies at the synodic period of Jupiter's rotation, 11.23 h at Europa. The varying part of the external field is dominated by the component along a radius vector from Jupiter's center. Equation (10.3) implies that a varying magnetic field generates an inductive electric field. If all of Europa's interior or even a shell, consists of conducting matter, the inductive electric field can drive current. In turn, the induced current generates a magnetic perturbation (see Eq. (10.4)) and for field variations in the direction radial from Jupiter, the induced dipole would have the form inferred from Galileo measurements.

A test was designed to support the speculation that the observed magnetic perturbation was the signature of induction. A planned future flyby was modified so that closest approach to Europa would occur at a rotation phase of Jupiter for which the external radial field component would be in the direction opposite to that of the first pass. The orientation of a permanent internal magnetic moment would not change from one pass to another but, if the dipole were generated by induction, the internal magnetic moment would reverse between the two passes. The internal moment was found to reverse, thereby demonstrating that the observed dipole field was induced.

The next problem was to identify the layer in which the induced current flowed. Gravity measurements supplemented by knowledge of average composition have been used to model Europa's interior as a dense metallic sphere with radius between 0.3 and 0.5 R_E (R_E is the radius of Europa) surrounded by a rocky shell and covered by an ice shell of roughly 100 km depth. The magnetic perturbations recorded on Galileo's first flyby of Europa were just what would be observed if the varying field of Jupiter's magnetosphere was inducing current in Europa's interior. The near-surface signature of currents flowing in Europa's conducting core, whose radius is 0.5 R_E or less, would be at most $0.5^3 = 0.12$ smaller than observed because of the rapid fall-off with distance of the amplitude of a dipole field. In order to account for the size of the induced dipole moment, the induced current must have been flowing very near the surface of the moon. The outermost layer of Europa was known to be composed of light material, mainly H_2O, with an icy surface. Ice is a poor conductor of electricity. However, liquid water containing dissolved electrolytes, something like Earth's ocean, is a reasonably good electrical conductor that could easily carry the required current density. Thus, it seems extremely probable that the conducting layer near the surface is a global-scale ocean buried beneath the icy surface, an ocean discovered through magnetic field measurements and logical deductions from those measurements. Missions are being planned to continue the investigation of this intriguing feature of Europa.

10.6 Moons without atmospheres

Generic properties of the flow onto a non-conducting body were discussed above and some features are illustrated in Fig. 10.2a. The sketch, although simplified, represents the most important features of the interaction of the solar wind with the Moon and of Saturn's magnetospheric plasma with Rhea. Because there are no current paths through the moon, there is no compressional wave propagating upstream, so to a good approximation, the upstream plasma flows directly onto the surface, whether the moon is in the super-magnetosonic solar wind or in the sub-magnetosonic plasma of a planetary magnetosphere. The theoretical arguments have been thoroughly tested by analysis of data acquired by the Artemis dual spacecraft mission to the Moon using data from time intervals when one spacecraft was in the undisturbed solar wind and the other was in the near vicinity of the Moon (Zhang *et al.*, 2014). Even small changes in plasma properties can be identified in this manner. As seen in Fig. 10.7, just downstream of the Moon, the wake is devoid of plasma and $|\mathbf{B}|$ is larger than in solar wind. Plasma refilling is largely governed by MHD slow-mode speed, which is very slow, so the wake is still depleted of plasma at distances of 10 lunar radii downstream. The disturbance in the downstream region propagates away from the wake center at the fast-mode speed. As

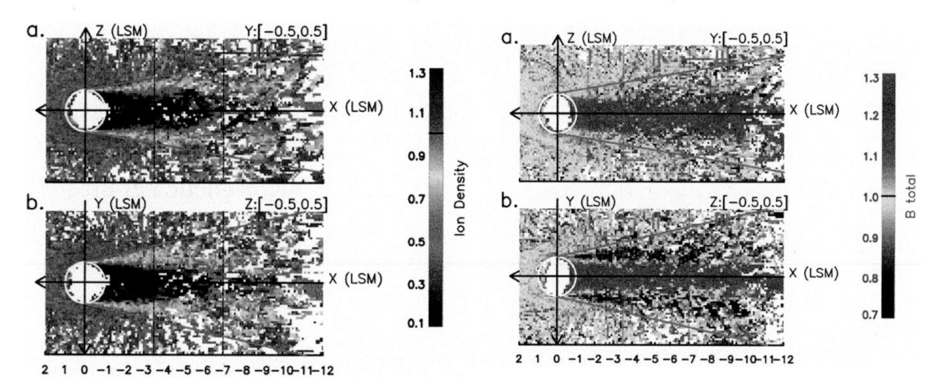

Fig. 10.7 Ion density (left) and magnetic field (right) in the vicinity of Earth's Moon from measurements by the *Artemis* spacecraft. The parameters represented by color are normalized by their values in the upstream solar wind. The x-axis is antiparallel to the solar wind flow. The data are plotted in the $x-z$ plane which is the plane of the solar wind field and the flow, and in the $x-y$ plane, perpendicular to this plane. The red lines diverging in the direction of negative x denote the wake boundary across which the density changes significantly. The divergence from the wake center is controlled by the propagation of fast mode waves. (From Zhang *et al.*, 2014.) A black and white version of this figure will appear in some formats. For the color version, please refer to the plate section.

anticipated for a non-conducting moon, no changes in plasma or field are evident upstream.

10.7 Moons with atmospheres or other sources of neutrals

The interaction between a moon and a flowing plasma differs greatly from the interaction described in Sect. 10.6 if a moon is enveloped in an atmosphere or another significant source of neutral gas. The upper levels of an atmosphere typically become ionized and capable of carrying current and we have already described how pickup ions and collisions with neutrals link allow currents to flow across the background field. Ionization of neutral gas scavenged from the upper atmosphere of the volcanic moon, Io, injects of order a ton of plasma per second into Jupiter's magnetosphere. Ionization of neutral gas from geyser-like plumes erupting from the surface of Enceladus injects tens of kilograms per second of plasma into Saturn's magnetosphere. Europa (at Jupiter) is an example of a moon with less intense sources of pickup ions, although there is some evidence for plumes arising from its surface. In all of these cases, the neutrals are central to the closure across the field of the Alfvén wing field-aligned currents that couple the moon with its parent planet.

Titan, Saturn's majestic moon that orbits at 20 R_S (Saturn's radius is 60 268 km) is shrouded in a very dense atmosphere. Its interaction region is exceptionally complex because as it moves around Saturn the field and plasma conditions change significantly. Sometimes Titan even leaves the magnetosphere and finds itself embedded in the solar wind. At Titan, the interaction with the atmosphere produces strong draping of the field, creating a highly localized region of intense interaction.

We know little about Neptune's large moon, Triton, but there is evidence suggesting that, like Enceladus, it may have geysers jetting vapor from its surface and these may also modify its magnetospheric surroundings.

10.8 Small bodies in the solar wind

The discussions of interaction in the previous sections have assumed that the scale of the obstacle to the plasma flow is large compared with the gyroradius of a solar-wind ion so that analysis of the interaction in the MHD limit is appropriate. However, some bodies are so small that MHD no longer applies, and the form of the plasma interaction changes.

10.8.1 Asteroids

Asteroids are widely distributed in the inner solar system although their orbits cluster between the orbits of Mars and Jupiter. They range in size from $\sim$1 km to $\sim$500 km, too small to be plausible sources of dynamo magnetic fields. Little is known about their interaction with the solar wind. Much of what we know about asteroids is based on studies of meteorites, small bits of asteroidal matter that reach Earth from outer space and may be found on its surface. We know that many meteorites contain iron and are magnetized. This magnetization, referred to as remanent magnetization, suggests that the parent body of the meteorite may have had an internal dynamo that impressed a field on surrounding iron-bearing matter prior to the collision that broke it to pieces.

Little is known about the magnetic fields of asteroids. The Galileo spacecraft flew by two asteroids: Gaspra (scale length of order 14 km) and Ida (scale length of order 30 km). In both cases, small-scale but distinctive magnetic fluctuations lasting 5–10 min near closest approach were present in the measured magnetic field. It is plausible that the fluctuations were, in both flybys, linked to the nearby asteroid. However, one must recognize that fluctuations in the solar wind take many forms and it is possible that the signatures attributed to these bodies were fortuitous solar-wind fluctuations. Assuming that the signatures were generated by a plasma–asteroid interaction, the signatures were analyzed using an approach that applies for cases in which length scales are small compared with the ion gyroradius but large compared with the electron gyroradius. In this parameter regime, no shock is expected to form upstream because the relevant waves propagate faster than MHD

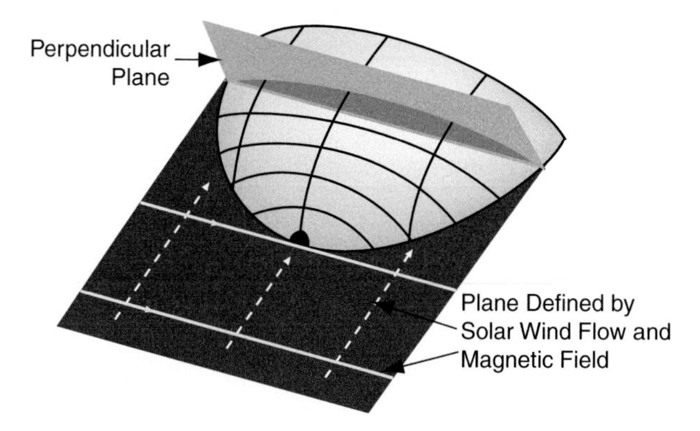

Fig. 10.8 Schematic of interaction region surrounding an asteroid of scale not large compared with ion gyroradii in the solar wind. Notice that the interaction region is extended along the field direction and compressed in the direction perpendicular to the field and the flow. (From Kivelson and Bagenal, 2007; credit: Steve Bartlett.)

waves. The interaction region is less cylindrical than that of a planet in the solar wind; it is compressed in the direction perpendicular to the field and the upstream flow as seen in the schematic illustration of Fig. 10.8. The duration of the signature associated with Gaspra required the asteroid to have a significant internal magnetic field corresponding to magnetization similar to that observed in some meteorites. This interpretation requires the parent body of Gaspra to have been magnetized before it broke up in some celestial collision. The Ida signature, also lacking an upstream shock, required that the body be conducting.

Magnetic-field perturbations near asteroid Braille are consistent with a specific moment (A m^2/kg) of the same order of magnitude as that inferred for Gaspra. By contrast, measurements taken by the NEAR-Shoemaker mission close to and on the surface of asteroid Eros indicate that its field is negligibly small, possibly because it is formed of magnetized rocks of random orientation. The DAWN spacecraft has spent months in the vicinity of asteroid Vesta and is off to an encounter with the largest asteroid, Ceres, but the spacecraft instrumentation does not include a magnetometer, so we have little information on the plasma interaction at these bodies. Other missions under discussion would add to our knowledge of asteroid magnetic properties.

10.8.2 Comets

Comets are small icy bodies that form in the outer solar system. The most spectacular ones are on highly elliptical orbits (on which they are likely to return to the inner solar system periodically) or sometimes on hyperbolic orbits (on which, in

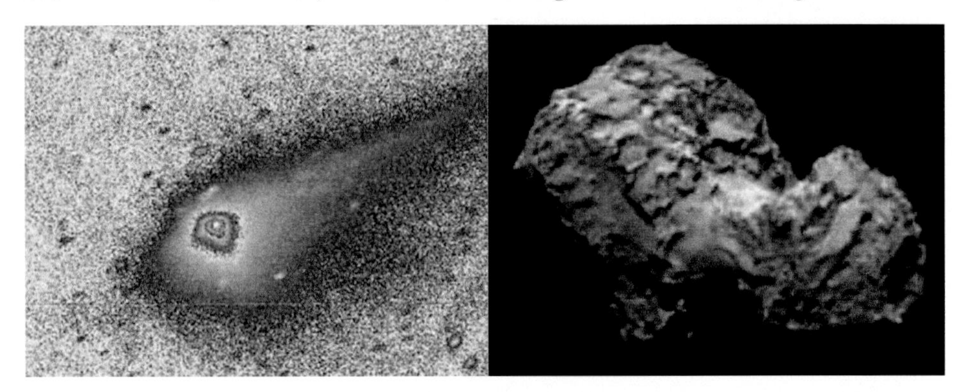

Fig. 10.9 Left: image of comet Churyumov–Gerasimenko from the 3.6 m telescope of the European Southern Observatory, La Silla/Chile. The comet was at 2.49 AU from the Sun when imaged on March 9, 2004. East is to the left and north is up. The nucleus is a source of asymmetric emissions. The scale is 70 000 km EW and 50 000 km NS. (From Glassmeier *et al.*, 2007.) Right: image of the nucleus acquired by Rosetta spacecraft on August 3, 2014. (Courtesy ESA/NASA – Rosetta spacecraft.)

the absence of significant perturbations, they are likely to escape the solar system). As these icy bodies approach the inner solar system, they are heated by the Sun and release neutral gas and dust, normally at rates that increase with approach to the Sun. The dust moves outward from the comet and, because it must conserve angular momentum, falls behind the comet increasingly as it moves farther outward, producing a tail that bends away from the direction radially outward from the Sun. The neutral gas can be ionized by solar ultraviolet radiation and the newly formed ions are picked up by the solar wind, as described in Sect. 10.5, forming a second tail directed radially outward.

Our understanding of cometary interactions will be greatly advanced by data acquired by the Rosetta spacecraft, which made measurements close to comet Churyumov–Gerasimenko, a 2-km object that reached perihelion in August of 2015. The image on the left of Fig. 10.9 shows the neutral and plasma environment, with a tail that extends tens of thousands of kilometers even though the comet is still far outside of Earth's orbit (2.49 AU). The tail is likely to extend over AU as the comet comes closer to the Sun. Comet Churyumov–Gerasimenko is a highly irregular object, heavily scarred by impacts as seen in an image taken as the spacecraft closed in on its target (Fig. 10.9 right). Often comets are described as dirty snowballs, but this comet, though dirty, is far from a ball-shaped object.

The solar wind flow is super-magnetosonic relative to the comet, but the flow slows as the wind approaches the comet because of its interaction with the cometary material. As its momentum decreases, the Mach number of the interaction also decreases. Because of the decreasing Mach number, shocks upstream of comets may be quite weak or possibly even disappear. As the slowed flow carries magnetic flux past the comet, the solar-wind magnetic field gets hung up on the region near the comet and the field drapes around the slowed region in the form of a hairpin. The oppositely directed draped fields on the downstream side of the comet can reconnect, and cause a portion of the ion tail to break off. For reviews of interaction of the solar wind with Comet 67P see Hansen *et al.* (2007) and Rubin *et al.* (2014).

10.8.3 Pluto

The New Horizons spacecraft is en route to Pluto, the most accessible dwarf planet. At this time, only computer simulations can hint at what kind of interaction will be observed. The spacecraft lacks a magnetometer, so there will be no direct measurement of the planetary magnetic field, but plasma measurements should reveal whether there is an upstream shock and the nature of the flow in the vicinity of Pluto (McNutt Jr. *et al.*, 2008; McComas *et al.*, 2008). The parameters of the plasma interaction anticipated lead us to expect a very asymmetric magnetosphere because the gyroradii of solar wind ions are not small compared with Pluto's radius, implying that an MHD analysis will not apply (e.g., Delamere, 2009). Pluto's orbit is very eccentric. Only near perihelion, when its orbit lies closest to the Sun, does Pluto have an atmosphere. The magnetosphere is affected by an atmosphere, so there may be different forms of the interaction at different orbital phases (Bagenal *et al.*, 1997).

10.9 Summary and expectations for other planetary systems

Plasma flowing onto the small bodies of the solar system creates interaction regions of many sorts, some very localized and others extended over many AU. Interpreting the response has required us to consider the scale of the body relative to fundamental length scales of the incident plasma, to note whether or not the body conducts electrical current, and to consider the neutral gas environment of the body. Dimensionless parameters of the incident plasma and the size of the body relative to the ion gyroradii of the incident plasma were found to be critical in determining the global structure of the interaction region and whether or not it was bounded by a shock on the upstream side.

Based on what we know from studies of bodies in the solar system, one can speculate on the nature of the interaction in systems that have not yet been explored.

Further from the Sun than the orbit of Neptune lie dwarf planets similar to Pluto but located in even more tenuous solar wind. The different plasma environment may modify the response in interesting ways.

Planets around stars other than the Sun are being discovered continually (see Ch. 6 for a discussion of exoplanet dynamos, and Ch. 5 of exoplanet discoveries). The conditions relevant to the formation of a magnetosphere may differ immensely if stellar winds differ substantially from the solar wind. One can be reasonably certain that there will be surprises whenever our measurements of other stellar systems begin to provide measurements of extrasolar magnetospheres.

11

Dusty plasmas

MIHÁLY HORÁNYI

11.1 Motivation

The study of dusty plasmas bridges a number of traditionally separate subjects, for example, celestial mechanics, mechanics of granular materials, and plasma physics. Dust particles, typically micron- and submicron-sized solid objects, immersed in plasmas and UV radiation collect electrostatic charges and respond to electromagnetic forces in addition to all the other forces acting on uncharged grains. Simultaneously, dust can alter its plasma environment by acting as a possible sink and/or source of electrons and ions. Dust particles in plasmas are unusual charge carriers. They are many orders of magnitude heavier than any other plasma particles, and they can have many orders of magnitude larger (negative or positive) time-dependent charges. Dust particles can communicate non-electromagnetic effects, including gravity, neutral gas and plasma drag, and radiation pressure to the plasma electrons and ions. Their presence can influence the collective plasma behavior by altering the traditional plasma wave modes and by triggering new types of waves and instabilities. Dusty plasmas represent the most general form of space, laboratory, and industrial plasmas. Interplanetary space, comets, planetary rings, asteroids, the Moon, and aerosols in the atmosphere, are all examples where electrons, ions, and dust particles coexist.

The observations of the inward transport of interstellar dust and the outflow of near-solar dust provide a unique opportunity to explore dusty plasma processes throughout the heliosphere. The flux, direction, and size distribution of interstellar dust can be used to test our models about the large-scale structure of the heliospheric magnetic fields, and its temporal variability with solar cycle. The measurements of the speed, composition, and size distribution of the recently discovered, solar-wind-entrained, nano-dust particles hold the key to understanding their effects on the dynamics and composition of the solar-wind plasma.

After its fly-by of Jupiter, the dust detector onboard the Ulysses spacecraft detected impacts of particles in the mass range of 10^{-14} to 10^{-11} g, predominantly

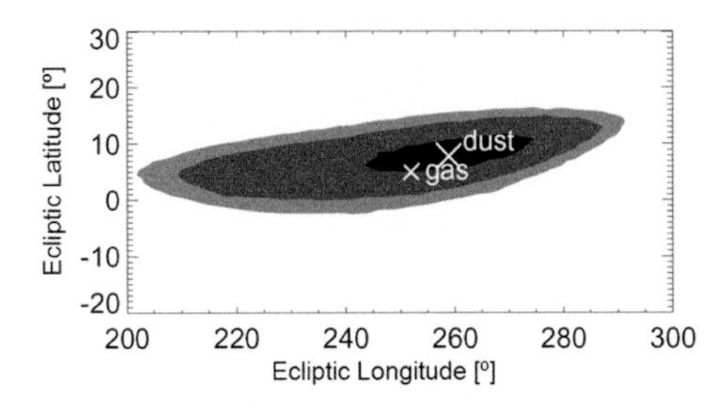

Fig. 11.1 The upstream direction of the interstellar dust flux observed by Ulysses is $\lambda = 259 \pm 20°$ and $\theta = 8 \pm 10°$ ecliptic longitude and latitude, respectively. The contour plot shows 1, 2, and 3 σ confidence levels as black, dark gray, and light gray. The helium upstream direction is $\lambda = 254°, \theta = 5.6°$. (From Frisch *et al.*, 1999.)

from a direction that was opposite to the expected impact direction of interplanetary dust grains. In addition, the impact velocities exceeded the local solar-system escape velocity (Grün *et al.*, 1993). Subsequent analysis showed that the motion of the interstellar grains through the solar system was approximately parallel to the flow of neutral interstellar hydrogen and helium gas (Fig. 11.1), both traveling at a speed of 26 km/s. The 20-year-long journey of these small dust particles through the heliosphere is strongly coupled to the large-scale structure of the heliospheric magnetic fields (IMF) through almost two full solar cycles. Hence, monitoring the variability of the flux of interstellar dust (ISD) grains provides a unique opportunity to test our models of the IMF through solar cycles (Sterken *et al.*, 2013).

The interaction of dust particles with the heliospheric magnetic fields depends on the grain's mass, charge, and material properties. The smallest interstellar grains (radii $<$10 nm) remain tied to the interstellar magnetic fields and are diverted around the heliosphere (Frisch *et al.*, 1999), only larger grains can penetrate the heliosphere. The Lorentz force created by the solar-wind magnetic field sweeping past the grains can both repel or focus the grains, depending on the configuration of the field and thus on the phase of the solar cycle (Grün *et al.*, 1994). In mid 1996, the flux of ISD grains was observed to decrease by a factor 3 as a result of the electromagnetic interaction of the grains with the magnetic field of the solar wind (Landgraf, 2000). The polarity of this magnetic field changes with the 22-year solar cycle. During the solar maximum in 1991, the field polarity became north-pointing. The azimuthal component of the expanding magnetic field deflected interstellar grains in the northern hemisphere to the north and grains in the southern hemisphere to the south. Thus, the net effect of the interplanetary magnetic field during the 1991 polarity cycle was to divert interstellar dust away from the ecliptic plane. During the 2000/2001 solar maximum the field polarity reversed. However, interstellar grains need about 20 years (almost 2 solar cycles) to traverse the distance from the heliospheric boundary (at about 100 AU) to the Sun, and therefore, the changes in the magnetic field configuration do not immediately affect the dust flow. For example, trajectory calculations indicate an approximately 6 year delay between solar magnetic field-reversals and the shift in the flow of ISD grains from focusing toward to defocusing away from the ecliptic plane (Sterken *et al.*, 2013). Ulysses measurements identified interstellar grains and determined their flow direction only by statistical means, and with large uncertainties. The full details of the influence of the heliosphere on the trajectories of ISD grains remains yet to be fully understood (Krüger and Grün, 2009).

Early dust instruments onboard Pioneer 8 and 9 and Helios spacecraft detected a flow of submicron-sized dust particles arriving from the solar direction. These particles originate in the inner solar system from comets, and from collisions among meteoroids, and move on hyperbolic orbits that leave the solar system under the prevailing radiation pressure force. The dust instrument onboard the Ulysses spacecraft observed escaping dust particles high above the solar poles, confirming the supposition that charged nanometer-sized dust grains are carried to high heliographic latitudes by electromagnetic interactions with the IMF. Recently, the STEREO WAVES instruments recorded a large number of intense voltage signals in the ecliptic plane at 1 AU, which were interpreted as impact ionization signals of streams of nanometer-sized particles striking the spacecraft at velocities of about solar-wind speed (Fig. 11.2). High fluxes of nanometer-sized dust grains at low heliographic latitudes, as well as strong spatial and temporal fluctuations of dust streams uncorrelated with solar-wind properties pose a mystery. The

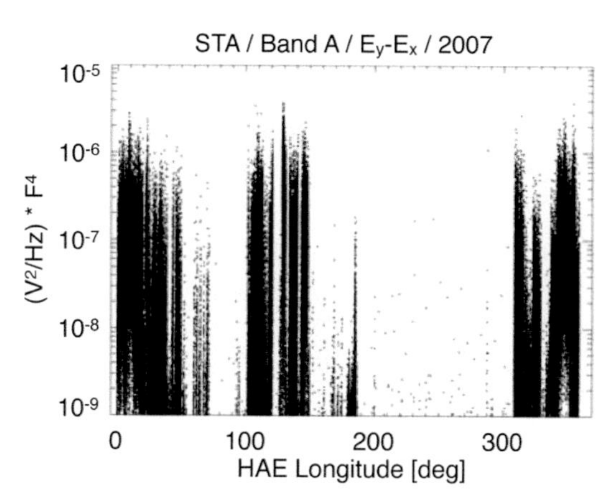

Fig. 11.2 The average power observed by the STEREO WAVES low-frequency receiver on STEREO A as a function of ecliptic longitude during 2007. The periods of high amplitudes have been suggested to be caused by high-speed nano-dust striking the spacecraft. (From Meyer-Vernet *et al.*, 2009.)

nano-dust, if real, represents a significant mass flux, which would require that the total collisional meteoroid debris inside 1 AU is cast in nanometer-sized fragments.

Dust adds mass to the solar wind as it is a source of atomic and molecular ions and neutrals, which are rapidly ionized. The charged ions and molecules are picked up by the solar wind magnetic field (Lemaire, 1990). However, the effects of mass loading by dust-generated species on the acceleration and dynamics of the solar wind near the Sun, including the position of the sonic point, where the subsonic solar wind plasma transitions to a supersonic flow, are not known and remain the subject of speculation. A slight increase in the average ion mass in the solar wind might, for example, generate shocks, at least in localized regions of high dust densities. These regions might also contribute to the energization of solar-wind ions via two-stream plasma instabilities (Shapiro *et al.*, 2005). Mass loading by dust has been invoked to explain enhancements observed in the IMF (Russell *et al.*, 1984). The exact nature, and the processes responsible for the generation of magnetic field enhancements remain an open issue.

Dust-generated neutrals and ions are thought to form an "inner source" of pickup ions (PUIs) whose velocity distribution differs from that of interstellar "outer source" PUIs. Both singly charged atomic ions and molecular ions (e.g., CH^+, NH^+, OH^+, and H_2O^+) have been identified in the inner source population (Gloeckler *et al.*, 2010). Processes proposed to explain the production of PUIs from the near-Sun dust include surface interactions and gas production by sublimation, sputtering, and collisional vaporization. Modeling indicates that collisional vaporization can account for the heavy elements in the inner source PUI population as well as molecular species, while surface interactions produce significant amounts of pickup protons (Mann and Czechowski, 2005). PUI production from dust sublimation or sputtering inside 0.1 AU is predicted to produce higher-charge-state PUIs; however, these processes have not yet been quantified (Mann, 2010). Solar-wind interaction with the near-Sun dust also contributes to the neutral solar wind and can generate energetic neutral atoms (ENAs) (Collier *et al.*, 2001) suggesting that dust interaction with solar energetic particles may be an additional inner heliospheric ENA source.

The dynamics of small charged dust particles in planetary magnetospheres can be surprisingly complex, possibly leading to levitation, rapid transport, energization and ejection, capture, and the formation of new planetary rings, for example. Dust particles immersed in plasmas are unusual charge carriers, and due to their very large mass compared to electrons and ions, they introduce new spatial and temporal scales. The Coulomb interaction between charged dust grains can lead to collective behavior, for example waves with unusually low frequencies and long wavelengths, making them visually observable in laboratory experiments. In this brief tutorial the discussion will be limited to dust charging, dust dynamics in planetary magnetospheres and in interplanetary space, and possible dusty plasma collective effects at comets. The material discussed is based on a collection of review papers (Horányi, 1996; Horányi *et al.*, 2004, Mendis and Horányi, 2013a). Comprehensive reviews on many other aspects of the physics of dusty plasmas are also readily available (Bliokh *et al.*, 1995, Shukla and Mamun, 2002, Tsytovich *et al.*, 2010). In addition, several modern general plasma physics text books now include sections on dusty plasmas (Bellan, 2006, Piel, 2010).

11.2 Dust charging

If a dust grain carries a charge, its motion will be influenced by electromagnetic forces in addition to all other forces acting on uncharged dust particles. The relative importance of the electromagnetic forces due to either the large-scale electric and magnetic fields or small-scale fields between charged grains, depend on the charge-to-mass ratio of a dust particle. A single isolated dust grain embedded in a plasma with a Maxwellian energy distribution and shielded from any radiation field energetic enough to cause photoelectric emission from the grain, will acquire a negative average charge, if the ion and electron number densities ($n_e = n_i$) as well as the ion and electron temperatures ($T = T_e = T_i$) are equal. While this is rarely the case, the assumptions are used here to simplify our discussion below. The negative average charge is a consequence of the higher thermal speed of the electrons due to their much smaller mass. In equilibrium the surface potential ϕ (with respect to $\phi = 0$ at infinity), and the grain charge q_d, will be given by

$$\phi = \frac{q_d}{4\pi \varepsilon_0 r_d} = -\alpha \frac{k_B T}{e} \tag{11.1}$$

where ε_0 is the vacuum permittivity, r_d is the radius of the dust grain (which we will assume to be spherical, hence its capacitance $C = 4\pi \varepsilon_0 r_d^2$), and k_B is Boltzmann's constant (Spitzer, 1978). The proportionality factor α is a function of the ion mass m_i, and for example it is approximately 2.5, 3.6, and 3.9 for a pure H^+, O^+, and S^+ plasma, respectively. In equilibrium, the rate at which electrons collide with the grain is equal to the rate at which positive ions strike the grain if each ion is singly ionized.

In addition to the collection of electron and ion fluxes (that by themselves can be further complicated if a grain is embedded in a multi-species plasma with non-Maxwellian energy distributions), grains can charge due to a number of other processes, including photo, secondary, and thermionic emissions (see Fig. 11.3) (Whipple, 1981, Mendis, 2002, Horányi, 1996).

Naturally, a dust grain charge is in equilibrium if the sum of all the charging currents is zero. However, due to the stochastic nature of these charging currents

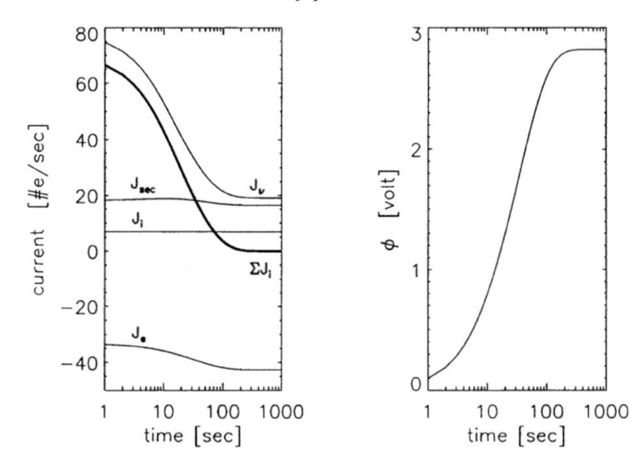

Fig. 11.3 The evolution of the electron and ion thermal currents, J_e, J_i; secondary and photoelectron currents J_{sec}, J_ν, and their sum $\sum J_k$ (left panel); and the electrostatic surface potential ϕ (right panel) of an initially uncharged dust particle with radius $a = 1$ μm at a heliocentric distance of 1 AU. (From Horányi, 1996.)

the charge of any dust particle is expected to fluctuate around its equilibrium (Gail and Sedlmayr, 1975). The presence of neutrals can also influence the charging processes. In a weakly ionized plasma the electron and ion fluxes will be limited to diffusion currents. However, in the solar wind, or in magnetospheric plasma environments discussed in this tutorial, the standard plasma Langmuir probe equations (Langmuir, 1923) used in the calculations of these currents give satisfactory results for regions where the dust density is much smaller than the plasma density $n_d \ll n_p$.

In these cases dust grains can be treated as test particles and their distribution in space has a negligible effect on the electric, magnetic, and gravitational fields. Dust-particle trajectories can be followed by simultaneously integrating their time dependent charge, and the dynamical effects of gravitational, electric, and magnetic forces, drag, and radiation pressure (Horányi, 1996). Following a large number of single particle trajectories, the size and spatial distribution of these grains can be calculated, including possible time dependence in their production (micrometeoroid bombardment of moon sources) and loss (sputtering and collisions).

In regions where the dust density is no longer negligible compared to the plasma density, the charge of a grain can be greatly reduced compared to an isolated particle in the same plasma environment, due to the depletion of the plasma density (Goertz and Ip, 1984, Havnes, 1984, Whipple *et al.*, 1985).

The influence of dust on collective effects (i.e., waves and instabilities) is particularly strong in plasmas in which the dust carries a significant fraction of either the negative or the positive charge. Havnes (1984) and Havnes *et al.* (1990) introduced the parameter

$$P \equiv \frac{4\pi\varepsilon_0 r_d}{e^2 n_0} n_{d0} k_B T \tag{11.2}$$

as an indicator of whether grains carry a significant fraction of the negative charge in a plasma in which photoelectric emission from the dust is unimportant. Here n_0 and n_{d0} are the number density of charges carried by ions and by dust, respectively. If $P \ll 1$, the dust will carry only a small fraction of the negative charge, and the magnitude of the average charge on a grain is well estimated by Eq. (11.1). If $P \gg 1$, the dust particles carry a large fraction of the negative charge and the average charge on a single dust particle is small compared to the result given by (11.1).

11.3 Dust in planetary magnetospheres

In order to follow the spatial and temporal evolution of dust density distributions in magnetized plasma environments, the equations describing the motion and the evolution of the charge of the particles have to be simultaneously integrated. The central body can be represented by multipole expansions of its gravitational and magnetic fields. The density and the temperature of the many-component plasma environment is to be predefined as a function of coordinates and, if necessary, the time as well. The charging currents are dependent not only on the instantaneous plasma parameters but on the velocity, as well as on the previous charging history of the dust grains. In Gaussian units the equation of motion of a dust particle with radius a, mass m, and charge Q, in an inertial coordinate system can be written as

$$\ddot{\mathbf{r}} = -\mu\nabla\left(\frac{1}{r} + \frac{R^2}{r^3}J_2P_2 + \frac{R^4}{r^5}J_4P_4\right) + \frac{Q}{m}\left(\frac{\dot{\mathbf{r}}}{c} \times \mathbf{B} + \mathbf{E_c}\right), \tag{11.3}$$

where $\mathbf{r}$ is the grain's position vector and an overdot signifies differentiation with respect to time. The first term on the right-hand side is the gravitational acceleration due to the planet with $\mu = GM$, the product of the gravitational constant G and the mass of the planet M, the planet's radius R, and the higher-order terms of its gravity are expressed in terms of Legendre polynomials P_2, and P_4 with coefficients J_2, and J_4, respectively, and possibly higher-order terms. The last term is the Lorentz acceleration where $\mathbf{B}$ is the local magnetic field and, assuming a rigidly co-rotating magnetosphere, the co-rotational electric field is

$$\mathbf{E_c} = (\mathbf{r} \times \boldsymbol{\Omega}) \times \mathbf{B}/c, \tag{11.4}$$

where $\boldsymbol{\Omega}$ is the angular velocity vector of the planet.

As grains traverse the various plasma regions their charge will not stay constant. A grain's charge can be followed via the current balance equation

$$\frac{dQ}{dt} = \sum_i I_i, \tag{11.5}$$

where I_i represent electron and ion thermal currents, and also the secondary and photoelectron emission currents. These are all functions of the plasma parameters, material properties, size, velocity, and also the instantaneous charge of a dust particle. This set of equations are already sufficient to find surprising outcomes: rapid dust transport, and even the energization and ejection of small grains from planetary magnetospheres.

11.3.1 Simplified dynamics

To gain insight into the dynamics of charged grains, we rewrite Eq. (11.3) for a grain moving in the equatorial plane of a spherical planet with a simple aligned and centered dipole magnetic field, using polar coordinates r, ϕ

$$\ddot{r} = r\dot{\phi}^2 + \frac{q}{r^2}(\dot{\phi} - \Omega) - \frac{\mu}{r^2} \tag{11.6}$$

$$\ddot{\phi} = -\frac{\dot{r}}{r}\left(\frac{q}{r^3} + 2\dot{\phi}\right). \tag{11.7}$$

We introduced $q = QB_0R^3/(mc)$, where B_0 is the magnetic field on the surface of the planet on its equator, and the combination $q/r^3 = \omega_g$, becomes the gyrofrequency of the dust particle (i.e., the angular rate the dust particle circles about magnetic field lines).

On a circular equilibrium orbit $\ddot{\phi} = \ddot{r} = \dot{r} = 0$, and $\dot{\phi} =$ constant $= \psi$. The differential equations in this case turn into an algebraic equation for the angular velocity ψ

$$\psi^2 + \omega_g\psi - \omega_g\Omega - \omega_K^2 = 0, \tag{11.8}$$

where $\omega_K = (\mu/r^3)^{1/2}$ is the Kepler angular rate of an uncharged particle on a circular orbit at a distance r from the planet. For big particles, terms that contain ω_g can be dropped and we recover $\psi = \pm\omega_K$, prograde or retrograde Kepler orbits. For very small particles, terms that are not multiplied with ω_g are to be dropped and $\psi = \Omega$. Very small grains are picked up by the magnetic field and co-rotate with the planet. These differential equations can be integrated to yield constants of the motion

$$\mathcal{E} = \frac{1}{2}(\dot{r}^2 + r^2\dot{\phi}^2) = \frac{\mu + q\Omega}{r}, \tag{11.9}$$

$$\mathcal{J} = r^2\dot{\phi} - \frac{q}{r}. \tag{11.10}$$

For large particles ($q \to 0$) these constants become the Kepler energy and angular momentum. The Jacobi constant, $\mathcal{H} = \mathcal{E} - \Omega\mathcal{J}$, remains a constant even if q changes with time (Northrop and Hill, 1983). For a fixed $\mathcal{J}$ the effective 1D potential (i.e., $\ddot{r} = -\partial U/\partial r$) becomes

$$U(r) = -\frac{\mu + q\Omega}{r} + \frac{\mathcal{J}^2}{2r^2} + \frac{q\mathcal{J}}{r^3} + \frac{q^2}{2r^4}. \tag{11.11}$$

Small dust particles are generated by interplanetary dust particle impacts, or by active volcanoes and plumes. In these cases the small particles have initial relative speeds that are small compared to the orbital speed of their parent body. For an initial Kepler orbit $\mathcal{J} = r^2(\omega_K - \omega_g)$. Figure 11.4 shows $U(r)$ for particles originating from Io ($r_0 = 5.9\ RJ$) at Jupiter for the two typical values -30 and +3 V for the surface potential $\phi = Q/a$ (Horányi *et al.*, 1993a,b). Particles with negative surface potentials remain confined in the vicinity of r_0. However, grains in a certain size range with positive charges are not confined as $U(r)$ shows no minima. In the case of positively charged grains the force due to the co-rotational electric field points radially out, opposing gravity. The upper limit in size for ejection a_{max}, is set by the condition $F_E/F_G > 1$. The lower limit in dust size for ejection results from the fact that very small grains behave like ions or electrons circling magnetic field lines. The gyroradius $r_g = |wmc/(QB)| = |w/\omega_g|$, where

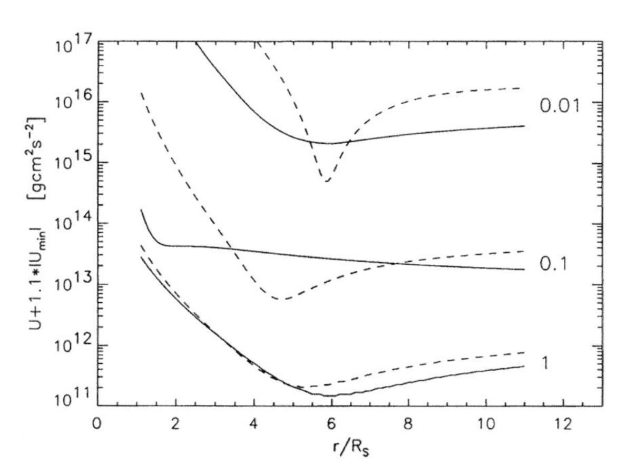

Fig. 11.4 The effective potential for dust grains with $a = 0.01, 0.1$, and 1 μm started from Io on circular Kepler orbits with $\phi = -30$ (dashed lines) and +3 V (continuous lines). To avoid the overlap of these curves, because only their shape is important, we have shifted them apart by plotting $U + 1.1|\text{minimum}(U)|$ instead of U itself. (From Horányi, 1996.)

w is the relative velocity between the co-rotating magnetic fields and the particle. For Kepler initial conditions $w = r(\Omega - \omega_K)$. The motion of these grains is well described by the guiding center approximation if the size of their orbit is smaller than the characteristic length scale for variations in the magnetic fields, $|r_g \nabla B/B| < 0.1$ ($|\nabla B/B| = 3/r$ in the equatorial plane of an aligned centered dipole). The upper limit of grain size satisfying this condition (i.e. the smallest grains that will be ejected) is

$$a^* \, [\mu\mathrm{m}] = \left(\frac{10^{-3} B_0 R^3 \phi}{4\pi r^2 \rho \omega c} \right)^{1/2} . \tag{11.12}$$

Grains in the size range $a^* < a < a_{max}$ will be ejected from the magnetosphere. As these positively charged grains move outward they gain energy from the co-rotational electric field

$$W = \int_{r_0}^{r_1} E Q dr = E_0 R^2 Q \left(\frac{1}{r_0} - \frac{1}{r_1} \right), \tag{11.13}$$

where the upper limit of the integration, r_1, is the characteristic size of the magnetosphere. This mechanism was suggested to explain the Ulysses and Galileo observations at Jupiter (Horányi *et al.*, 1993a,b) and the Cassini measurements at Saturn (Horányi, 2000; Kempf *et al.*, 2005) of small dust grains streaming away from these planets.

11.3.2 Stream particles at Jupiter and Saturn

Small grains ejected from planetary magnetospheres represent a surprising example of the possible complexity of the dynamics of charged dust particles. Electrons and ions with their large charge-to-mass ratios follow adiabatic motion and remain confined by magnetic fields. Large dust particles with vanishingly small charge-to-mass ratios follow Kepler orbits and remain bound by gravity. When the magnitude of the electromagnetic and gravitational forces are comparable the outcome can be quite surprising.

Jupiter was first recognized as a source of dust particles during Ulysses' encounter with the planet in 1992, as high speed intermittent streams of small grains were discovered (Grün *et al.*, 1993). The first estimates put the mass of the stream particles in the range of $1.6 \times 10^{-16} < m < 1.1 \times 10^{-14}$ g and their velocity in the range of $20 < v < 56$ km/s. Assuming an average density $\rho \simeq 1 \text{ g cm}^{-3}$, the radii of these grains were estimated in the range of $0.03 < a < 0.1$ μm. However, these estimates remained uncertain, because the detector was not calibrated in this size and velocity range. Similar fluxes were seen with the identical dust detector onboard the Galileo spacecraft as it first approached Jupiter in 1995 (Grün *et al.*,

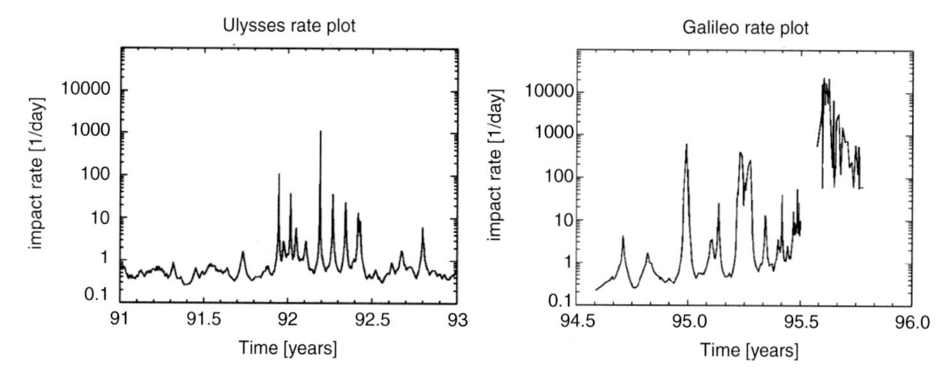

Fig. 11.5 Dust impact rates measured by *Ulysses* (left) and by *Galileo* (right) at Jupiter. (From Horányi, 1996.)

1997). The discovery observation from Ulysses and the data from Galileo's initial approach to Jupiter are shown in Fig. 11.5.

High-velocity streams of nanometer-sized dust particles originating from the inner Saturnian system were observed during Cassini's approach to Saturn (Kempf *et al.*, 2005, Hsu *et al.*, 2010). The Cosmic Dust Analyzer (CDA) often registered impact signals that were most likely caused by particles moving faster than 70 km/s, the fastest impact speed for which CDA was calibrated at the Heidelberg dust accelerator facility. Using Eq. (11.13) to calculate the work done by the co-rotating electric field and the expected charges of the grains, a simple order-of-magnitude relationship can be derived to show that the expected escape velocity is inversely proportional to the size of the particles (Horányi and Juhász, 2000):

$$v_{escape} = \frac{3}{a\ [\mu\text{m}]}\ \text{km/s} \quad \text{for Jupiter, and} \tag{11.14}$$

$$v_{escape} = \frac{0.6}{a\ [\mu\text{m}]}\ \text{km/s} \quad \text{for Saturn.} \tag{11.15}$$

These expressions assume that Jovian dust particles start at Io, and that Saturn's grains are accelerated outward from Dione's distance, because that is where the grains' charge becomes positive in Saturn's magnetosphere. The faster speeds at Jupiter result mainly from its stronger magnetic field. Beyond Saturn's magnetosphere, the dynamics of the stream particles is governed by interactions with the interplanetary magnetic field convected by the solar wind.

11.3.3 Spokes in Saturn's rings

Spokes are intermittently appearing, approximately radial markings on Saturn's B ring (Fig. 11.6), consisting of small charged dust particles lofted from their parent

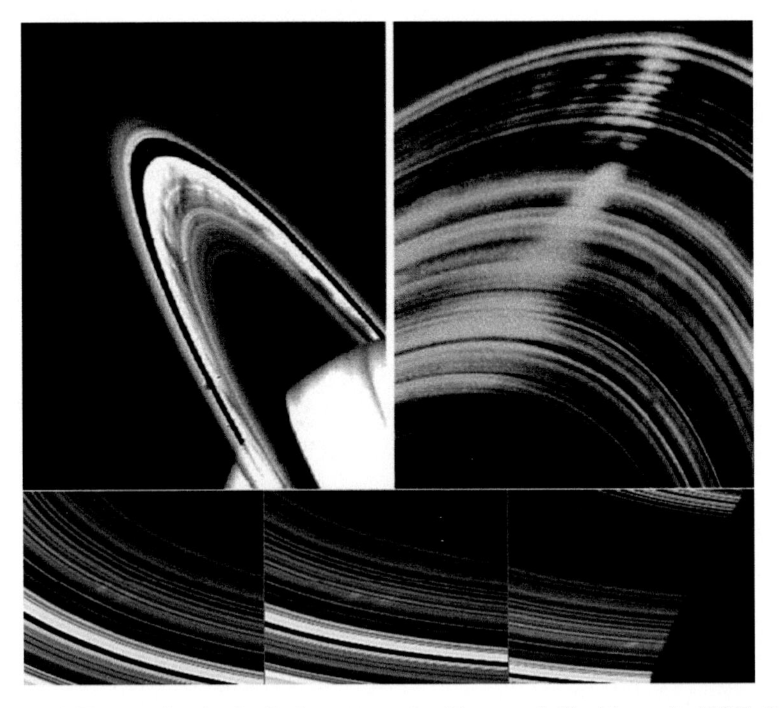

Fig. 11.6 Top: spokes in the B ring as seen by Voyager 2 (Smith *et al.*, 1982). The left-hand image was captured in back-scattered light before closest encounter, with the spokes appearing as dark radial features across the ring's center. The right-hand image was taken in forward-scattered light after the spacecraft crossed the ring plane, and was looking back towards the Sun; the spokes now occur as bright markings. Typical dimensions of these spokes are 10 000 km in length and 2000 km in width. The nature of the changing brightness indicates that spokes consist of small submicron-sized grains, i.e., that are comparable in size to the wavelength of visible light. At the time these images were taken, the rings' opening angle to the Sun was $B_0 \simeq 8°$. Bottom: the initial spoke observations taken by Cassini on September 5, 2005 ($B_0 \simeq 20°$), over a span of 27 min. These faint and narrow spokes were seen from the un-illuminated side of the B ring. These first spokes are $\simeq$ 3500 km long and 100 km wide, smaller than the average spokes seen by Voyager. These images were taken with a resolution of 17 km per pixel at a phase angle of 145° when Cassini was 135° above the unlit side of the rings as the spokes were about to enter Saturn's shadow (Mitchell *et al.*, 2006).

ring bodies owing to electrostatic repulsion. While they were first recognized in images taken by Voyagers 1 and 2 (Smith *et al.*, 1981, 1982), spokes were possibly noticed even earlier in ground-based telescopic observations (Robinson, 1980). These perplexing features have attracted great attention; following their appearance in Voyager 1 images, Voyager 2, during its approach to Saturn, dedicated sequences to spoke observations, providing an invaluable dataset. The spokes' characteristics have been derived with an increasing level of sophistication since their acquisition in these images. The desire to provide an explanation for the spokes has played a large role in the emergence of the research field of Dusty Plasmas.

The key physical characteristics of spokes based on the Voyager (Porco and Danielson, 1982; Grun *et al.*, 1983; Grün *et al.*, 1992; Eplee and Smith, 1984; Doyle and Grun, 1990) and Cassini (Mitchell *et al.*, 2013) data can be summarized as follows: These features are generally most common near the morning terminator, (i.e., the ring extremity (or ansa) on the dawn side), and seem to form primarily in that region. Spokes develop on a time scale of minutes, and can become more intense over a period of a few hours. Appearing at radial distances that are near to, or straddle, synchronous orbit, they move around the ring nearly co-rotating with the planet. During increases in spoke intensity, these features extend forwards and backwards from the synchronously moving longitude inside and outside the co-rotation distance, respectively, while their central regions intensify; this indicates that the spoke material follows Keplerian trajectories, in broad terms at least. Once spokes no longer intensify, they fade while traveling around the dayside of the rings. Newly formed spokes in the Voyager data often coincided with the positions of older spokes that seem to have survived an entire revolution around Saturn. The periodicity of spoke formation is close to the observed period of a strong low-frequency radio emission, called the Saturn kilometric radiation (SKR) measured by the Voyagers (Porco and Danielson, 1982), suggesting a formation trigger that is linked to Saturn's magnetic field.

Numerous formation theories were proposed to explain the spokes' existence, but none could be definitively tested without further observations. The Hubble Space Telescope (HST) monitored spoke activity from 1995 until 1998, when HST no longer detected spokes due to Saturn's equinox (when the solar illumination hit the rings edge-on). McGhee *et al.* (2005) proposed that spokes, while always present, can be detectable only when the observer lay close to the ring plane. It was therefore anticipated that Cassini would detect spokes on its 2004 arrival at Saturn and that its observations would finally decide which, if any, of the competing theories were correct (Horányi *et al.*, 2004). However, contrary to predictions, Cassini did not observe spokes, even when close to the ring plane, until September 2005 (Fig. 11.6). The variability in spoke occurrence in HST data was therefore not an observational effect: spokes are indeed a seasonal phenomenon, and their formation can be suspended for extended periods. This seasonal variation of spoke activity may be a consequence of the variable plasma density near the ring. The plasma density is a function of the solar elevation angle $B_{\odot}$, measured from the ring plane, because it is generated mainly from the rings by photoelectron production and by photo-sputtering of neutrals that are subsequently ionized (Mitchell *et al.*, 2006; Farrell *et al.*, 2006).

Spokes are comprised of dust particles in a narrow size distribution centered at about a radius $s \simeq 0.6$ μm (Doyle and Grun, 1990). It is generally believed that spoke formation involves charging and thus electric fields acting on these small grains, but this process requires, as we show below, a much higher plasma density than is commonly expected near the rings (Hill and Mendis, 1982; Goertz and Morfill, 1983). When formed, spokes initially cover an approximately radial strip with an area of $A \simeq 10^3 \times 10^4$ km^2, with a characteristic optical depth of $\tau \simeq 0.01$. The total number of elevated grains can be estimated to be on the order of $N_d \simeq A\tau/(\pi s^2) \simeq 10^{23}$. If the grains are released approximately at the same time and carry just a single electron when released from their parent bodies, the formation of the spoke cloud requires a minimum surface charge density (measured in units of electron charges e) $\sigma_e^* = N_d/A \simeq 10^6$ cm^{-2}, orders of magnitude higher than the charge density, σ_0, expected from the nominal plasma conditions in the B ring.

The nominal plasma environment near the optically thick B ring is set by the competing electron and ion fluxes to and from the ring due to photoelectron production from the ring (as well as the ionosphere) and the photoionization of the rings' neutral atmosphere that is maintained by photo-sputtering. All of these are expected to show a seasonal modulation with the ring's opening angle with respect to the Sun, $B_{\odot}$. The characteristic energy for photoelectrons is $T_e \simeq 2$ eV, and the plasma density is expected to be $n \sim 0.1$–1 cm^3 (Waite *et al.*, 2005). The characteristic plasma shielding distance is $\lambda_D = 740 \times (T_e/n)^{1/2} \simeq 1$–$3 \times 10^3$ cm, larger than the average distance between the cm–m sized objects in the B ring, which has a comparable vertical thickness, $h \simeq 10$ m. Hence, it is reasonable to treat the B ring as a simple sheet of material (Goertz and Morfill, 1983). The nominal surface potential, including its possible seasonal variations, is expected to be in the range of $-5 < \phi_R < 5$ V. The surface charge density can be estimated from Gauss's law,

$$\sigma_0 \simeq \frac{\phi_R}{4\pi\lambda_D} \simeq 2.5\phi_R \left(\frac{m}{T_e}\right)^{1/2} < 1\text{–}3 \times 10^3 \text{cm}^{-2}. \tag{11.16}$$

Because $\sigma_0 \ll \sigma_e^*$, the formation of a spoke requires higher than normal plasma densities.

Several spoke formation theories have been put forward (McGhee *et al.*, 2005). Of these, the proposed spoke formation trigger theories that arguably have been most widely accepted are those of meteoroid impacts onto the rings (Goertz and Morfill, 1983) and field-aligned electron beams originating from the auroral regions of Saturn (Hill and Mendis, 1982). Both of these could transiently increase the plasma density above a critical threshold, and trigger the formation of spokes.

A meteoroid impact-produced plasma cloud was shown to expand, cool and recombine as it rapidly propagates in the radial direction, possibly explaining many of the observed spoke characteristics (Goertz and Morfill, 1983). However, the estimated propagation speed of such a cloud remains difficult to estimate (Farmer and Goldreich, 2005; Morfill and Thomas, 2005). The electron-beam mechanism has been suggested to loft small particles instantaneously along the entire radial extent of a spoke (Hill and Mendis, 1982). Other spoke formation ideas include dusty plasma waves (Tagger *et al.*, 1991; Yaroshenko *et al.*, 2008) and impact-induced avalanches of small charged dust particles (Hamilton, 2006).

During the first four years (2004–2008) of Cassini observations spokes remained a high priority. For most of this interval, spokes were much fainter and less frequent than those seen by the Voyagers. By late 2008, $B_\odot$ had reached values similar to those during the Voyager encounters, and spoke activity was indeed approaching – if not matching – the activity observed by the Voyagers (Mitchell *et al.*, 2008). Based on the increase in spokes at the time of writing, it is anticipated that Cassini should answer key questions regarding the nature of these perplexing ring features by the end of its mission in 2017.

11.4 Waves in dusty plasmas: possible role in comets

The presence of charged dust can alter collective behavior of the plasma as manifested by altered dispersion relationships of the customary plasma waves, and the emergence of new plasma waves and instabilities (Shukla and Mamun, 2002; Bliokh *et al.*, 1995; Tsytovich *et al.*, 2010). In the following we describe dusty plasma wave modes that possibly have relevance to comets.

The so-called "Dust Ion Acoustic" (DIA) dusty plasma wave was predicted by Shukla and Silin (1992) and it was first observed in the laboratory by Barkan *et al.* (1996). Here the charged dust does not participate in the wave dynamics; it simply modifies the wave mode via the quasi-neutrality condition in the plasma. If we consider the simple case of dust particles (all of the same size) in a singly ionized two-component (electron and ion) thermal plasma, where the only charging currents to the grains are electron and ion collection, the grains become negatively charged (as discussed in Sect. 11.2). If each dust grain carries an excess of Z electrons on its surface, the quasi-neutrality condition in the undisturbed plasma is:

$$n_{i0} = n_{e0} + Zn_{d0}, \tag{11.17}$$

where $n_{\alpha 0}$ refers to the ion, electron, and dust number densities in the undisturbed plasma. Note that this leads to a depletion in the electron density relative to the ion density. This increases the phase velocity of a DIA wave above the usual ion acoustic wave in a dust-free plasma. With the approximation of negligible electron inertia and immobile dust grains, the dispersion relation for the DIA wave is given by (Shukla and Silin, 1992)

$$\omega^2 = \frac{\delta k^2 c_s^2}{1 + k^2 \lambda_{De}^2}, \tag{11.18}$$

where $c_s = (k_B T_e / m_i)^{1/2}$ is the usual ion acoustic speed, λ_{De} is electron Debye shielding length, and $\delta = (n_{i0}/n_{e0})$. Hence, in a dust-free charge-neutral plasma $\delta = 1$, and the DIA dispersion relationship is simply that of an ion-acoustic wave (Chen, 1974). In the long-wavelength regime ($k\lambda_{De} \ll 1$), Eq. (11.18) reduces to

$$\omega = k\delta^{1/2} c_s. \tag{11.19}$$

Because $\delta^{1/2}$ could be $\gg 1$ if the electron depletion due to dust is sufficiently large, the phase velocity (ω/k) of this wave could be $\gg c_s$, and, consequently, Landau damping could become negligible. This is in contrast to a pure electron–ion plasma, where $T_e \gg T_i$ is required for the acoustic wave to remain undamped. The implication of this in various cosmic dusty plasma environments was discussed by Mendis and Rosenberg (1994).

Several authors have considered the excitation of the dust ion acoustic instability in the Saturnian magnetosphere, due to the relative motion of the co-rotating plasma and the charged dust particles moving with speeds intermediate between co-rotation and the local Kepler velocity (Rosenberg, 1993; Winske *et al.*, 1995). Both the linear and the non-linear properties of this instability have been investigated, including its saturation due to the trapping of plasma ions. It has been identified as a possible explanation for the temperature increase of O^+ ions in the region 4–8 R_S from about 40 to 200 eV, first observed by the Voyager mission at Saturn (Richardson and Sittler, 1990). There is also relative motion between the charged dust and the plasma throughout the cometary dusty plasma environment, in both the cometary head and the tail. Perhaps, a good candidate location is the region behind the nucleus where the already high-speed ions forming the ion tail diverge from the flow of dust into the dust tail of a comet.

The second dusty plasma wave mode to consider is the so-called "Dust Acoustic" (DA) mode, whose existence was predicted by Rao *et al.* (1990), and confirmed in the laboratory by Barkan *et al.* (1995). In this case, contrary to the DIA mode, the charged dust grains also participate in the wave dynamics, in addition to modifying the usual quasi-neutrality condition. The laboratory experiment, shown in Fig. 11.7, used a potassium plasma with $k_B T_e \simeq 3$ eV, $k_B T_i \simeq 0.2$ eV, micron-sized dust grains of equal mass $m \simeq 10^{-15}$ kg, a typical dust charge of $Z_d \simeq 2 \times 10^3$e, and had $n_d/n_i \simeq 5 \times 10^{-4}$. In this setup a slowly propagating ($v_\varphi \simeq 9$ cm/s), long-wavelength ($\lambda \simeq 0.6$ cm), low-frequency ($\omega \simeq 15$ Hz), longitudinal wave of significant amplitude ($A = |\Delta n_d/n_{d0}| \simeq 1$) was observed by laser scattering. With these conditions (i.e., $k_B T_e \gg kT_i$), in the long-wavelength regime ($k\lambda_{De} \ll 1$) the dispersion relation of the DA mode reduces to (Rao *et al.*, 1990)

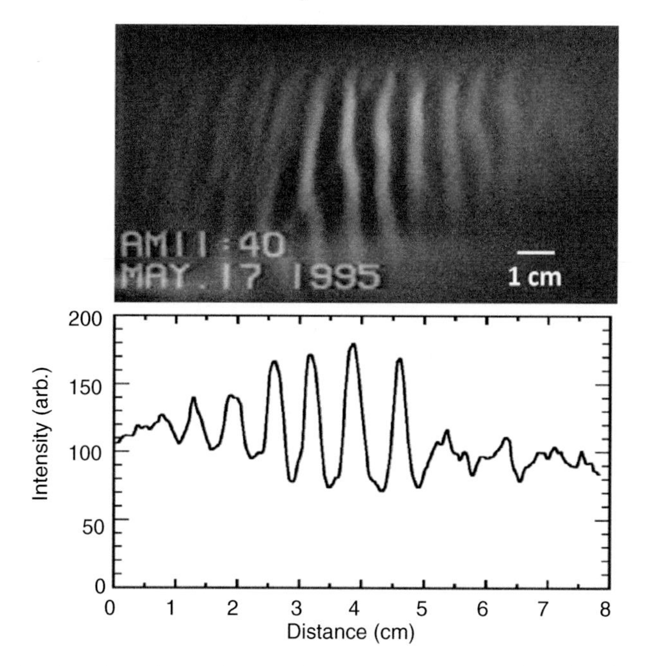

Fig. 11.7 Single video frame image of a Dust Acoustic Wave (DAW) observed in the laboratory. The bright vertical bands correspond to the wave crests (dust compressions). The wavelength can be seen to be $\lambda \simeq 0.6$ cm. By following the wave fronts from frame to frame of the video, the wave speed was measured $v_\varphi \simeq 9$ cm/s, so that the frequency $f = v_\varphi/\lambda \simeq 15$ Hz (Thompson *et al.*, 1999).

$$\omega^2 = Z_d^2 k^2 \left(\frac{k_B T_i}{m_d} \frac{n_{d0}}{n_{i0}} \right). \tag{11.20}$$

Substituting the experimental values in Eq. (11.20) $v_\varphi = \omega/k \simeq 10$ cm/s, results in a good agreement with the observed value.

The possibility of observing such waves in the dusty plasma environment of comets has not been discussed in the literature, and owing to their small spatial scales, could not have been possibly observed during the fly-by missions to date. The high-resolution cameras onboard Rosetta might be able to capture the propagating small-scale structures of possible DA waves. Perhaps, new dusty plasma phenomena will be found by Rosetta that could lead to the inference of the excitation of DIA and DA waves and instabilities (Mendis and Horányi, 2013a,b).

11.5 Summary and conclusions

In most solar-system plasma environments the dynamics of small charged particles is strongly influenced, if not dominated, by electromagnetic forces acting simultaneously with gravity, drag, and radiation pressure. Dust particles traversing various regimes adjust their electrostatic charges as dictated by the changing plasma conditions, and in fact they act as active electrostatic probes, continuously adjusting their surface potential towards the local equilibrium value. The fields and particle environment can uniquely shape the size and the spatial distribution of the dust grains. Dust particles are unusual charge carriers, as compared to electron or ions, as their charge-to-mass ratio is not fixed. Micron-sized dust particles can have thousands of missing or extra electrons. Because of their huge mass, they enable physical processes on very different spatial and temporal scales compared to plasmas composed of electrons and ions only. Dusty plasma waves, for example, can show wavelengths and frequencies that enable them to be visually observed and easily recognized.

Studies of the motion of charged dust particles connect a number of observations that are often thought to be unrelated. Space missions, in general, are designed to make simultaneous in-situ and remote observations with some combination of the following experiments.

- New sophisticated dust detectors (Grün *et al.*, 2003, Fiege *et al.*, 2014) can provide in-situ measurements of the mass, velocity vector, charge, chemical, and possibly even the isotopic composition of the dust grains.
- Plasma instruments provide the composition, density, and energy distributions of the electrons and ions. These data are used to calculate the charging currents of the grains and to learn whether grains are in charge equilibrium or will have significant charge variations due to fluctuations and/or gradients in composition and/or density and/or temperature of the plasma. Apparent lack of charge balance between electrons and ions can result from the presence of sufficiently large dust densities (Ye *et al.*, 2014b; Morooka *et al.*, 2011).
- The plasma-wave and radio-science experiments onboard Voyager detected broad-band noise passing through the ring planes at Saturn, Uranus, and Neptune. This noise is believed to be caused by small dust grains bombarding the body of the spacecraft. The few kilometer per second relative velocity between the spacecraft and the dust grains is sufficient to fully vaporize the impacting grains and in part ionize the produced gas. The expanding plasma cloud causes the detected noise. This phenomenon led to the recognition that all giant planets are surrounded by vast tenuous sheets of small grains that could not have been discovered via imaging alone (Gurnett *et al.*, 1983; Ye *et al.*, 2014a).
- A magnetometer provides in-situ field measurements that can be used to describe the global structure of magnetic fields (Behannon *et al.*, 1977; Dougherty *et al.*, 2004). These data are essential to calculate the trajectories of charged dust particles.

- An imaging experiment can supply images taken through filters at various phase angles to show the spatial and size distribution of the dust particles. Ultimately, the spatial distributions of the fine dust can be independently modeled based on the transport processes at work and compared with the images (Juhász *et al.*, 2007).

It is a unique and powerful consistency test if our models describing dust transport, based on particles and field data, match the observations of the dust detectors and the images. However, without in-situ data on particles and fields, images showing the spatial distribution of small dust grains can be used to infer the plasma conditions. In addition to planetary magnetospheres and comets included in this tutorial, there are many other examples where dust and plasma interactions might be important. These include the propagation of interstellar dust particles through the heliosphere, noctilucent clouds in the Earth's mesosphere, or the surfaces of all airless planetary bodies.

12

Energetic-particle environments in the solar system

NORBERT KRUPP

Energetic particles in our heliosphere are key in the understanding of the evolution and the current status of our solar system. The energy distributions of ions and electrons are widely used to understand acceleration phenomena in the heliosphere and in the vicinity of planets and moons. Knowing the energy spectra of those particles can tremendously help in the characterization of plasma sources and sinks of many different environments in the Milky Way and beyond. The knowledge of those energetic particle distributions in fact is essential in many fundamental plasma physics problems. They are a very nice tool to investigate acceleration mechanisms, geochemistry and solar-system evolution, atmospheric composition and solar-system evolution, and last but not least energetic particles are important in studying the configuration and dynamics of planetary magnetospheres and the interaction of magnetospheric plasma with their moons, rings, neutral gas, or dust clouds.

Energetic particles in the energy range between 1 eV to 10^{20} eV can be found everywhere in our solar system as sketched in Fig. 12.1. Their sources can be either outside our solar system from galactic or extra-galactic interstellar space or inside our solar system from the Sun or the planets or created in various acceleration processes in interplanetary space or inside planetary magnetospheres, i.e., at interplanetary shocks, corotating interaction regions, planetary bow shocks or at the termination shock of the heliosphere. Types of energetic particles range from electrons to charged atoms and molecules to neutral atoms and molecules as well as dust particles. Figure 12.2 shows the particle intensity versus energy spectra of various types of energetic particles (left) and for cosmic rays (right).

This chapter focuses on the description of charged energetic particle populations from the Sun and particle distributions inside planetary magnetospheres (including galactic cosmic rays (GCRs) as a potential source of charged particles in planetary radiation belts). Energetic particles from acceleration processes at shocks are reviewed in Ch. 7 of Vol. II.

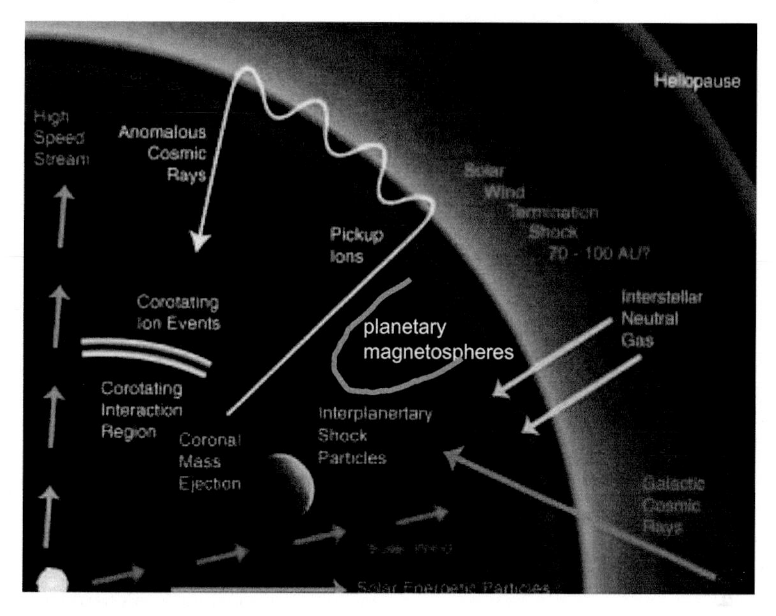

Fig. 12.1 Sources of energetic particles in the heliosphere. (Adapted from www.issibern.ch/teams/Suprat.)

In order to set the scene for later sections and without going into much detail we introduce the motion of charged particles in a magnetic field. For further details and equation derivations the reader is referred to books like Roederer (1970) or Walt (1994). A excellent text used for this chapter is Kallenbach *et al.* (2006).

Starting from the fundamental equation of charged particle motion the Lorentz force $\vec{F}$ is given by

$$\vec{F} = \frac{d\vec{p}}{dt} = q(\vec{v} \times \vec{B} + \vec{E}), \tag{12.1}$$

with $\vec{v}$, $\vec{p}$, $\vec{B}$, $\vec{E}$ being the particle velocity, particle momentum, magnetic field, and electric field vectors, and q being the elementary charge. In a uniform magnetic field the momentum component parallel to the magnetic field is constant and its derivative is equal to zero. Centrifugal force balances the Lorentz force and the resulting motion of the particle is a circle when projected into a plane perpendicular to the magnetic field direction with the cyclotron or gyroradius r_G given by

$$r_G = \frac{mv_\perp}{|q|B} = \frac{v_\perp}{|\omega_G|}, \tag{12.2}$$

where ω_G is the gyrofrequency. The gyromotion direction is given by the sign of the charge of the particle. In addition to the motion perpendicular to the magnetic field the particle can also move parallel to it. The pitch angle $\alpha = \tan^{-1}(\frac{v_\perp}{v_\parallel})$ is the

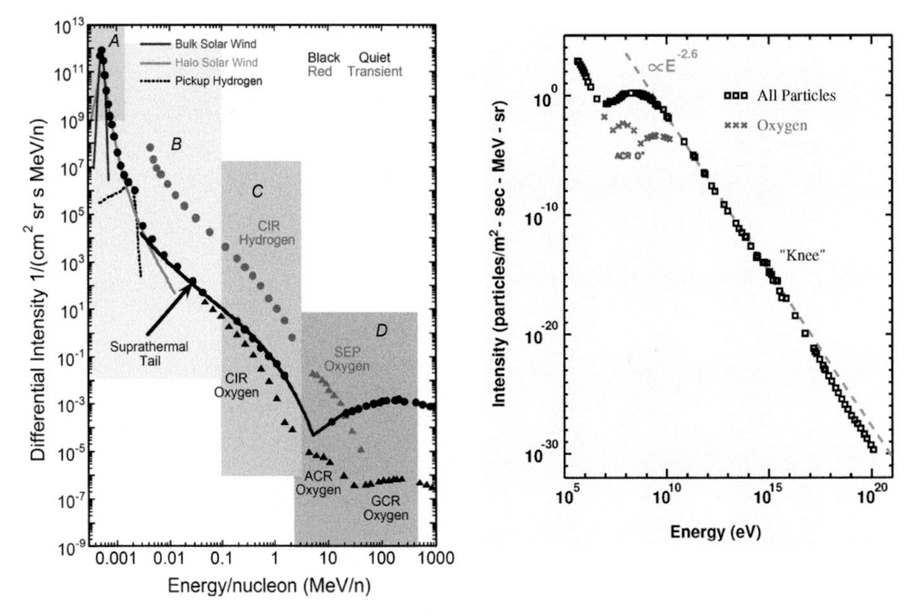

Fig. 12.2 Energy spectra of energetic particles in the heliosphere (left) and for cosmic rays (right). The curves illustrate the energy spectra during quiet time and disturbed solar wind conditions. The dots and triangles represent the suprathermal part of the spectrum and the particles accelerated at corotating interaction regions (CIRs), galactic cosmic rays (GCRs), and the anomalous cosmic rays (ACRs) together with Solar Energetic Particles (SEPs). The right figure shows the high-energy part of the GCR energy spectrum to TeV energies. Note the characteristic peak at about 10 MeV and the $E^{-2.6}$power-law dependence for energies above the peak. (Figure adapted from Ch. 3 in Vol. II.)

angle between the particle velocity vector and the direction of the magnetic field. The gyration creates a current I, a magnetic flux ϕ, and a magnetic moment μ.

The magnetic moment is defined as

$$\vec{\mu_G} = \frac{mv_\perp^2}{2B^2}\vec{B}|\mu| = \frac{W_\perp}{B}, \tag{12.3}$$

where $W_\perp$ is the particle energy perpendicular to the magnetic field. The first adiabatic invariant of a particle is defined as $\mu =$ constant, which means that a charged particle moving in a changing magnetic field conserves its magnetic moment and, as a consequence, changes its energy component perpendicular to the magnetic field.

The second adiabatic invariant J of a charged particle describes the periodic parallel motion or bounce motion back and forth along the magnetic field:

$$J = \oint mv_\parallel ds = \frac{v_\perp d\alpha}{\omega_G}. \tag{12.4}$$

From μ = constant it follows that a charged particle with a pitch angle α_1 in a magnetic field B_1 moves up and down the field line until it reaches a point (mirror point) where the magnetic field is $B_m = B_1/\sin\alpha_1^2$ and is reflected by the related mirror force $-\mu\nabla B$.

Finally the third adiabatic invariant for a charged particle describes approximately the magnetic flux conservation enclosed by the full drift motion around a planet for example.

$$\phi = \frac{2\pi m}{q^2} M = \text{constant}, \tag{12.5}$$

with M being the magnetic moment of the enclosed magnetic field. For details see Northrop (1963).

The presence of a uniform magnetic and electric field perpendicular to each other results in a drift of particles perpendicular to $\vec{B}$ and $\vec{E}$:

$$\vec{V}_E = \frac{\vec{E} \times \vec{B}}{B^2}. \tag{12.6}$$

In the presence of an inhomogeneous magnetic field the gradient in B and the curvature of the field lines cause deviations which can be described by the gradient drift

$$\vec{V}_G = \frac{mv_\perp^2}{2qB^3}(\vec{B} \times \nabla B) \tag{12.7}$$

and the curvature drift

$$\vec{V}_C = \frac{mv_\parallel^2}{qB^3}(\vec{B} \times \nabla B). \tag{12.8}$$

The total drift velocity of a particle is a combination of all three:

$$\vec{V}_\perp = \vec{V}_E + \vec{V}_G + \vec{V}_C = \frac{\vec{E} \times \vec{B}}{B^2} + \frac{m}{qB^3}\left(\frac{v_\perp^2}{2} + v_\parallel^2\right)\vec{B} \times \nabla B. \tag{12.9}$$

As an example of drift values the bounce-averaged drift velocity in Saturn's magnetosphere is shown together with equatorial gyroradius and bounce period as a function of dipole L in Fig. 12.3.

The detection of energetic charged particles goes back to the pioneers in space instrumentation like James van Allen in the late 1950s, when the first rocket-based "Geiger counters" measured huge intensity increases and dropouts at several thousand kilometers altitude above the Earth. The historic discovery of the Earth's radiation belts was made by measuring the charged particle environment of the Earth. In the beginning, relatively simple counters were used, followed by more advanced semiconductor detector devices and electrostatic analyzers. A real breakthrough was the possibility to distinguish between different energies, different

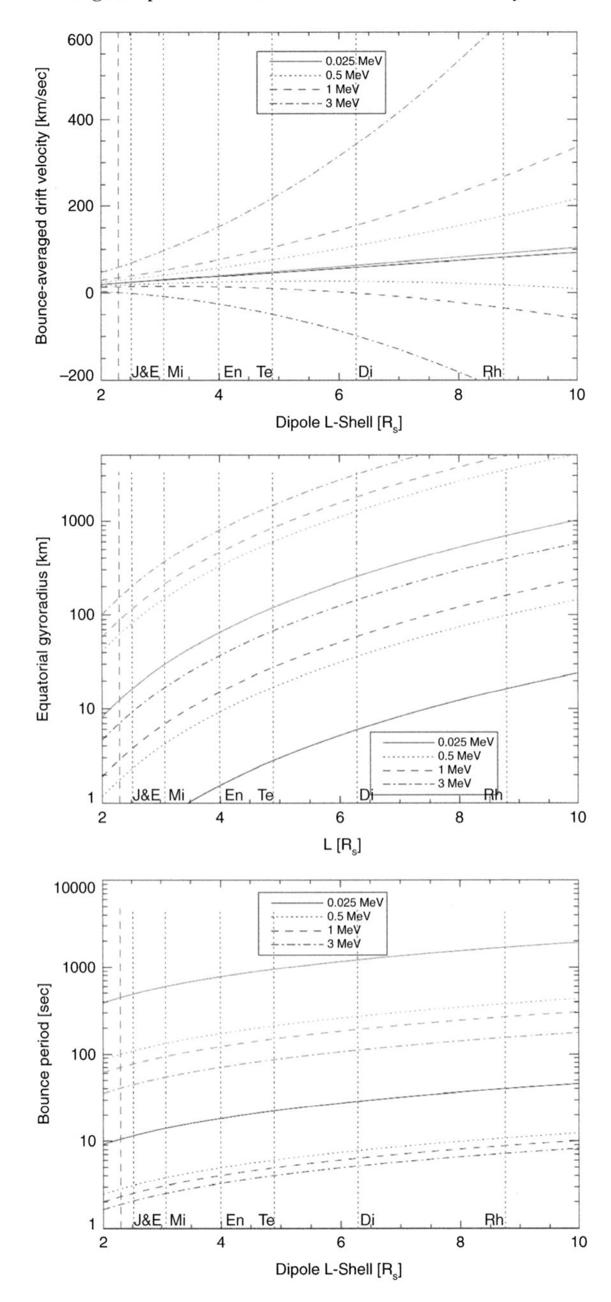

Fig. 12.3 Top: bounce-averaged drift velocities at Saturn as a function of energy and L-shell (lower four curves for electrons and upper four curves for ions). All values are for equatorial particles (90 degree pitch angle) assuming rigid corotation in an inertial, Saturn-centered coordinate system. Middle: same plot for the gyroradius of equatorial electrons and ions. Bottom: bounce period of electrons and ions with a pitch angle of 45 degrees. In all panels, the location of the various icy moon L-shells are indicated with dotted lines. The dashed line indicates the L-shell of the main ring's outer edge. (From Roussos, 2008.)

species, and different directions using combinations of so called time-of-flight systems and post-acceleration subunits. Today's technology makes it possible to "visualize" the invisible particle environment. Particle detectors with imaging capabilities measure charged and even neutral particles at energies between eV and GeV. A nicely summarized description of particle detector types can be found in Gloeckler (2010).

12.1 Energetic particles from the Sun

The first detection of solar energetic particles goes back to 1946 when Scott Forbush related the source of high-energy-particle intensity increases to the Sun and not to cosmic rays (Forbush, 1946). Since then solar energetic particles have been measured in a wide range of energies between keV and GeV. They are observed on time scales between several hours up to several days. Their intensities increase above average by several orders of magnitude. Details are found in Kallenrode (2003), Reames *et al.* (2013), Reames (2013), Kahler (2013a,b), and Kahler and Vourlidas (2013). It is currently accepted that the occurrence of energetic particles is directly related to either Solar Flares or to Coronal Mass Ejections (CMEs). The ultimate energy source of those particles is the magnetic energy of the Sun. As nicely illustrated in Fig. 12.4 from Kallenrode (2003), the idea is that an instability triggers a flare or a CME with a number of associated acceleration mechanisms

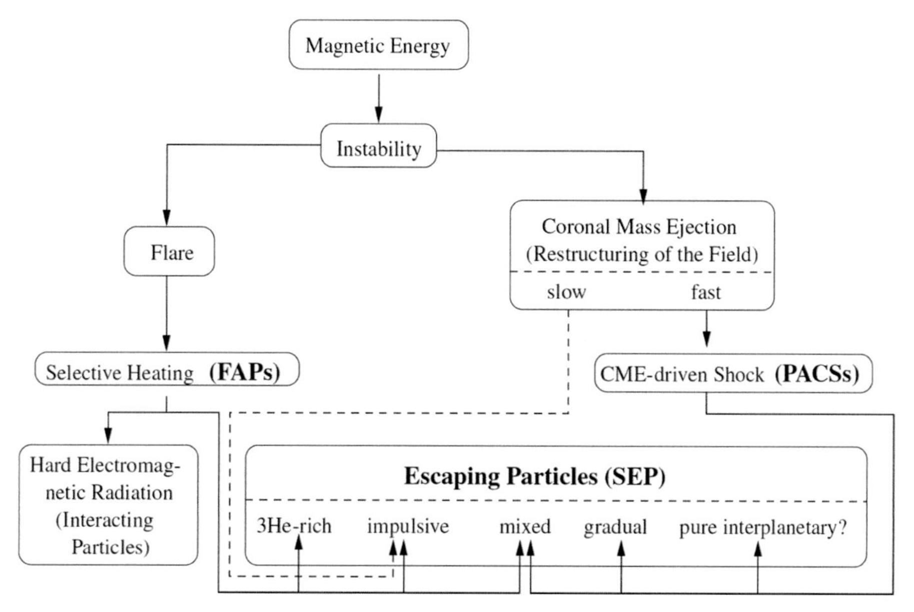

Fig. 12.4 Classification of solar energetic-particle events. (From Kallenrode, 2003.)

and/or compositional variations of Sun material, i.e., enrichment of ^{3}He relative to ^{4}He, etc. For a full description the reader is referred to the original literature starting from Kallenrode (2003) and earlier references therein.

The classification of solar energetic particle events into "^{3}He-rich", "impulsive", and "gradual" is based on a number of parameters such as relative abundance ratios of ^{3}He/^{4}He, H/He, Fe/O to name a few (after Reames, 1995). Solar energetic particle events are also in good correlation to the observation of type II or type III radio bursts on the Sun (Kane *et al.*, 1974).

12.1.1 Energetic particles from Coronal Mass Ejections (CMEs)

Coronal Mass Ejections are events originating at the Sun where particles are trapped in closed magnetic loop structures. Their footpoints are still connected to the Sun's surface and the entire structure is moving outwards. CMEs are well-known features, since their first discovery in the 1970s with coronagraph instruments onboard spacecraft. Data of the LASCO coronagraph, an instrument onboard the ESA/NASA spacecraft SOHO (launched in 1996), could show that CME characteristics vary from event to event. The CMEs move at speeds between 50 km/s and more than 2000 km/s with large angular widths which could cover the entire sky. During CMEs billions of tons of material are ejected.

Figure 12.5 shows the spatial structure of a CME as observed by the LASCO instrument onboard SOHO. It is clearly visible that CMEs consists of three different clearly distinguished spatial regions: a leading edge, a cavity, and a core.

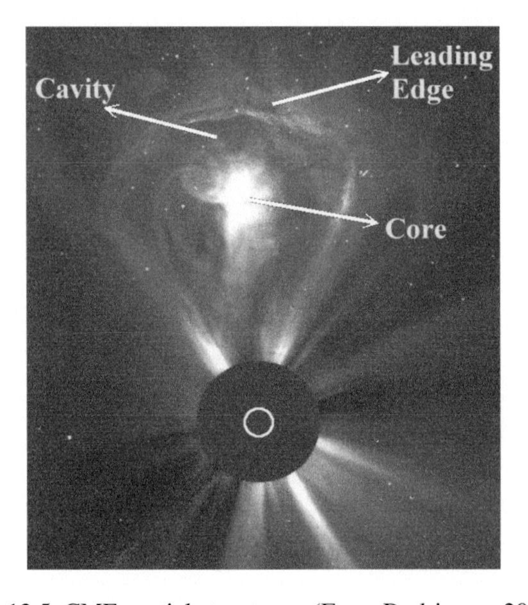

Fig. 12.5 CME spatial structures. (From Rodriguez, 2005.)

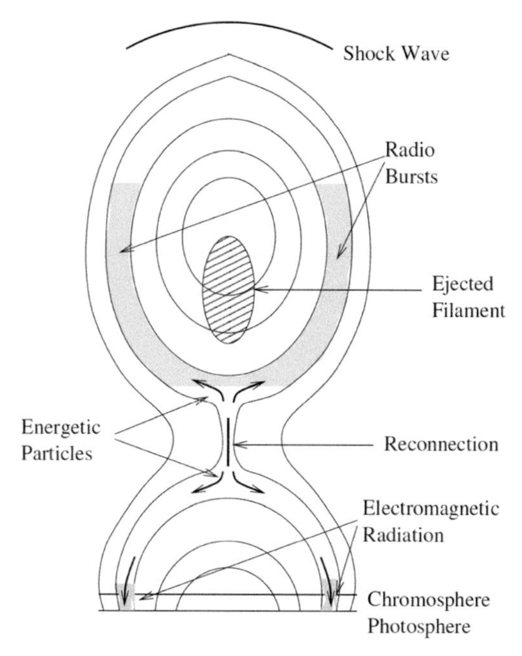

Fig. 12.6 Concepts of particle acceleration in a solar gradual event. (Adapted from Kallenrode, 2003.)

Gradual events are those particles observed in correlation to fast shock waves driven by CMEs in the corona of the Sun or in interplanetary space. The acceleration process itself in those events could be either shock drift acceleration, diffusive shock acceleration (Fermi 1 acceleration) or stochastic acceleration (Fermi 2 acceleration). For further discussion of these processes, the reader is referred to Ch. 7 in Vol. II.

The principal scenario of particle acceleration in those gradual events is sketched in Fig. 12.6 (Kallenrode, 2003). Energetic particles are accelerated close to the magnetic reconnection region where they move bi-directionally along the field lines. As pointed out by Reames *et al.* (2013) the acceleration on the shock wave occurs on open magnetic field lines in contrast to the closed magnetic loop picture.

The intensities of those gradual events observed at different solar longitudes relative to the nose of the shock front are shown in Fig. 12.7. The intensity–time profiles are quite different from each other and depend on the location of the observer relative to the nose of the shock front.

Near the nose of shock front they display a four-stage intensity–time profile as shown in Fig. 12.8 from Reames (2013): onset (velocity-dispersed with highest velocity ions arriving first), plateau (generated Alfvén waves scatter outward streaming particles up to a "streaming limit"), shock peak (enhancement typical for shock acceleration), and reservoir (slow time decay of intensities). Only the large events create high-energy particles.

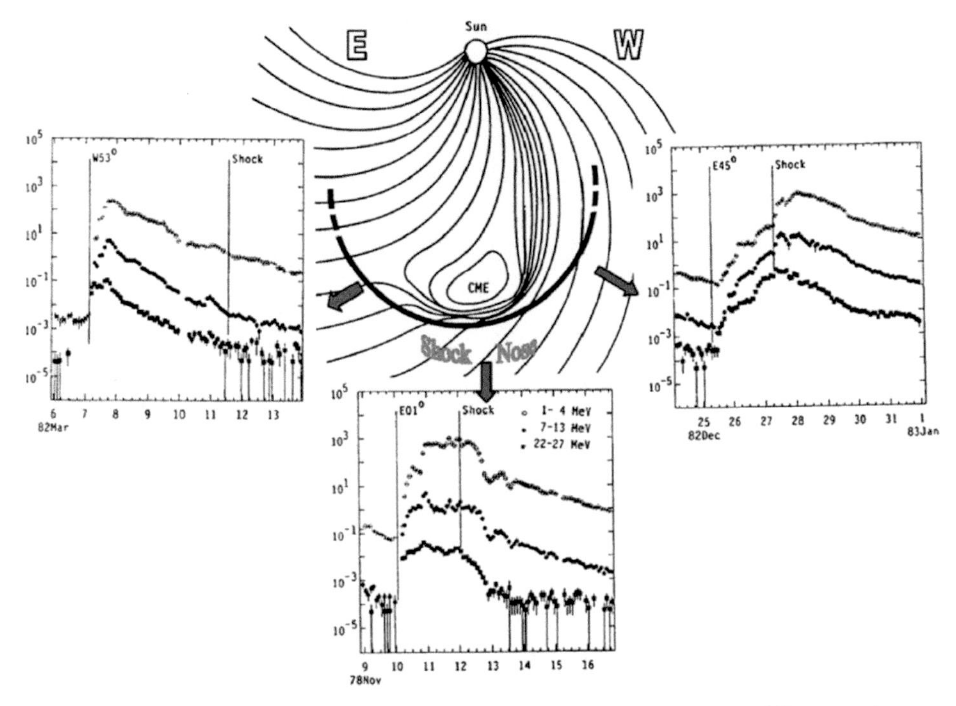

Fig. 12.7 Intensity–time plots of particle fluxes ejected from three different solar longitudes with respect to the nose of the shock front. (From Reames, 2013.)

12.1.2 Flares

Impulsive events originate in impulsive jets or flares on the Sun and are observed in correlation with type III radio bursts from outward streaming electrons with energies of about 40 keV (Reames, 2013). Resonant stochastic acceleration caused by magnetic reconnection additionally can also enhance the ^{3}He/^{4}He- and Fe/O-ratio. Figure 12.9 illustrates the scenario.

12.2 Energetic particles in planetary magnetospheres

Planetary magnetospheres represent huge and unique plasma laboratories with sources and sinks of neutral and charged particles in partly rapidly rotating magnetic fields. As described in the introduction section of this chapter the particles in dipole-like magnetic fields gyrate, bounce, and drift in those environments. The adiabatic motion of particles in a rotating magnetosphere is mathematically described in detail by Northrop and Birmingham (1982) and Roederer (1970). The understanding and characterization of transport and motion of those particles are key to understanding astrophysical phenomena in general.

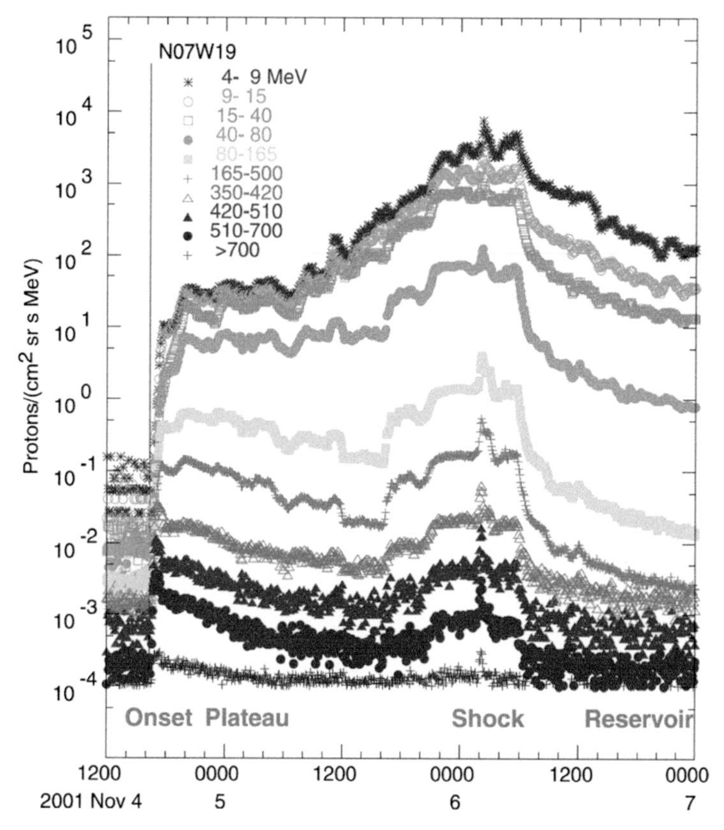

Fig. 12.8 The four phases of a large gradual CME event compared to energy-dependent energetic particle fluxes. This event was observed by NOAA/GOES spacecraft at 1 AU. Different gray shades indicate the intensities of protons for different energies. (From Reames, 2013.)

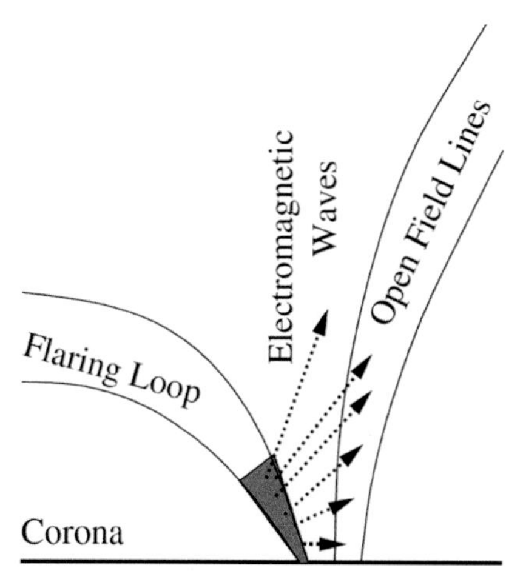

Fig. 12.9 Concept of particle acceleration in an impulsive flare. (Adapted from Kallenrode, 2003.)

Energetic particles are found in planetary magnetospheres in their radiation belts where they are trapped in a strong, basically dipolar magnetic field, bouncing back and forth between the northern and southern hemispheres. Correlated with the high energy and high intensities are radio emissions up to synchrotron radiation when relativistic MeV electrons are present. Energetic particles are also present in the magnetodisk and/or magnetotail regions of planetary magnetospheres where they bounce and drift around the planet. In the Jovian and Saturnian magnetosphere, i.e., the keV–MeV particle population start as neutral particles from their sources Io or Enceladus, get charged by solar UV and are transported radially outwards through the centrifugal forces in the rapidly rotating magnetic fields of the planets. Several hundred kg/s of heavy ions are released into the magnetospheres. As a consequence the magnetic field lines mass-loaded with this material stretch significantly predominantly in the equatorial plane into a magnetodisk configuration until the magnetic stress is too strong and reconnection occurs. Reconnection in the magnetotail, wave–particle interactions, and radial inward diffusion are prime candidates of processes to accelerate those particles from eV to MeV energies.

Energetic particles also play a major role to study planetary aurorae and other high-latitude polar magnetospheric processes. As an example it should be mentioned that hundreds of keV electrons accelerated into Jupiter's atmosphere are mainly responsible for the main auroral oval at Jupiter.

Particle sources can be either the atmosphere or ionosphere of the planet itself, or the sources can be deep inside the magnetospheres as in the cases for Jupiter and Saturn with the moons Io and Enceladus, respectively, or they can be outside the magnetospheres (solar wind or GCRs), being trapped and/or accelerated inside the magnetosphere. Particles are accelerated, at least partially, in these processes. It is currently believed that most of the energetic particles in the outer planet magnetospheres gain their energies up to MeV in the outer magnetosphere. From phase-space density calculations (under conservation of first and second adiabatic invariant) it is currently accepted that most of the particles in the keV–MeV range observed in the middle to inner magnetosphere are diffusing radially inward, gaining energy on the stronger magnetic field. However, especially in the case of Saturn, the inner icy moons orbiting the planet prevent this radially inward motion and most of the particles are lost onto the moons. Therefore the charged particle populations of the radiation belts must have another source: galactic cosmic rays (GCRs).

12.2.1 Galactic cosmic rays as sources of radiation belts

It is currently believed that GCRs originate from supernova explosion remnants. The particles are accelerated at the expanding shock fronts. They enter into our

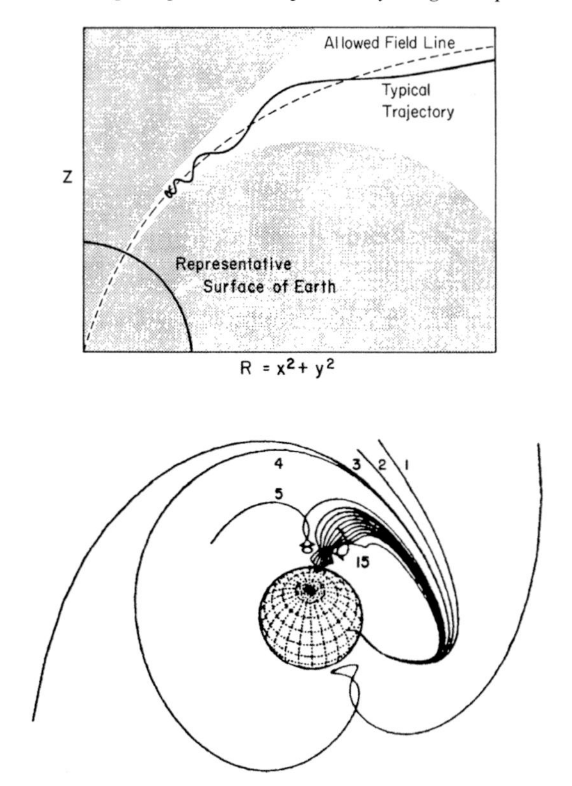

Fig. 12.10 Trajectories of cosmic rays in a planetary dipole field. (Adapted from Smart *et al.*, 2000.)

solar system and can penetrate into planetary magnetospheres on trajectories as shown in Fig. 12.10.

Incoming cosmic rays with GeV energies hit either the atmosphere of a planet or its moons and rings and interact with the material there. As an example Fig. 12.11 shows the simulated spectra of particles produced by 30-GeV protons passing a 10-mm-thick slab of water.

Secondary particles are produced in a wide variety of energies and particle types. For the radiation belts of planetary magnetospheres one of the most important processes is the "cosmic ray albedo neutron decay" (CRAND) in which cosmic rays produce neutrons via knockoff of nucleons in the target material. Most of the neutrons escape from the system, but some of the neutrons decay in the β-process creating protons with lower energy. This is the process to explain MeV protons in inner radiation belts of the Earth or Saturn. The CRAND process is responsible for the peak at about 10 MeV in the energy spectrum of cosmic rays shown in Fig. 12.2. Jupiter's radiation belts are the most intense in the solar system. Figure 12.12 shows

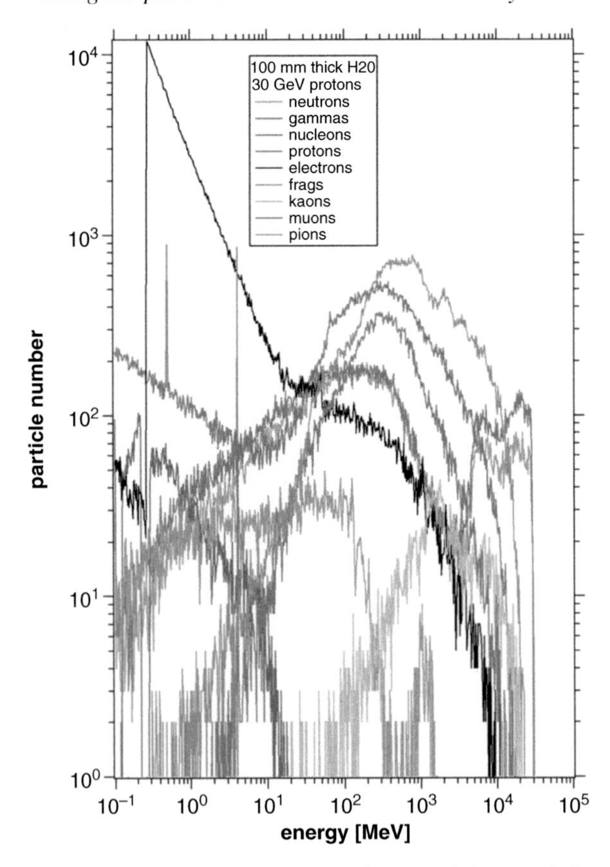

Fig. 12.11 Galactic cosmic-ray spectra for various particle populations. (Diagram provided by D. Haggerty, JHUAPL.) A black and white version of this figure will appear in some formats. For the color version, please refer to the plate section.

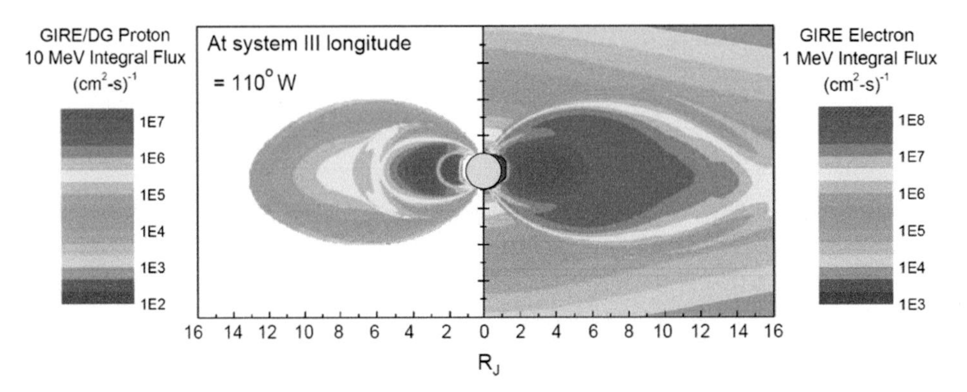

Fig. 12.12 Two-dimensional integral flux distributions for electrons (right section) and protons (left section) at Jupiter based on the Divine-Garrett/GIRE radiation models. (From Paranicas *et al.*, 2009.) A black and white version of this figure will appear in some formats. For the color version, please refer to the plate section.

the modeled integral fluxes of >1 MeV electrons and >10 MeV protons in Jupiter's radiation belts.

MeV electrons with high intensities are found out to at least 10–15 Jovian radii. These relativistic electrons can cause damage in electronics of instruments. Therefore the detailed knowledge of their spatial and energy distribution are crucial to investigate for future missions to Jupiter.

12.2.2 Energetic particles as a useful tool to study magnetospheres

Energetic-particle distributions measured inside planetary magnetospheres are very useful tools to investigate plasma parameters of transport and magnetospheric dynamics. Many new findings are related to the analysis of energetic particles which by far are too many to list them all here in this chapter. They range from the discovery of unknown objects such as moons, plumes, rings, ring arcs, neutral clouds, and tori to the determination of transport parameters, electric fields, flow velocities, open–closed field-line boundaries, surface weathering of moons, remote sensing of surfaces, global imaging of magnetospheric dynamics. Below we describe three examples.

12.2.2.1 Moon absorption signatures in energetic-particle intensities and the determination of diffusion coefficients

The absorption signatures of moons in planetary magnetospheres are distinguished into macrosignatures and microsignatures (Paranicas and Cheng, 1997). Macrosignatures are regions of decreased count rates, where such absorptions are always present at all longitudes along the object's orbit with decreased count rates. As an example Fig. 12.13 shows the macrosignatures of energetic ions (caused by the moons Janus, Mimas, Enceladus, Tethys, and Dione in the Kronian magnetosphere). It is clearly visible that along the L-shells of the corresponding moons and at all latitudes covered by the Cassini measurements a significant depletion is observable.

Microsignatures are small-scale absorptions in the particle distributions highly dependent on the relative position of the causing obstacle and the observer onboard a spacecraft. Moons, planetary rings, or ring arc material as well as dust particles orbiting a planet inside its magnetosphere are subject to heavy "bombardment" from energetic charged particles gyrating, bouncing, and drifting on planetary magnetic field lines. When they hit the object they are lost leaving a gap in the distribution function. Owing to diffusion processes this gap in filled up again slowly. Using the diffusion equation from van Allen *et al.* (1980):

$$f = 1 - 0.5[erf(\frac{1 - x/R}{\sqrt{\tau}}) + erf(\frac{1 + x/R}{\sqrt{\tau}})]\tau = 4D_{LL}t_{rk}/R^2 D_{LL} = D_0 L^n, \quad (12.10)$$

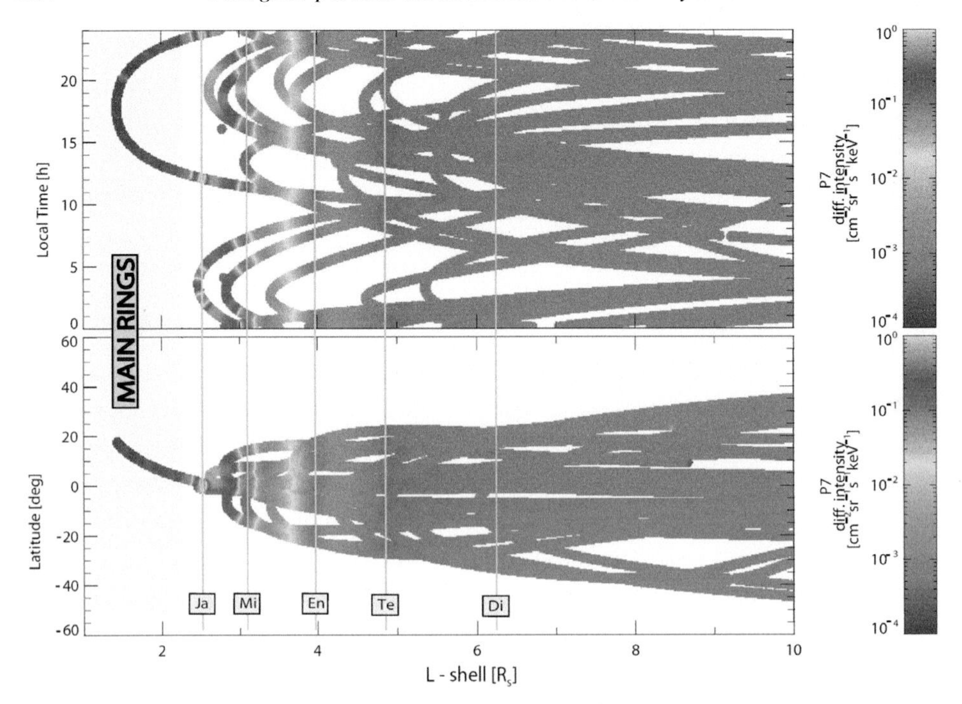

Fig. 12.13 Macrosignatures of energetic ions in the inner magnetosphere of Saturn as a function of L-shell and either local time (upper panel) or latitude (lower panel). Color-coded are the differential intensities of ions ($>$ 10 MeV/nucleon) as measured between 2004 and 2007 by the Low Energy Magnetospheric Measurement System LEMMS onboard the Cassini spacecraft. (From Roussos, 2008.) A black and white version of this figure will appear in some formats. For the color version, please refer to the plate section.

with t_{rk} being the time an electron needs to travel through the absorption region, D_{LL} is the radial diffusion coefficient, R is the radius of the object causing the absorption. Therefore it is possible to calculate radial diffusion coefficients by studying the absorption signature profiles as a function of azimuthal separation from the object at the appropriate radial position, the magnetic flux shell or L-shell, connected to the object. By studying the same processes at different objects (or at different L-shells) a radial diffusion coefficient D_{LL} as a function of distance (or L-shell) can be determined as indicated in Fig. 12.14 from Roussos *et al.* (2007, 2008).

The same method was used to discover a neutral gas torus around the orbit of the Jovian moon Europa (Lagg *et al.*, 2003) or to identify new ring arcs (Roussos *et al.*, 2008), and to discover an additional noon-midnight electric field in Saturn's magnetosphere (Andriopoulou *et al.*, 2012).

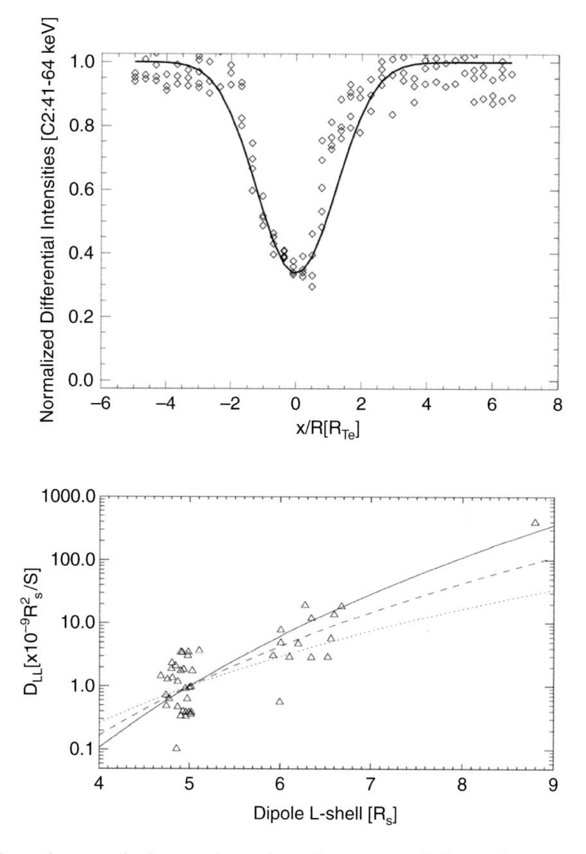

Fig. 12.14 Top: the symbols are the microsignature of Saturn's moon Tethys, the solid curve is the best fit. Bottom: L-dependence of radial diffusion coefficient D_{LL}. (From Roussos *et al.*, 2007.)

12.2.2.2 Determination of flow velocities

Another very important quantity that can be derived from energetic particle distributions is the calculation of particle flow velocities. Using the directional information and energy spectra from energetic-particle measurements it is possible to derive the particle anisotropies (Krupp *et al.*, 2001). Under the assumption that the anisotropy A_F is mainly due to flow and with the assumption that the energy spectrum can be described by a power law with exponent γ the flow velocities can be determined as

$$v_F = \frac{A_F v_{ion}}{2(\gamma + 1)}, \tag{12.11}$$

with v_{ion} being the particle velocity. As an example the derived global flow patterns in Jupiter's equatorial plane are shown in Fig. 12.15 as derived from energetic-particle measurements onboard the Galileo spacecraft. With these measurements it

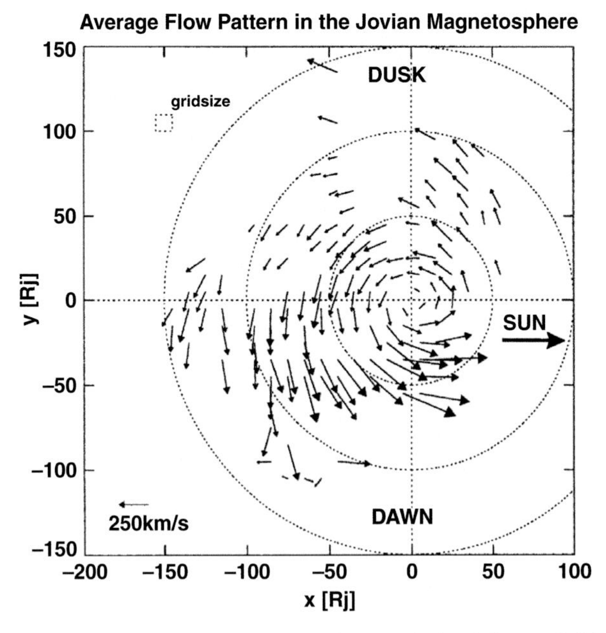

Fig. 12.15 Global flow patterns around Jupiter as derived from *Galileo* energetic particle measurements binned in 10 × 10 R_J bins. (From Woch *et al.*, 2004.)

is clear that plasma and energetic particles (sub-)co-rotate with the planet and that a strong dawn–dusk asymmetry in the flow patterns exists.

12.2.2.3 Determination of open–closed field-line boundaries

Another application of energetic particle distributions as a tool to investigate key magnetospheric regions is the determination of the boundary between open and closed field lines in the auroral zones of the Earth or Saturn. The principle is quite simple and requires only a good pitch angle coverage of the particle sensor, making it able to measure charged particles at a particular energy, but at both directions parallel and anti-parallel to the local magnetic field. Knowing that energetic particles bounce along closed field lines between their mirror points the expectation is that the ratio of particles with pitch angle 0 and 180 degrees is close to unity if the field lines are closed and substantially differ from unity if the spacecraft was connected to an open field line. Monitoring this ratio as a function of time therefore is used to determine the open–closed field-line boundary. An example from measurements inside Saturn's magnetosphere is shown in Fig. 12.16.

Marked in yellow is the section where the ratio between the two opposite channels parallel and anti-parallel to the local magnetic-field direction are very different, and an indication of open field lines. In the cases of Earth and Saturn, this boundary is related to the region in the plane's ionosphere of the main auroral oval.

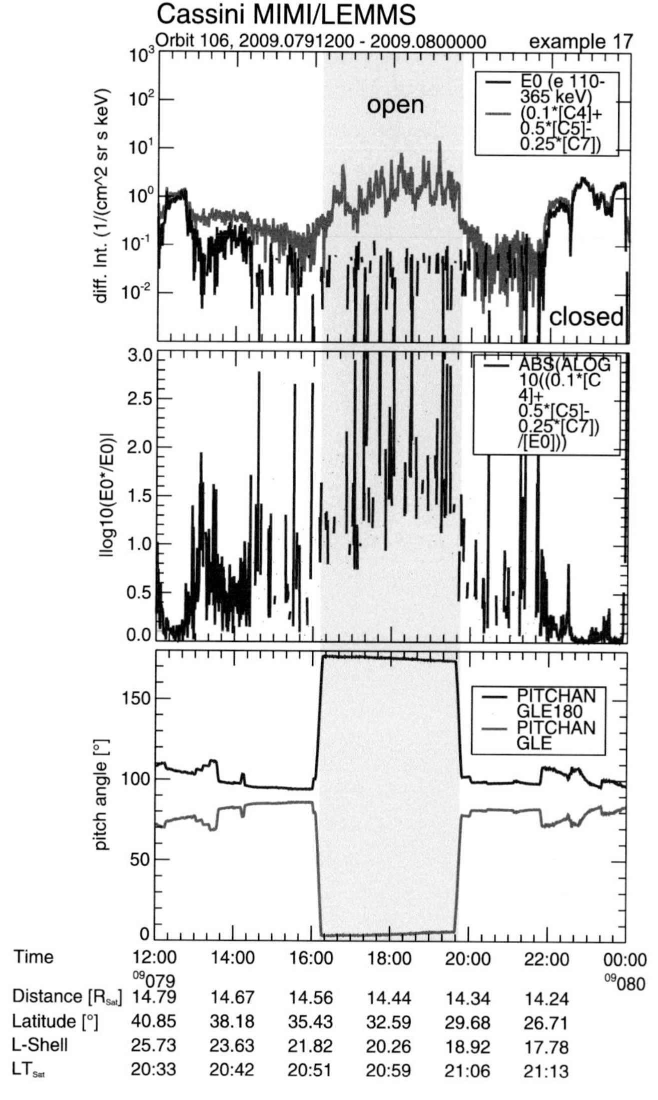

Fig. 12.16 Intensity of electrons along and against the magnetic-field direction inside Saturn's magnetosphere. A black and white version of this figure will appear in some formats. For the color version, please refer to the plate section.

12.3 Summary

In this chapter we have revisited the topic of energetic particles originating from the Sun and in planetary magnetospheres have been revisited. Covering energies up to hundreds of MeV energetic particle characteristics (energy spectra, directional information, intensities, ion composition, etc.) can be used to investigate acceleration mechanisms as well as global configuration and dynamics both in the interplanetary space and inside planetary magnetospheres. They can be used as a tool to study plasma parameters nearly impossible to measure otherwise. Knowing their characteristics in our local environment in our solar system (distribution functions, sources and sinks, etc.) can help to better understand in general astrophysical phenomena within and outside our galaxy.

13

Heliophysics with radio scintillation and occultation

MARIO M. BISI

There are multiple techniques that can be employed to study the Sun and solar wind. Remote-sensing observations include radio and visible/white-light observations of the Sun from the ground, and observations in the ultraviolet (UV) and X-ray bands from space along with any other parts of the electromagnetic (EM) spectrum. The density irregularities in the solar wind can also be observed using ground-based or space-based coronagraphs for the inner solar wind as it emerges from the Sun's corona, by space-based white-light heliospheric imagers, and by radio measurements of distant compact radio sources to indirectly observe the solar wind through scintillation in the radio signal received from distant astronomical or artificial radio sources.

In-situ measurements of solar-wind velocity, density, magnetic field, temperature, and other plasma and field parameters are also of high importance. An advantage of in-situ measurements is that the physical parameters recorded are measurements of the primary solar wind parameters at that point in space and time. However, the obvious disadvantage to this is that measurements can be made only at that particular position of the spacecraft at certain times. Some regions are only accessed rarely, if ever; for example, in-situ measurements inside of 0.3 AU (the perihelion of the orbit of the planet Mercury) or outside of the ecliptic plane of the solar system are not available at present. Only the Ulysses spacecraft data set is available for past years outside of the ecliptic (1990 launch, mission data from 1994 to 2009) and some measurements are available from the twin Helios spacecraft both outside of the ecliptic and just inside the orbit of Mercury (1974 and 1976 launches, mission data from 1975 to 1985). Remote-sensing observations have the advantage of covering the solar wind over a wide range of heliocentric distances and all heliographic latitudes (including out of the ecliptic plane) and at almost any time; this is especially true of radio observations. The disadvantage to remote-sensing observations is that the primary parameters are not measured but are inferred from the observations.

13.1 Observing radio waves

The radio spectrum itself is broken down into various segments as defined through radio astronomy (see Table 13.1). It is important to remember that frequencies below about 10 MHz (sometimes as high as below 30 MHz, depending on ionospheric conditions) can only be observed outside of the Earth's ionosphere because these lower radio frequencies are reflected by the ionosphere and hence cannot propagate through from space to the ground.

13.1.1 The Sun and radio waves

James Clerk Maxwell (1831–1879) was the first to theoretically predict the existence of EM waves by the derivation of his equations that describe mathematically how electric induction occurs from an oscillating magnetic field and showed that they traveled at the speed of light and that light, too, was just another form of EM radiation (Maxwell, 1873). The original paper on the theory behind his equations was read before the Royal Society on 8 December 1864. Heinrich Hertz (1847–1894) confirmed during the years 1885–1889 that radio waves existed and was the first to send and receive them. In 1901, Gugliemo Marconi (1874–1937) was the first to send a radio message across the Atlantic Ocean from England to Canada using the Earth's ionosphere to reflect the radio waves (although the nature of this reflectivity was not known at the time).

The "great storms" of radio emission from the Sun in February of 1942 were the beginning of the modern development of radio astronomy. The 1942 detection occurred when British military radar stations that were operating at wavelengths of the order of a few meters experienced some form of jamming in late February of that year. An investigation described by Hey (1983) led to the conclusion that radio waves of very high intensity were being emitted from the Sun and were apparently due to a very large presence of sunspots that were active on the solar disk. Later that same year, Southworth (in the USA) discovered that the quiet Sun also emitted solar radio waves; this was the normal radio emission from the Sun at centimeter wavelengths (Hey, 1983).

In 1951, when observations of the outer solar corona were underway, it was noticed that radio waves were coming from the Crab Nebula. These radio waves were refracted due to the electron-density irregularities in the solar corona which caused the apparent size of the Crab Nebula to increase as observations of the radio waves moved in closer to the Sun where the Crab Nebula was going behind the Sun on the sky (Hewish, 1955). Interplanetary scintillation (IPS) of the radio waves from distant compact radio sources produced by density variations in the

Table 13.1 *A breakdown of the radio spectrum. N.B. Frequencies below about 10 MHz and sometimes as high as below 30 MHz can only be observed outside of the Earth's ionosphere.*

Frequency band definition	Frequency range	Wavelength range	Band designations
ELF – Extremely Low Frequency	<300 Hz	>1000 km	—
ULF – Ultra Low Frequency	300 Hz–3 kHz	1000 km–100 km	—
VLF – Very Low Frequency	3 kHz–30 kHz	100 km–10 km	—
LF – Low Frequency	30 kHz–300 kHz	10 km–1 km	—
MF – Medium Frequency	300 kHz–3 MHz	1 km–100 m	—
HF – High Frequency	3 MHz–30 MHz	100 m–10 m	—
VHF – Very High Frequency	30 MHz–300 MHz	10 m–1 m	—
UHF – Ultra High Frequency	300 MHz–3 GHz	1 m–100 mm	P (sometimes) = 300 MHz–1 GHz
			L = 1 GHz–2 GHz
			S = 2 GHz–4 GHz
			C = 4 GHz–8 GHz
SHF – Super High Frequency	3 GHz–30 GHz	100 mm–10 mm	X = 8 GHz–12 GHz
			Ku = 12 GHz–18 GHz
			K = 18 GHz–26 GHz
			Ka = 26 GHz–40 GHz
			Q = 30 GHz–50 GHz
			U = 40 GHz–60 GHz
EHF – Extremely High Frequency	30 GHz–300 GHz	10 mm–1 mm	V = 50 GHz–75 GHz
			E = 60 GHz–90 GHz
			W = 75 GHz–110 GHz
			F = 90 GHz–140 GHz
			D = 110 GHz–170 GHz

solar wind further out from the Sun were discovered soon afterwards (Hewish *et al.*, 1964; Cohen *et al.*, 1967) and have now been in almost constant use in an effort to determine some of the properties of the solar wind via radio remote sensing.

In all, there are four primary methods of observing the Sun by radio waves: radio emissions both from the Sun itself and from the solar wind (such as from the electron beams responsible for solar radio bursts, as well as both coherent and incoherent emission); radar echoes where radio waves sent to the Sun reflect off the solar atmosphere; the Faraday rotation of polarized spacecraft beacons or astronomical radio sources (such as by pulsars or the background galactic synchrotron radio emission) due to changes in the magnetic field in the solar corona (and the solar wind); and IPS with compact astronomical radio sources (such as quasars, or occasionally with spacecraft beacons). A broader introduction to the radio aspects of the Sun and inner heliosphere not covered in this chapter can be found in Ch. 4 of Vol. II.

13.2 Astronomical radio sources and spacecraft beacons

Quasars, otherwise known as quasi-stellar radio sources, are most typically used for observations of IPS. They are compact, broadband radio emitters. Quasars are extremely luminous and the most-energetic and distant type of phenomena classified as Active Galactic Nuclei (AGN). AGN quasars can be 100 times more luminous in radio light than the whole Milky Way galaxy, and are extremely compact on the sky plane.

Fast-rotating neutron stars that make pulsars are created as a result of Type II supernovae of stars between about eight and 25 solar masses. These too are compact, broadband radio emitters on the sky. They are very small objects being some tens of kilometers in diameter between about 1.5 and five solar masses that rotate very fast. The pulse comes from a beam of high-intensity radiation being emitted from the pulsar's magnetic axis which spins with the neutron star where the magnetic axis determines the direction of the electron beam, but is not necessarily aligned with the rotational axis of the neutron star; hence the pulsing observations as viewed from the Earth.

Spacecraft radio beacons emit radio waves at one or more known frequencies that for use in heliophysics are usually in the S or X band of the radio spectrum (but also at much lower frequencies depending on the spacecraft beacon design specifications). They are normally wholly circularly polarized in either a left-handed or right-handed sense, but will have a small polarization leakage into the other polarization. They are typically used to track the spacecraft position, but can be used to obtain properties of the solar corona, solar wind in the heliosphere, and even of planetary atmospheric properties.

13.3 Radio occultation

An occultation occurs when one object is hidden by another that passes between it and the observer; i.e., when an apparently larger body passes in front of an apparently smaller one. Radio occultation is another remote-sensing technique, but is generally used for determining the physical parameters of planetary atmospheres, although a similar technique can be used for observations of spacecraft beacons as spacecraft move behind the Sun. It is the result of refraction of a radio wave by a planetary or the solar atmosphere.

For planetary occultation, the bending of the radio waves is a result of the different properties in the planetary atmosphere/ionosphere, and thus the amount of bending is determined by these properties. At radio frequencies the amount of bending cannot be measured directly but can be calculated using the Doppler shift of the signal given the geometry of the radio emitter and the radio receiver. In the case of neutral atmosphere properties, those below the ionosphere in particular, properties such as the water vapor content, atmospheric pressure, and the gas temperature can be derived, so that radio occultation has some direct applications for the meteorology aspects of heliophysics.

13.4 Radio scintillation

The term scintillation, known more commonly as twinkling, is the generic term for variations in apparent brightness or position of a distant luminous object viewed through some medium. There are several types of radio scintillation that can occur, each with different causes, and each somewhat dependent on the source size on the sky, on where from the Earth it is observed, and also on the frequency that it is observed at.

For the most extended astronomical radio sources (as well as for spacecraft and satellite communications/beacons in general) the only real contributor for scintillation is that of the Earth's ionosphere. Ionospheric scintillation is caused by turbulent structures in the ionosphere that scatter the incoming radio waves forming an interference diffraction pattern by varying the path length along which the waves reach a telescope. Ionospheric scintillation can cause the biggest concern for global navigation systems, for example, because the scintillation can cause an error in the timing from the satellite and hence can give an inaccurate position measurement. Ionospheric scintillation is generally predominant at lower radio frequencies and also at high Earth magnetic latitudes, but during high solar activity and resulting space weather at the Earth the effects can also be seen at much lower latitudes (and even at the higher observing frequencies). Thus, ionospheric scintillation can also cause problems for radio astronomy and can sometimes dominate over other aspects of a radio signal and over other types of scintillation that are of scientific interest.

If you were to use "point-like" compact astronomical radio sources, then for milli-arcsecond source sizes on the sky, you can also get IPS in your radio signal under the right conditions. This form of scintillation is a result of density inhomogeneities and turbulence in the solar-wind plasma outflow from the Sun. This also results in the scattering of the radio waves, but for a sensible scintillation pattern to be formed which can then be used for the study of the solar-wind plasma, the source ideally needs to be simple and not complex in nature. This means that the radio emission needs to be coming from a single point. The sources are usually within 90° elongation from the Sun (although IPS can occur for signals received from radio sources beyond 90° elongation, i.e., radio sources on the sky in the anti-solar direction). In addition, throughout the late 1960s, IPS was also used extensively for the study of the source structure of quasars before the advent of very-long-baseline interferometry (VLBI), which combines multiple telescopes together to form a single, much-larger overall collecting area, and much-enhanced imaging resolution.

In an even-more "point-like" compact sense, micro-arcsecond source sizes on the sky can be used for interstellar scintillation or ISS (for example, a 10 μarcsec source size in radio would be about 10 000 times finer in terms of the resolution needed than is achieved by the Hubble Space Telescope operating at its shortest wavelength). ISS can provide information on the speed and direction of flows in the interstellar medium (ISM), just as IPS does for the interplanetary medium, but only if the radio source is that much more compact.

13.5 Radio occultation, with a focus on planetary occultations

Planetary occultation is detected through the effects on the radio-wave propagation resulting in changes in the velocity of the radio waves from that in free space. In addition, changes in the frequency, phase, and amplitude of the signal, and also changes in the refractive index and hence the amount of bending of the radio waves caused as a result of the planet occulting the spacecraft. Similar effects are also observed with solar occultations, but for different parameters and methods, one of which being the detection of Faraday rotation (FR) which is covered in Sect. 13.7 as well as for higher-frequency observations of IPS as covered by Sect. 13.6.

The technique of planetary occultation has been proven to be important for the study of planetary atmospheres (e.g., Eshleman, 1973). However, as with all observing techniques, there are uncertainties involved with the experiment, and these are well documented in the literature (e.g., Withers, 2010).

The effects of planetary occultation on radio signals come from many aspects of a planetary atmospheric–ionospheric environment. Thus, radio occultations probe both lower atmospheric properties (neutrals/molecules) as well as ionospheric electron densities.

The first use of a coherent radio transmission for investigating planetary atmospheres/ionospheres other than that of the Earth was on 15 July 1965 with the Mariner IV spacecraft as it disappeared behind the limb of Mars. The spacecraft remained in occultation for around 54 min and, prior to the start and immediately following the occulation, the spacecraft's S-band radio signal passed through both the atmosphere and the ionosphere of Mars. This first is well documented by Kliore *et al.* (1965) where Doppler tracking was also used for this experiment. Planetary occultation relies on the interaction of the radio signal being transmitted with the planetary atmosphere for any effects in the radio-wave propagation to be recorded and for the experiment to work (e.g., Withers, 2010).

Details on the atmosphere of Venus have been discovered through several occulation experiments from multiple spacecraft whereby the temperatures, densities, and cloud layers have been determined and investigated with height above the planet's surface (e.g., Imamura *et al.*, 2011, and references therein). Figure 2 of Imamura *et al.* (2011) provides an excellent image example of how the radio signal is refracted due to a planet's atmosphere/ionosphere, along with a full description and set of equations explaining such.

A list of key references of giant-planet occultation observations is given in Table 8.1, along with an extensive list of the basic planetary and upper-atmospheric properties that can be gleaned from the use of radio-occultation measurements due to planetary atmospheres. In addition, Fig. 8.4 provides examples of ionospheric electron-density profiles as derived from spacecraft radio occultation experiments carried out at Jupiter, Saturn, Uranus, and Neptune, and their respective altitudes refer to the radial distance above the one bar pressure level of atmosphere. Despite these experiments, any detailed knowledge of ionospheric structure at the giant planets is still relatively unconstrained by observation. This is because the radio-occultation experiments at these planets remain to be few with only nine published experiments for Jupiter, 31 for Saturn, two for Uranus, and two for Neptune. So even less is known about the ice giants than the gas giants.

Saturn has had the greatest amount of scientific investigation carried out via radio-occulation measurements, and this is mostly due to the Cassini spacecraft being in orbit there since 1 July 2004. This far-increased number of radio-occultation measurements has allowed for the identification of two main global ionospheric structures. These are in the form of dawn/dusk asymmetries and an unexpected behavior in how the electron density varies with changing latitude in that the electron densities are actually smallest at the equator. Further details of these findings can be found in Ch. 8, and also in Nagy *et al.* (2006, 2009), and Kliore *et al.* (2009).

Coming back to Mars, dust storms provide an episodic forcing on the upper atmosphere and ionosphere of Mars where, thus far, there has not been found an analog at Earth and thus this is of high interest since it is something not able to be investigated at Earth. The dust storms can be regional or global in nature, and are also very much unpredictable in their occurrence and ferocity. There appears to be a time where dust storms are more prevalent on Mars, and this has been termed the "dust storm season", and it encompasses the second half of the year on Mars (Zurek *et al.*, 1992; Kass *et al.*, 2014). Dust storms have the effect of changing the normal radio-occulation recording parameters for experiments conducted through Mars' upper atmosphere, and further details of these can be found in Ch. 9. As an example, Fig. 9.15 clearly illustrates that for an event which took place in 2005 on Mars, the dayside heights of the atmosphere rose by 5 km as a result of the dust storm taking place. This and other effects have been measured through the radio-occulation technique. Profiles for the Mars atmospheric environment can also be found in, e.g., Tellmann *et al.* (2013), and references therein, as well as the detection of gravity waves, which is of a high importance since it is thought that gravity waves have a significant influence on the atmospheres of the other terrestrial planets, including the Earth (e.g., Tellmann *et al.*, 2013).

A detailed discussion of the Earth's ionosphere, where planetary occultation methods are still used, as well as space-weather effects on and as a result of the ionosphere, are well documented and described in the chapter by Norbert Jakowski in Vol. V (cf., Table 1.2). In addition, the latest advances in ionospheric scintillation where the secondary power spectrum of an observation can provide a distance to the scattering screen responsible for the scintillation (see Sect. 13.6 for the description in relation to IPS), i.e., the height of the ionosphere's scattering screen are given in Fallows *et al.* (2014).

13.6 Interplanetary scintillation in the context of heliophysics

13.6.1 A brief history of IPS and IPS-capable receivers

In 1951, Vitevitch in Russia, and both Machin and Smith at Cambridge, independently suggested the possibility of studying the outer solar corona by using observations of distant radio sources. This arose from observations of the radio component of the Crab Nebula whereby its apparent size seemed to change and broaden as a result of the scattering of the radio waves by variations/inhomogeneities in the coronal electron density (described in Hewish, 1955; Brandt, 1970; and Hey, 1983). An extension of this was that of interplanetary scintillation; this technique was developed by Hewish from 1962 onwards (Hewish *et al.*, 1964) following the serendipitous discovery of unusual fluctuations in source signal strength during a program to accurately map the position of a large number of radio sources on the sky. These astronomical radio sources were subsequently determined to be "point like" in nature (P. J. S. Williams, private communication to A. R. Breen, 2002, as cited in Canals, 2002; Bisi, 2006). It was thus hypothesized that these same density irregularities causing the broadening of the Crab Nebula radio-source size were also responsible for the scintillation in the radio waves from the distant point-like sources, and also that the solar corona extended out into the interplanetary medium (subsequently coined as the solar wind). It was then recognized that IPS could be a powerful tool in probing the solar wind and also for measuring the finer structure of the radio sources themselves (Hewish *et al.*, 1964; Hewish and Okoye, 1965).

The amount of scintillation detected at Earth was suggested to be highly dependent on the elongation angle of the source from the Sun (i.e., the distance from the point of closest approach, the P-point, of the Earth-to-source line of sight (LOS) to the Sun), and that, beyond some critical elongation angle, the scintillation would no longer be detectable by the receiver. This critical elongation angle was described as "the effective radius of the corona for occultation", which was proposed to be dependent on the effective wavelength being observed (for more details, see Machin and Smith, 1952). Hence, the possibility was then suggested that you could estimate the electron density of the corona by measuring the occultation radius over a range of different wavelengths.

Hewish's first observations of IPS gave solar wind velocity estimates of between $200\,\mathrm{km\,s^{-1}}$ and $400\,\mathrm{km\,s^{-1}}$. It was IPS that first indicated that the velocity of solar wind was greater coming from above the polar regions of the Sun than was observed in the plane of the ecliptic (Dennison and Hewish, 1967; Coles *et al.*, 1980). This was due to IPS being capable of observing at all heliographic latitudes wherever there are suitable radio sources on the sky. This discovering of faster wind over the Sun's polar regions was then only later confirmed by the Ulysses polar pass in-situ measurements many years later (Phillips *et al.*, 1994). Similarly, Houminer (1971) found features in IPS that are transient and those that co-rotate.

Multi-site observations of IPS were subsequently extensively undertaken from the early 1970s onwards via the use of several different systems around the world over the years. For example, up until 1987, scientists at the University of California, San Diego (UCSD) described in a series of papers (Armstrong and Coles, 1972; Armstrong *et al.*, 1972; Coles and Rickett, 1976; Coles *et al.*, 1980; Rickett and Coles, 1991; Rickett, 1992) observations that were undertaken at a frequency of 74 MHz with a maximum baseline of 94 km using a three-antenna system. These observations of IPS provided results that gave an overview of changes in the solar wind over a whole solar cycle covering all heliographic latitudes. These results showed a large change in the nature of the solar-wind structure throughout the solar cycle and also followed the magnetic evolution of the inner corona (Rickett and Coles, 1991). The declining phase of that cycle, which lasted for seven years, saw the high-latitude heliosphere dominated by near-uniform fast flow occupying approximately half of the heliosphere in all. This was the same during the solar minimum itself. The low helio-latitudes were dominated by a low-velocity flow from around the heliospheric current sheet. A latitudinal gradient could be seen at solar minimum which then vanished at solar maximum, with slow flow extending to all latitudes during the period of, and around, the maximum activity (Coles and Rickett, 1976; Coles *et al.*, 1980). In addition, over in Japan (and covering the same time period), observations of IPS were conducted by scientists at the Solar Terrestrial Environment Laboratory (STEL/STELab) at Nagoya University (Kojima and Kakinuma, 1987), now the Institute for Space-Earth Environment (ISEE). These observations continue even today at a higher frequency of 327 MHz, and thus were used and are being used to probe closer to the Sun than the UCSD system (the reasons for this are discussed later in this chapter). Observations of IPS at a higher frequency of 933.5 MHz began in 1982 at the European Incoherent SCATter (EISCAT) radar facility based across northern Scandinavia (Bourgois *et al.*, 1985) which allowed for the determination of solar wind parameters from around 20 $R_{\odot}$ to over 75 $R_{\odot}$; these were the closest-in multi-site observations of IPS at the time.

The resolving ability of such multi-site observations of IPS was also improved using EISCAT due to the much larger baselines of up to 390 km. These observations were able to show simultaneously from individual observations that there were two clear solar wind components ahead of any confirmation by the Ulysses spacecraft. EISCAT (Rishbeth and Williams, 1985) has been used for IPS-observing campaigns regularly since 1985 and more extensively from 1991. A unique opportunity arose in 1994 and 1995 for a comparison with the in-situ measurements from 1.4 AU and outward of the Ulysses spacecraft as it passed over the poles of the Sun reaching latitudinal extents of 80.2° South and North. This bi-modal nature of the solar wind was determined with the fast stream of around 800 km s^{-1} and slow stream around 400 km s^{-1} (Breen *et al.* 1996a, b). These then agreed with data taken using the Ulysses spacecraft (Phillips *et al.*, 1994, 1995). Further analysis of these data sets have revealed a possible two-mode structure of the inner fast solar wind (Bisi, 2006; Bisi *et al.*, 2007a) and thus this warrants further investigation. EISCAT was later enhanced with the building of the EISCAT Svalbard Radar (ESR) observing at 500 MHz (Wannberg *et al.*, 1997), and EISCAT then upgraded to 1420 MHz as an alternative observing frequency on two of its three sites on the mainland (Wannberg *et al.*, 2002). In addition, these same two mainland sites have had the UHF ~930 MHz receivers replaced with VHF 224 MHz receivers in 2012–2013.

In 2002, 2004, 2005, and again in 2006, observations were made using the Multi-Element Radio-Linked Interferometer Network (MERLIN) facility (Thomasson, 1986) with IPS observing frequencies of either 5 GHz or 6 GHz as well as at

1420 MHz. Canals (2002) states that the Very-Long-Baseline Array (VLBA) has also been used at 22 GHz for observations of IPS. There have been combined programs of MERLIN and EISCAT to probe the acceleration region of the solar wind and also to look in more detail into the large-scale structure of the solar wind and in particular, the direction of flow of the solar wind where non-radial flows have been detected (e.g., Moran, 1998; Moran *et al.*, 1998; Breen *et al.*, 2006; Bisi, 2006; Bisi *et al.*, 2007b; Breen *et al.*, 2008; Dorrian *et al.*, 2013) and the possible two-mode structure of the inner polar fast solar wind (Bisi, 2006; Bisi *et al.*, 2007a). The VLBA at 22 GHz has the capability of observing the solar wind as close as 2.5 $R_\odot$ (Klinglesmith, 1997), and MERLIN as close as 5 $R_\odot$ (Canals, 2002) at 5 GHz.

IPS is also very capable of detecting the passage of a coronal mass ejection (CME) through the inner heliosphere (e.g., Gothoskar and Pramesh Rao, 1996; Ananthakrishnan *et al.*, 1999; Jones *et al.*, 2007; Tokumaru *et al.*, 2007; Bisi *et al.*, 2009b; Tappin and Howard, 2010; Glyantsev *et al.*, 2015; and references therein). Indeed, Jackson (1984) was one of the first papers linking CME signatures in IPS data with the white-light CMEs as seen in coronagraphs at the time.

Multiple IPS-capable systems are still in use and under development around the world today over half a century after its first use as a method of probing inner-heliospheric outflows and structure. Single-site systems include the Ootacamund (Ooty) Radio Telescope (ORT) in India, the Big Scanning Array (BSA) at Pushchino in Russia, the Ukrainian T-shaped Radio telescope – second modification (official abbreviation UTR-2) near Hrakovo in Ukraine, the MEXican Array Radio Telescope (MEXART) in Coeneo, Michoacán, in Mexico, the Korean Space Weather Center (KSWC) IPS array on Jeju Island in the Democratic Republic of Korea (South Korea), and the Miyun Synthesis Radio Telescope (MSRT), Urumqui radio telescope, and the Miyun radio telescope, all located in China. Multi-site systems include the aforementioned EISCAT/ESR joint system (four sites in all), the aforementioned 327-MHz ISEE IPS arrays in Japan (originally four sites, now three), and the LOw Frequency ARray (LOFAR) centered in the Netherlands with international sites across central and western Europe (many Netherlands sites with, presently, nine international stations and three more to start construction). In addition, as a single-site system based on LOFAR technology and as a test bed for the future EISCAT_3D planned system, there is the Kilpisjärvi Atmospheric Imaging Receiver Array (KAIRA) which can be considered as a reconfigured LOFAR Netherlands non-core station. There are systems with other *ad hoc* IPS capabilities such as the Giant Metrewave Radio Telescope (GMRT) near Pune in India, the VLBA across the USA and the US Virgin Islands, the European Very Long Baseline Interferometer (EVLBI) Network (EVN), and the Millstone Hill Steerable Antenna (MISA) located at the MIT Haystack Observatory in Westford Massachusetts in the USA for example. Finally, the Murchison Widefield Array (MWA) in Western Australia has, serendipitously, recorded its first observation of IPS with planning underway for some short, dedicated IPS test experiments as well as IPS capabilities to be tested with the Long Wavelength Array (LWA) in the USA (S. M. White, private communication, 2015).

13.6.2 An overview of IPS theory

Variations in the local refractive index of the interplanetary medium are the foundations for the IPS detected in the radio signals. These inhomogeneities have a characteristic scale size of tens to a few hundreds of kilometers. The irregular changes of refraction bend the radio waves in varying directions and so they then reach the observer from different directions. The refracted waves then combine with others causing constructive and destructive interference in the form of amplitude variations in the received signal (amplitude scintillation). The frequency of the fluctuations is about 0.5 Hz when detected from the ground (Hewish, 1989). Only if the patterns have a sufficient coherence (i.e., when the angular diameters are less than around 0.5 seconds of arc) will the scattered waves then combine to form a diffraction pattern. This diffraction pattern then drifts over the ground to give rise to the fluctuations in intensity (Ekers and Little, 1971; Coles and Harmon, 1978) as can be seen in Fig. 13.1.

Solar-wind parameters can be obtained from single-site analyses of a power spectrum of an observation of IPS (e.g., Manoharan and Ananthakrishnan, 1990; Aguilar-Rodriguez *et al.*, 2014), but more-robust methods are generally based on the cross-correlation of two simultaneous observations of the same radio source (e.g., Coles, 1996; Breen *et al.*, 1996a; Fallows *et al.*, 2008; Bisi *et al.*, 2010b; Fallows *et al.*, 2013). Figure 13.2 provides a simplified overview of a multi-site IPS experiment. The two observations are typically at the same observing frequency from both receivers; however, Bisi (2006) and Fallows *et al.* (2006) have shown that multi-frequency cross-correlations are also possible, and much science can still be achieved through such.

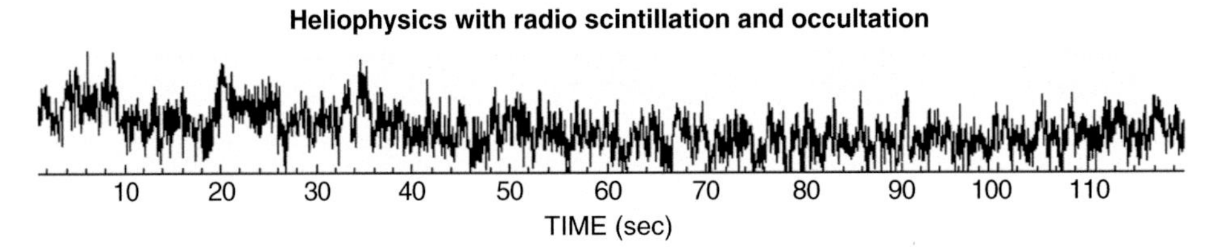

Fig. 13.1 Intensity fluctuation seen in a single observation of IPS using the Sodankylä site during an EISCAT observation on 05 September 2004 with radio source J1256-057 (3C279). (Adapted from Bisi, 2006.)

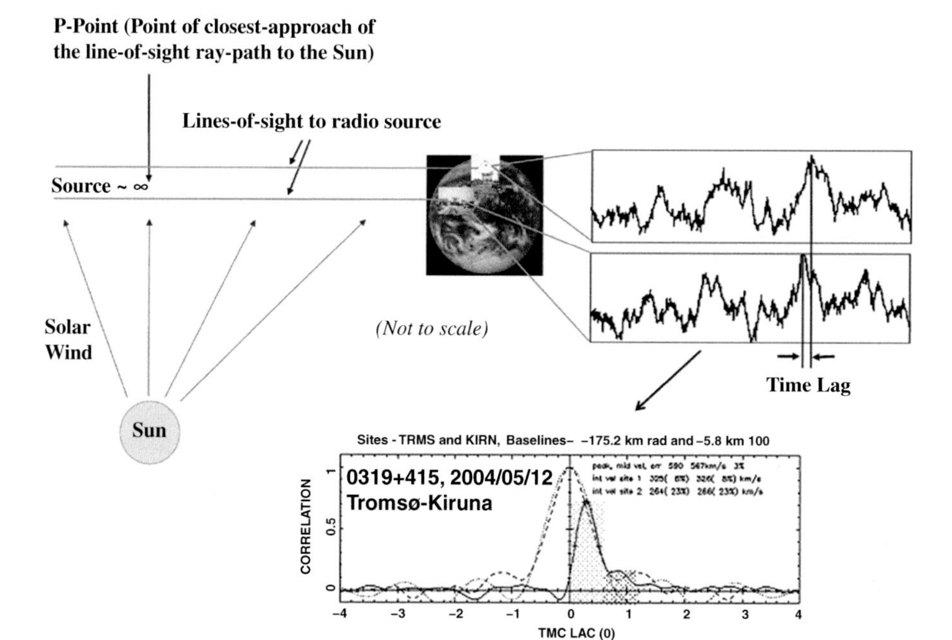

Fig. 13.2 The basic principles of multi-site IPS (in this case two EISCAT radar antennas, Tromsø and Kiruna) through the simultaneous observation of a single radio source as described in the text. The signal's amplitude variation is directly related to turbulence and density variation in the solar-wind outflow crossing the line of sight (LOS). A cross-correlation analysis of the two simultaneous signals received yields what is known as a cross-correlation function (CCF) which can be used as a first estimate of the velocities crossing the LOS. (Adapted from Bisi *et al.*, 2010b.)

The variations in the received signal at Earth depend on the structure of the solar-wind plasma through which it traverses. The IPS of the radio waves causes a 2D (two-dimensional) interference pattern across the Earth's orbit. As the irregularities casting the intensity pattern are moving out in the solar wind away from the Sun, a receiver on Earth will see the apparent intensity of the source varying on time scales of between ~0.1 s and 10 s. Fluctuations can become smooth due to minima from one part of the source overlapping with minima from another part. This depends on frequency and on distance from the Sun because the irregularities are larger further out from the Sun because of their natural expansion and the approximate decrease in density squared with increasing distance from the Sun. A simple formula for working out if a radio source is too large, is whether or not a source subtends an angle $\theta_s \geq L/D$, where L is the average size of the irregularity of solar wind, and D is the distance from the observer. If the radio source size on the sky is larger than $\sim\theta_s$, then the amplitude fluctuations will be smoothed out and no IPS will occur.

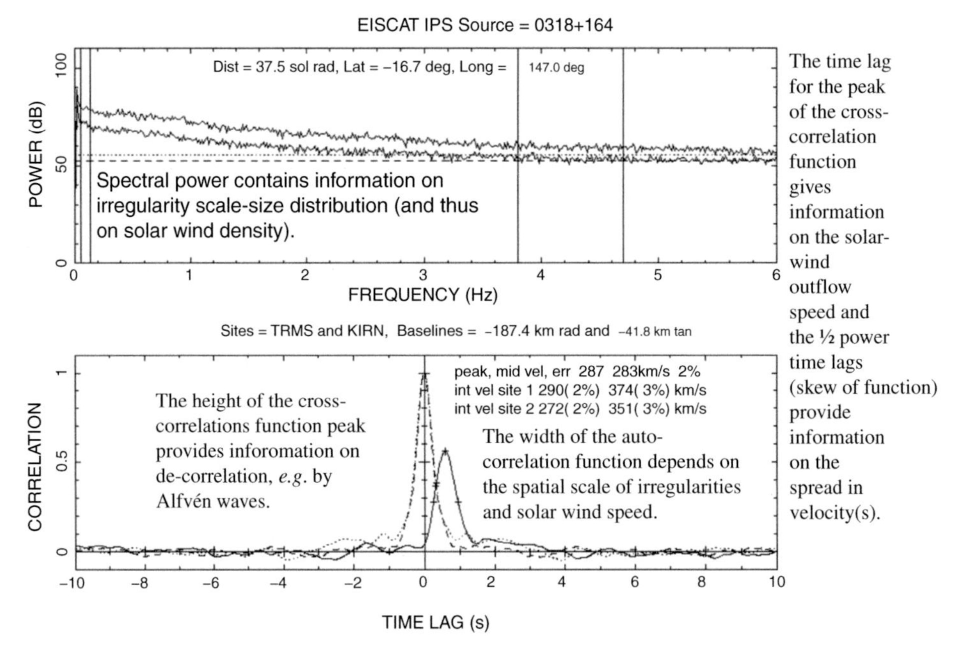

Fig. 13.3 The top plot displays an example Fourier-transformed spectrum from the two time series of a multi-site observation of IPS with EISCAT. The bottom plot shows a CCF of this observation. The vertical lines on the power spectrum to the left are the high-pass filters for each of the two sites and similarly to the right are the low-pass filters for each of the two sites. (Courtesy of R.A. Fallows.)

Figure 13.3 provides an example of the Fourier-transformed IPS power spectrum on a log–linear scale as well as a cross-correlation function (CCF) from an EISCAT observation of IPS with radio source J0318+164 (CTA21). Figure 13.4 provides an additional example of an IPS power spectrum on a log–log scale where features are more-clearly marked.

Ionospheric scintillation is also possible as already noted. This generally occurs with larger source sizes than IPS (because D is much smaller, so θ_s larger). Typical IPS time scales of 0.5 Hz are significantly faster than those of ionospheric scintillation. In addition, the diffraction of about 200 km of ionosphere is considerably large (likely due to the larger electron column density) so that high-frequency IPS is not normally confused with any effects that could come from the ionosphere (Hewish, 1989) unless there are very rapid flows in the ionosphere (private communication to A. R. Breen, 1995, as cited in Canals, 2002, and Bisi, 2006) where ionospheric scintillation may rise above a frequency of 0.2 Hz as seen in the power spectrum (A. R. Breen, private communication, 2005). In these cases, the ionospheric contribution may be seen as an extra bulge in the spectrum close to the low-frequency cut-off end of the IPS spectrum where there is a high-pass filter

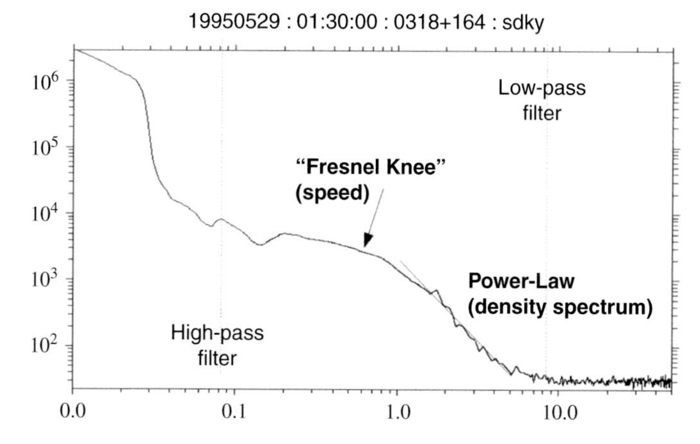

Fig. 13.4 An example IPS power spectrum on a log–log scale. The aforementioned low- and high-pass filters are more-clearly illustrated and individually labeled. The Fresnel knee is related to the speed/velocity of the solar-wind outflow, and the power law is related to the density. (Courtesy of R.A. Fallows.)

placed in the analyses. A low-pass filter in the analyses removes the background white noise from the higher-frequency end of the spectrum. At lower observing frequencies, the ionosphere can become more dominant and great care then needs to be taken to adequately filter out the ionospheric scintillation (where possible) or to not trust the validity of that observation, particularly for receivers at the Earth's high geomagnetic latitudes where ionospheric turbulence and variations in density are much greater. This has become prevalent in test observations of IPS using both KAIRA and the two new EISCAT 224 MHz remote sites.

The variations in the local refractive index are directly proportional to the density of electrons provided the variation in the electron density is small enough (Hewish, 1989). Using the notation of Uscinski (1977) and adapted from Fallows (2001), Canals (2002), and Bisi (2006), the refractive index, $n(x, y, z)$, can be described by the following set of equations starting with Eq. (13.1), where n_0 is the mean refractive index, $n'(x, y, z)$ is the variation of refractive index about the mean (above and below the mean refractive index), μ is the standard deviation of $n'(x, y, z)$, and $n_1(x, y, z)$ is the scaled form of $n'(x, y, z)$ with unit standard deviation:

$$n'(x, y, z) = \mu n_1(x, y, z) = n(x, y, z) - n_0. \tag{13.1}$$

The variation in refractive index is determined by the relative electron permittivity, ϵ, of the medium, i.e., the solar wind in this case. In the absence of particle collisions (the solar wind is a collisionless plasma), and if the magnetic field is of no concern (a simplification), this can then be represented by

$$\epsilon = n^2 = 1 - \frac{f_p^2}{f^2}, \tag{13.2}$$

where f_p is the plasma frequency (in Hz) and f is the observing frequency of the observation of IPS:

$$f_p^2 = \frac{N_e e^2}{4\pi^2 \epsilon_0 m_e}, \tag{13.3}$$

where N_e is the electron number density (in $\mathrm{m^{-3}}$), ϵ_0 is the permittivity of free space ($8.854 \times 10^{-12}\,\mathrm{F\,m^{-1}}$), e is the electric charge of the electron (-1.602×10^{-19} C), and m_e is the mass of the electron (9.109×10^{-31} kg).

Because $f_p \ll f$, the refractive index can be written using a Taylor expansion, neglecting terms beyond first order:

$$n = 1 - \frac{1}{2}\frac{f_p^2}{f^2}, \tag{13.4}$$

so that

$$n' = \left[1 - \frac{N_e e^2}{8\pi^2 \epsilon_0 m_e f^2}\right] - \left[1 - \frac{\langle N_e \rangle e^2}{8\pi^2 \epsilon_0 m_e f^2}\right] = \frac{\delta N_e e^2}{8\pi^2 \epsilon_0 m_e f^2}, \tag{13.5}$$

where $\langle N_e \rangle$ is the average electron density at the P-point (the point of closest approach of the LOS to the Sun). The weighting contributions about the P-point along with the weighting function are illustrated in Fig. 13.5 and further description can be found later in this sub-section and in the next.

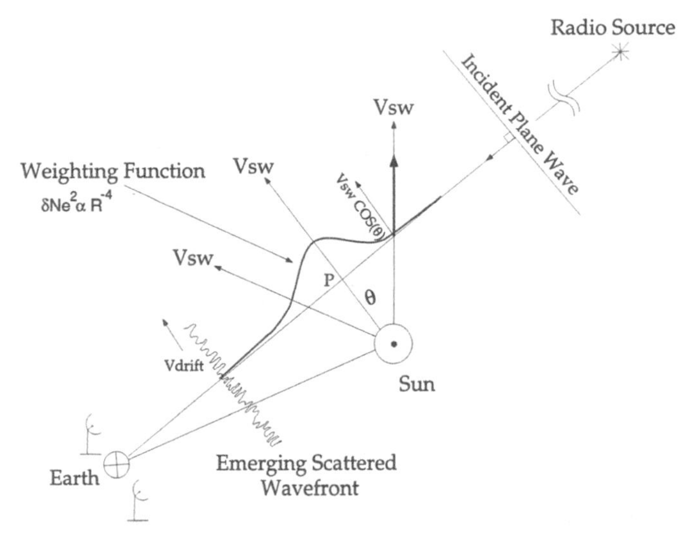

Fig. 13.5 The geometry of observations of IPS and how the weighting of scintillation potential along the LOS from the radio telescopes to the source varies due to the solar wind flowing across the LOS at different angles. (From Grall, 1995.)

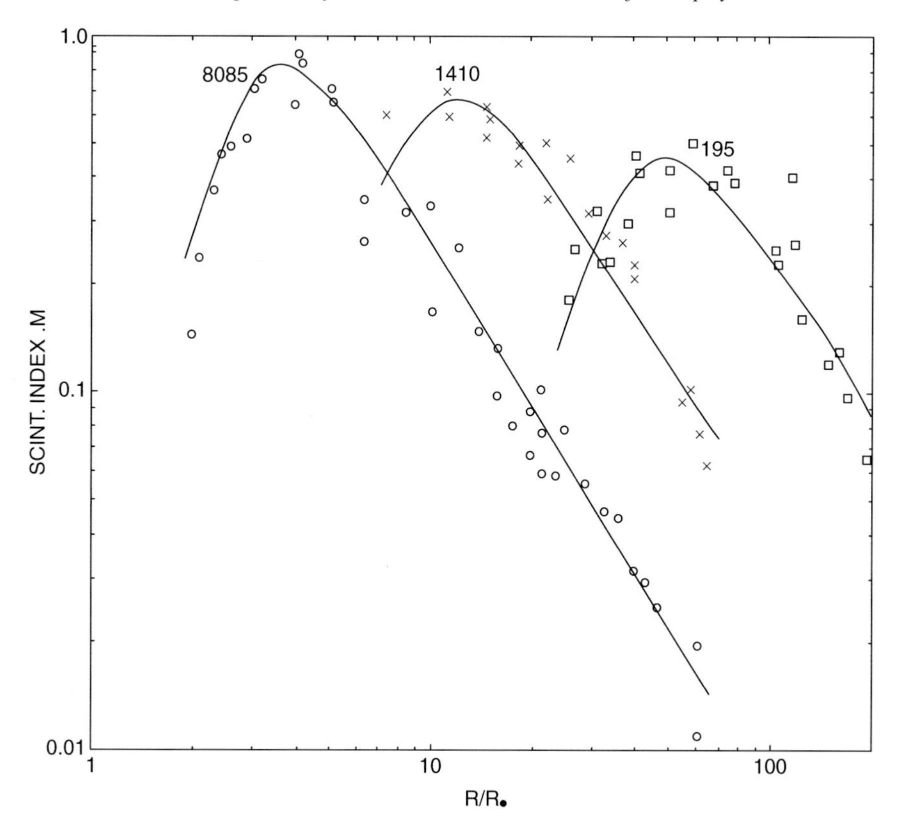

Fig. 13.6 A log–log plot showing for three different IPS observing frequencies of 8085 MHz, 1410 MHz, and 195 MHz, the scintillation index (m) response with distance from the Sun (in units of solar radius R_{Sun}). The scintillation index is the square of the scintillation power received at the antenna while observing a particular radio source. As illustrated by the figure, the higher the observing frequency, the closer-in to the Sun the peak in the scintillation index occurs and hence higher observing frequencies can be used to enable observations in the weak-scattering regime closer-in to the Sun as described by the earlier equations. (Reproduced from Coles, 1978.)

For observations closer-in to the Sun, N_e is larger and it is therefore assumed that δN_e is also increased and also the scale sizes of the irregularities are decreased (e.g., Fallows, 2001, and references therein). Hence, n' increases as the strong scattering regime is reached. Because n' is inversely proportional to the observing frequency squared, a higher observing frequency allows observations in the weak scattering regime to be made closer-in to the Sun than that of a lower observing frequency, as seen in Eq. (13.5) and Fig. 13.6. In contrast, as n' decreases with distance from the Sun, so does the scintillation level (amplitude) of the observation and a point is reached where the IPS amplitude variations become swamped with noise from the observing system. IPS is a capable method of observing the solar wind over a very wide range of distances, especially if multiple frequencies are at the observer's disposal.

In weak scattering, when the variation between the phase changes is much less than one radian, the detected scintillation pattern can be treated as a linear sum from all the scattering events along the LOS from the source to the receiver. Diffraction of the radio waves can be thought of as occurring at a series of "thin screens" and the phase variations that are introduced by diffraction at any single "screen" build up into amplitude variations over a distance which is then dependent on the actual size of the density variations in that particular "screen". In strong scattering, however, this is not at all the case. Although analysis of the IPS results is still possible – as demonstrated by Imamura *et al.* (2014) and references therein – the amount of reliable information that can be extracted from the results is considered to be vastly reduced. The variation in refractive index (and thus for phase change) is shown by Eq. (13.5). It shows that the variation in phase change will be reduced if the observing frequency is increased and that the transition from weak to strong scattering therefore occurs closer to the Sun at greater observing frequencies (Fallows, 2001; Fallows *et al.*, 2006; Bisi, 2006). If the strong scattering regime is reached, then a sudden drop in the scintillation index occurs as is demonstrated by Fig. 13.6 where, for each of the three example frequencies shown respectively, as the P-point gets closer-in to the Sun, the scintillation index increases to a maximum before turning over and decreasing again, signifying the change of regime from weak to strong scattering for each example frequency.

13.6.2.1 Details of multi-site IPS and further IPS theory

Figure 13.5 (from Grall, 1995) shows the IPS scintillation potential and observing geometry along the LOS from the Earth to the distant point-like radio source. This figure takes into account the cos θ effect for the foreshortening of the velocity crossing the LOS (because IPS is sensitive to the perpendicular component of velocity across the entire LOS).

Multi-site IPS is capable of providing a much more accurate determination of the solar-wind velocity (and other parameters) in general compared to a single-site observation of IPS. As already described, the radio waves scintillate due to the solar wind and cause a 2D pattern that drifts across the Earth due to the density irregularities in the solar wind flowing outwards from the Sun. Only if the alignment of the radio source, solar wind, and telescopes is such that the two receivers lie in the same plane which passes through the center of the Sun can the time lag between the signals recorded be used to derive the velocity, or velocities (depending on whether there are one or more streams crossing the LOS), flowing over the LOS. Observations of IPS are henceforth used as a tracer of the solar-wind outflow, and can be used to study the evolution of the solar wind structure throughout interplanetary space (e.g., Canals, 2002; Bisi, 2006).

One can also use IPS to determine the direction of outflow because of the nature of where the maximum peak in the CCF occurs compared with the radial direction of flow, and it can indicate non-radial flows in the meridional direction (e.g., Moran, 1998; Moran *et al.*, 1998; Bisi *et al.*, 2005; Breen *et al.*, 2006; Bisi, 2006; Breen *et al.*, 2008; Bisi *et al.*, 2010a). The time lag, δt, can be used at maximum correlation along with the information on the component of the baseline length measured parallel, the parallel baseline (B_{Par}), to the Sun–Earth line to give, to a first approximation, the velocity of the primary/dominant stream of solar wind flowing across the lines of sight. As the Earth rotates, the geometric baseline between the two observing sites projected onto the sky also rotates making the long side of a right-angled triangle (from the point of view of the radio source). The outflow of plasma makes one side of the triangle, B_{Par} between the two observing antennas lies along this outflow direction, and finally, the third side is as a result of the angle between B_{Par} and the projection of the geometrical baseline completing the right-handed triangle, is the perpendicular baseline (B_{Perp}) (which for a cross-correlation is usually significantly smaller than the length of B_{Par}). The initial velocity estimate is accomplished by using Eq. (13.6), where the term $Distance$ in the equation refers to the length of B_{Par} (which, for completely radial outflow, would result in maximum cross-correlation when B_{Par} lies completely along the radial direction with no perpendicular component):

$$v_{\mathrm{SW}} = \frac{Distance}{\delta t}. \tag{13.6}$$

This equation can also be used for the time lag of other peaks (if any) that appear in the CCF. Observations of IPS contain contributions from the whole of the LOS, but are only sensitive to the perpendicular component of velocity across the LOS as already noted. The aforementioned cos θ effect will therefore cause a broadening of the CCF and will lead to an underestimation of the velocity of the solar-wind stream(s) detected crossing the LOS. The maximum scattering will occur in a region along the LOS around the P-point of the LOS (again, see Fig. 13.5) because the scattering potential of the solar wind decreases with increasing distance from the Sun for a given frequency within the weak-scattering regime (again, see Fig. 13.6) as well as the perpendicular component across the LOS becoming progressively smaller. To a first approximation, this is $\propto 1/R^4$, which shows that the majority of the scattering occurs at the P-point (where $\cos\theta \simeq 1$). Hence, by a rough correction, the IPS results can be corrected by a multiplication of 1.18 (Breen *et al.*, 1996b), although more thorough methods of velocity calculation are usually undertaken (e.g., Coles, 1996; Fallows, 2001; Bisi, 2006; Fallows *et al.*,

2006; Jones, 2007; Jones *et al.*, 2007; Fallows *et al.*, 2008; Bisi *et al.*, 2010a,b; Fallows *et al.*, 2013).

Equation (13.8) shows the temporal power spectrum under weak scattering conditions, where r_e is the classical electron radius

$$r_e = \frac{1}{4\pi\epsilon_0}\frac{e^2}{m_e c^2} = 2.818 \times 10^{-15}\text{m}, \tag{13.7}$$

λ is the observing wavelength, α is the power-law exponent (e.g. Kolmogorov), v_p is the component of solar wind velocity perpendicular to the line of sight, q is the 2D spatial wavenumber (normally in $x-y$ coordinates), q_i is the inner-scale for turbulence (the scale at which the turbulence dissipates), z is the distance from Earth to the scattering "screen", θ_0 is the diameter of the source in radians, $v(q, z, \theta_0)$ is the visibility function of a radio source, ϵ_0 is the permittivity of free space, e is the charge of the electron, m_e is the mass of the electron, and c is the speed of light in a vacuum (2.998×10^8 m s^{-1}). It is described in its various forms by Salpeter (1967), Scott *et al.* (1983), Klinglesmith (1997), Moran (1998), Bisi (2006), Fallows *et al.* (2006), and references therein:

$$P(f) = 2\pi r_e^2\lambda^2 \int_0^\infty \frac{2\pi}{v_p(z)} \int_{-\infty}^\infty 4\sin^2\left(\frac{q^2\lambda z}{4\pi}\right) |v(q, z, \theta_0)|^2 q^{-\alpha} \exp - \left(\frac{q}{q_i}\right)^2 R^{-4} dq_y dz. \tag{13.8}$$

The full development and derivation of the underlying steps to get to this equation can be found in Salpeter (1967).

The Fresnel filter, described by Eq. (13.9) acts as a high-pass filter attenuating wavenumbers below the Fresnel spatial frequency, q_f, as given in Eq. (13.10). The Fresnel filter is necessary because the wavefronts need a certain distance from the point of scattering to develop into a scintillation pattern – this is known as the Fresnel distance – and it is defined as the distance at which these wavefronts are again in phase and the amplitude variations first become "fully developed". Because of this, the diffraction of radio waves that occur very close to the observer (near the Earth) would thus not have time to fully develop from phase changes into amplitude changes and therefore would not contribute very much to the overall scintillation pattern received at the Earth from along the IPS LOS. This is also illustrated by the weighting along the IPS LOS as seen earlier in Fig. 13.5. This means that the cause of the scintillation along the LOS is slightly biased to the source side of the P-point along the LOS and not the Earth side, although the majority of the scintillation comes from around the P-point as already described. Additional details and exploration of the scintillation potential along the IPS LOS are described by Fallows (2001).

$$Fresnel\ Filter = 4\sin^2\left(\frac{q^2\lambda z}{4\pi}\right), \tag{13.9}$$

$$q_f = \sqrt{\frac{4\pi}{\lambda z}}. \tag{13.10}$$

The Fresnel radius, defined by Eq. (13.11) is that of the first Fresnel zone:

$$r_f = \sqrt{\lambda z}. \tag{13.11}$$

This gives a maximum scale size of the irregularities for which amplitude scintillation can occur; it is dependent on the observing frequency and the distance to the scattering "screen". For example, for an observing wavelength (frequency) of ~21 cm (1420 MHz), the maximum scale-size of irregularity at a "thin screen" of scattering at 1 AU is ~177 km; at ~32 cm (928 MHz), this is ~219 km; and at ~60 cm (500 MHz), it is ~300 km.

The $\exp[-(q/q_i)^2]$ term in Eq. (13.8) describes the turbulent dissipation towards smaller irregularity scales and it also attenuates the scintillation power spectrum at wavenumbers higher than q_i. The source visibility function also acts as a low-pass filter attenuating wavenumbers above:

$$q_s = \frac{1}{z\theta_0}. \tag{13.12}$$

13.6.2.2 Increasing baseline length for multi-site observations of IPS

The ability of observations of IPS to resolve streams of solar wind with different velocities increases as the radial separation in the plane of sky of the lines of sight increases (Klinglesmith, 1997; Moran *et al.*, 1998; Bisi *et al.*, 2005; Breen *et al.*, 2006; Bisi *et al.*, 2010b; Fallows *et al.*, 2013). However, the time lag between the two sites is increased by the increase in B_{Par} between the two observing sites and as the irregularity pattern is evolving in time; any increase in B_{Par} will thus lead to a decrease in the degree of correlation, although the length of time that coherency can be maintained is still yet to be studied in any detail (Klinglesmith, 1997; Moran *et al.*, 1998; Bisi *et al.*, 2005; Bisi, 2006; Breen *et al.*, 2006). Thus far, correlations over periods of around eight seconds have been recorded (Bisi *et al.*, 2005; Bisi, 2006; Breen *et al.*, 2006).

The effect of increasing B_{Par} for a particular observation can be seen in Fig. 13.7. As B_{Par} is increased, the correlation of the signal is decreased at the primary peak (the dominant solar wind flow velocity) and is transferred elsewhere within the CCF. A second peak (i.e., a second solar-wind flow) can be resolved if one is present when B_{Par} is long enough. Indeed, Bisi *et al.* (2007a), Breen *et al.* (2008), Fallows *et al.* (2008), and Bisi *et al.* (2010a) have shown that up to three streams

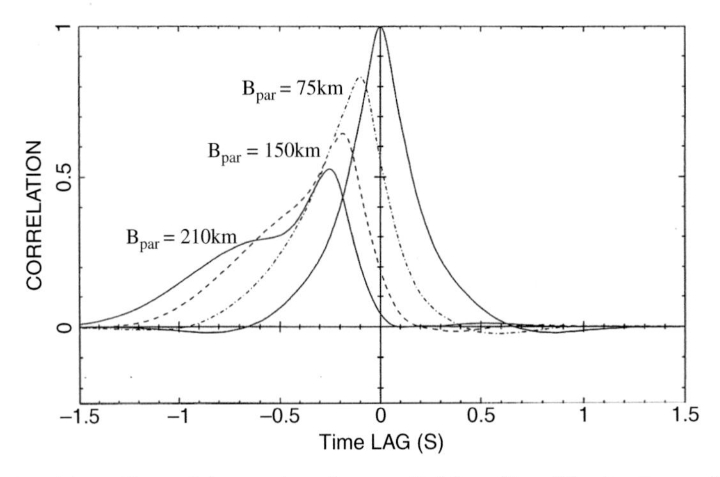

Fig. 13.7 The effect of increasing the parallel baseline (B_{Par}) of a multi-site observation of IPS on the CCF. (From Klinglesmith, 1997.)

can be resolved and subsequently determined in terms of their parameters and relative weighting factors between the differing streams present crossing the LOS.

The general approach of estimating the velocity of solar wind from the time lag does have some limitations, however. The most important of these is that a finite spread in solar-wind speeds across the stream considered will introduce a skew of the correlation function to shorter time lags, this time, leading in turn to an overestimate of the plane-of-sky speed (e.g., Breen *et al.*, 1996a).

The larger B_{Par}, the larger the B_{Perp} for a given angle relative to the solar wind outflow (θ):

$$\tan\theta = \frac{B_{\mathrm{Perp}}}{B_{\mathrm{Par}}}. \tag{13.13}$$

Thus, the likelihood of detecting any non-radial flow is increased if you use extremely-long-baseline (ELB) IPS (e.g., Bisi *et al.*, 2005; Breen *et al.*, 2006; Bisi, 2006; Bisi *et al.*, 2007b; Bisi *et al.*, 2010a) over shorter-baseline multi-site IPS alone.

13.6.2.3 Dual-frequency IPS

Equation (13.8) is suitable only for representing the CCFs for a single observing frequency. However, relatively simple modifications can be made to the equation to allow the cross-correlation of two different frequencies: one at each of the receivers used (Salpeter, 1967; Fallows *et al.*, 2006), or from the same site when observing at multiple frequencies or over a wide bandwidth. It is the terms involving λ^2 that split into two single λ_i terms (as per Fallows *et al.*, 2006, and also Bisi, 2006). The resulting Eq. (13.14) is taken directly from Fallows *et al.* (2006):

$$P(f) = 8\pi^2 r_e^2 \lambda_1 \lambda_2 \int_0^\infty \frac{2\pi}{v_p(z)} \int_{-\infty}^{\infty} \sin\left(\frac{q^2\lambda_1 z}{4\pi}\right) \sin\left(\frac{q^2\lambda_2 z}{4\pi}\right)$$

$$|v(q, z, \theta_0)|^2 q^{-\alpha} \exp - \left(\frac{q}{q_i}\right)^2 R^{-4} dq_y dz. \qquad (13.14)$$

Observations of IPS are sensitive to density scales that can overlap, and so still provide meaningful correlation when the two different observing frequencies are not too far apart and while observing at both still remains in the weak-scattering regime for both observing frequencies. Fallows *et al.* (2006) have shown that observations of IPS at frequencies of 500 MHz at one receiver and 1420 MHz at another still provide meaningful IPS results – suggesting that different scale sizes of micro-structure within the solar-wind outflow move outward with the same velocity and hence are in line with the bulk velocity as measured in situ. The maximum scale size of a density irregularity of a range of scale sizes is more-specifically determined by the observing wavelength (or frequency) as defined earlier in Eq. (13.11). These multi-frequency correlations are still obtainable even over the very long baselines such as from the ESR to the EISCAT mainland sites, and also with the ELB observations between EISCAT and MERLIN sites using the Tromsø receiver recording at 928.5 MHz and all others at 1420 MHz (e.g., Bisi, 2006; Jones, 2007).

13.6.3 Three-dimensional computer-assisted tomography (CAT)

Many observations of IPS over several weeks or months will allow for the three-dimensional (3D) computer-assisted tomography (CAT) reconstruction of the detailed structure of solar-wind outflow throughout the inner heliosphere. Since the 1990s, a set of CAT techniques have been developed at UCSD. Here, we focus on the UCSD techniques that are used for both detailed science investigations as well as for real-time space-weather forecasting (see, e.g., http://ips.ucsd.edu/, http://www.spaceweather.go.kr/models/ips, and http://helioweather.net/models/ipsbd/vel3r1e1b/index.html).

The UCSD 3D CAT reconstructions use perspective views of solar co-rotating plasma (e.g., Jackson *et al.*, 1998, 2011b) and of outward-flowing solar wind and transient features (e.g., Jackson and Hick, 2005; Jackson *et al.*, 2011b) crossing the IPS observing lines of sight, and also with the inclusion of various near-Earth in-situ measurements directly into the time-dependent reconstructions (Jackson *et al.*, 2010b, 2013). Both velocity and density are obtained through the use of a kinematic solar-wind model based on the conservation of mass and mass flux in the heliosphere as structure propagates outwards with a radial-flow assumption from a model source surface at 15 $R_\odot$ out to 3 AU (or about 650$R_\odot$). Schematics showing

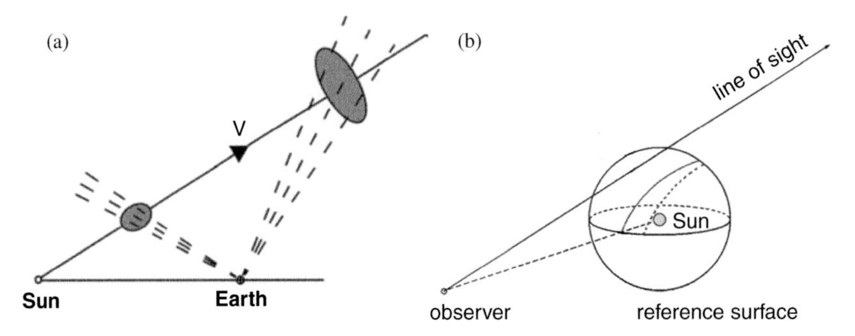

Fig. 13.8 The left-hand image (a) shows a schematic of the outward-flowing solar-wind structure which we know follows very specific physics as propagation away from the Sun takes place. By observing large areas of the sky with time, this allows individual features to be observed several times over time and this has the same effect as a moving detector in a comparable instantaneous time such as CAT scans used in medical imaging where sensors rotate around the body and not the body moving through the sensors in time. The right-hand image (b) depicts a schematic of how an individual IPS LOS maps/projects onto the source (reference) surface for use in the UCSD 3D CAT; the solid line is the immediate projection of the LOS to this surface and the dashed line is the projected location that takes into account the solar wind speed. (Adapted and combined from Jackson *et al.*, 2010a, 2011b.)

separately the overall perspective view as well as then how the projection of each LOS maps onto a source surface above the Sun are shown in Fig. 13.8.

The IPS lines of sight (or portions of each line of sight) are mapped onto the source surface for a particular region of interest, usually a Carrington rotation (CR) in length, before then being fitted iteratively to a best-fit solution. A display of the lines of sight onto two CR maps is shown in Fig. 13.9.

Using the IPS scintillation level converted to the disturbance factor level or normalized scintillation level, otherwise known as the g-level, as a proxy, the solar wind density can be inferred from observations of IPS in the 3D CAT reconstructions (e.g., Hick and Jackson, 2004; Jackson and Hick, 2005; Bisi *et al.*, 2007c; Breen *et al.*, 2008; Bisi *et al.*, 2009b; Jackson *et al.*, 2013). The 3D velocity reconstructions can take place directly from the IPS velocity determinations, however (e.g., Bisi *et al.*, 2007b, 2010c). For density, the g-level is defined by:

$$g = m/\langle m \rangle, \tag{13.15}$$

where, m is the observed scintillation level and $\langle m \rangle$ is the modeled mean level of $\Delta I/I$ for the source at its elongation and gain calibration at the time of its observation, for source-intensity variation, ΔI, and the measured signal intensity, I. As described in detail by Jackson and Hick (2005) and taken from Bisi *et al.* (2008c),

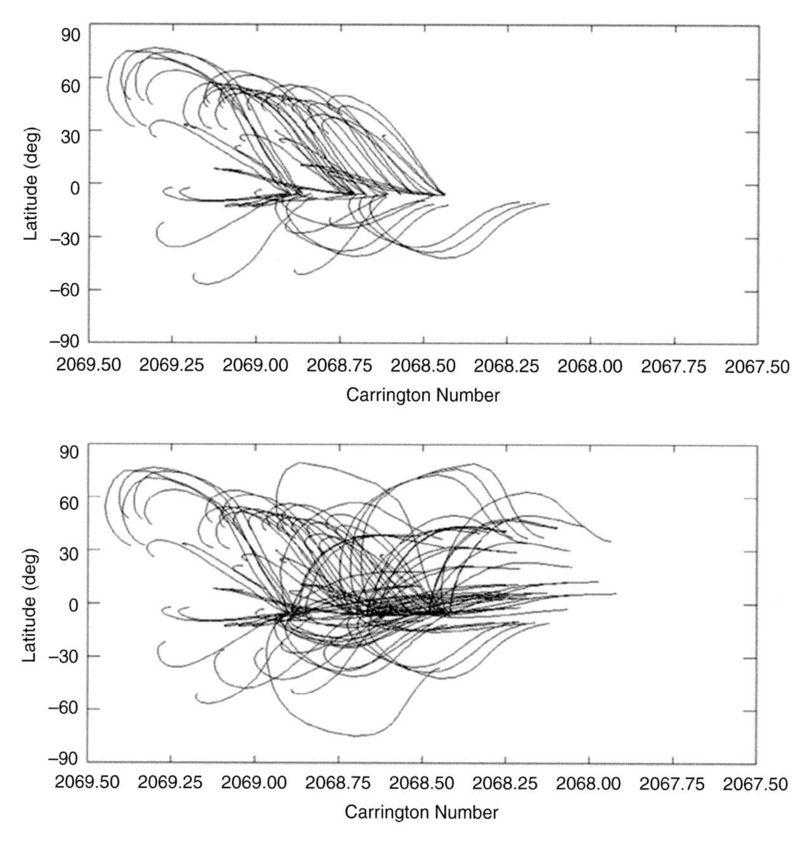

Fig. 13.9 The STELab IPS lines of sight used by the UCSD 3D CAT from g-level (top) and velocity (bottom) determinations from observations of IPS using the STELab system mapped back to the 15 $R_\odot$ source surface (lower boundary) in the UCSD 3D CAT. The display is over two Carrington rotations (CRs), and as can be seen, the velocity coverage is far greater onto these two CR maps used for the tomography (CR2067.50 to CR2069.50) than that of the g-level coverage in this particular example. The numbers of data points (lines of sight used) for velocity and g-level need not be the same and indeed often differ. (From Bisi *et al.*, 2009a.)

scintillation-level measurements have been available from STELab (now ISEE) since 1997 for example, but velocity determinations from much earlier.

The g-level proxy for density uses Eq. (13.16) because density values along a LOS are not known a priori but are assumed for small-scale variations with a power-law scaling of heliospheric density:

$$\Delta N_e = A_c R^\alpha N_e^\beta. \tag{13.16}$$

Here, A_c is a proportionality constant, R is the radial distance from the Sun, α is a power of the radial falloff, and β is the power of the density. The parameters A_c,

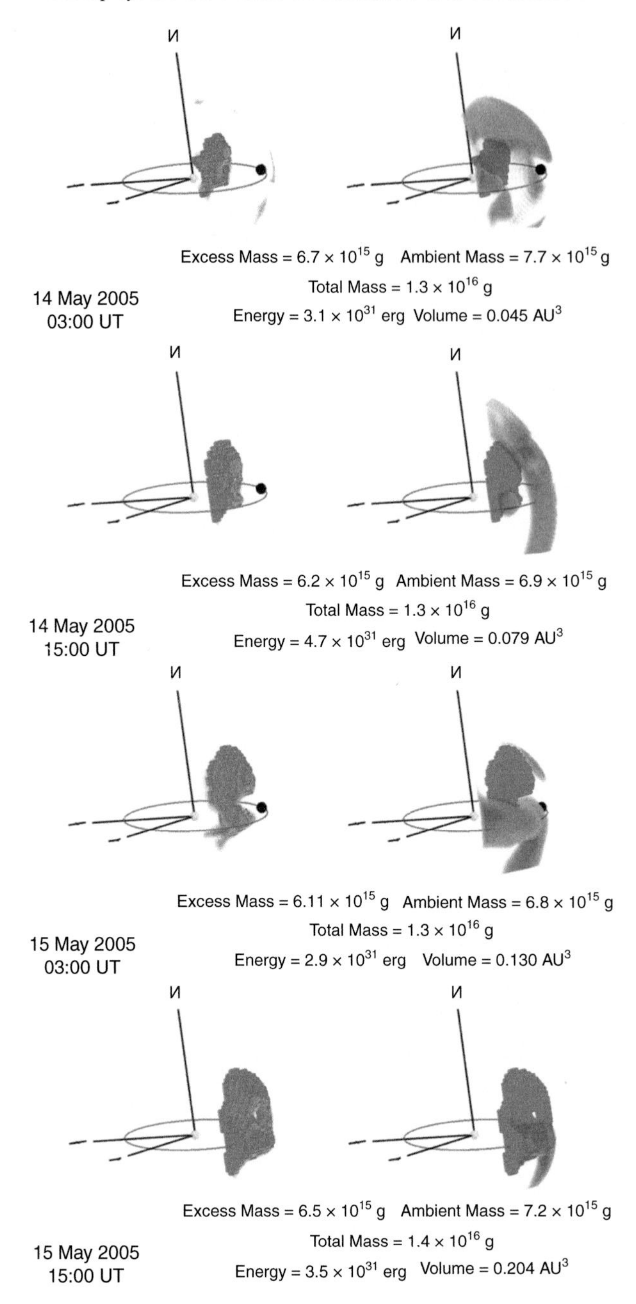

Fig. 13.10 The 3D CAT reconstructed visualization of the distribution of solar-wind density upwards of $8\,e^{-}\,cm^{-3}$ (brighter colors toward yellow mean increasing density) on the left-hand side and high-velocity portions (blue) on the right-hand side showing the developing and changing reconstructed structure of the 13–15 May 2005 coronal mass ejection (CME) event sequence. The left-hand density images are highlighted with green cubes to encompass the reconstructed volume of the mass portion of the CME. This same highlighted volume is depicted

α, and β are determined using best-fit comparisons with in-situ measurements at 1 AU. The values shown here are as previously used for 327 MHz IPS observations: A_c is set equal to 1, α is determined as -3.5, and β determined to be 0.7 (see Jackson *et al.*, 2003, for further details).

The UCSD 3D CAT reconstruction techniques can also be applied to the Thomson-scattered white-light observations from the Solar Mass Ejection Imager (SMEI) (Eyles *et al.*, 2003; Jackson *et al.*, 2004) with much-improved spatial and temporal digital resolution in density alone (e.g., Jackson *et al.*, 2008; Bisi *et al.*, 2008a,c; Jackson, 2011; Jackson *et al.*, 2011a,b; Yu *et al.*, 2014), and to IPS data from the ORT at an intermediate-level resolution for both density and velocity (e.g., Bisi *et al.*, 2009b). Similar much-improved-quality 3D IPS CAT reconstructions are expected from LOFAR and MWA (when the MWA becomes fully IPS capable), and eventually with the Square Kilometre Array (SKA) also. Plans are underway for the use of Solar TErrestrial RElations Observatory (STEREO) Heliospheric Imager (HI) data for input to the UCSD CAT to enable higher-resolution density reconstructions confined to a relatively-small volume along the Sun–Earth line. IPS can also be included with SMEI (or HI) data to allow for velocity to also be reconstructed alongside the density from SMEI (or HI). However, both the radio source size (angular size on the sky) and the observing frequency determine a weighting factor needed for the distribution of IPS contributions along each LOS. Generally, however, the greater the number of LOSs as input to the tomography, the higher the digital resolution and temporal cadence that can be achieved. In addition, these 3D reconstructions can be used to better constrain the analyses of the greater-sensitivity ELB observations of IPS by providing accurate contextual information on where streams/structures are located crossing the LOS (e.g., Breen *et al.*, 2006; Bisi *et al.*, 2007b; Breen *et al.*, 2008; Bisi *et al.*, 2010a).

Figure 13.10 (from Bisi *et al.*, 2010a; further details of this event and this set of images are given there) shows an example of some of the physical parameters that are able to be obtained from low-resolution UCSD CAT reconstructions (in this

Caption for 13.10 (cont.)

on the right-hand velocity reconstructions for illustrative purposes. Each image is labeled with the masses, volume, and energy values on each date and time as shown. All non-CME-related features have been removed for clarity of viewing when displaying the 3D volume. The axes are heliographic coordinates with X-axis direction pointing toward the vernal equinox, and Z-axis directed toward solar heliographic North. An r^{-2} density increase has been added to better-show structures further out from the Sun (the central sphere) to the Earth (the blue sphere) along with the Earth orbit (ellipse). (From Bisi *et al.*, 2010a.) A black and white version of this figure will appear in some formats. For the color version, please refer to the plate section.

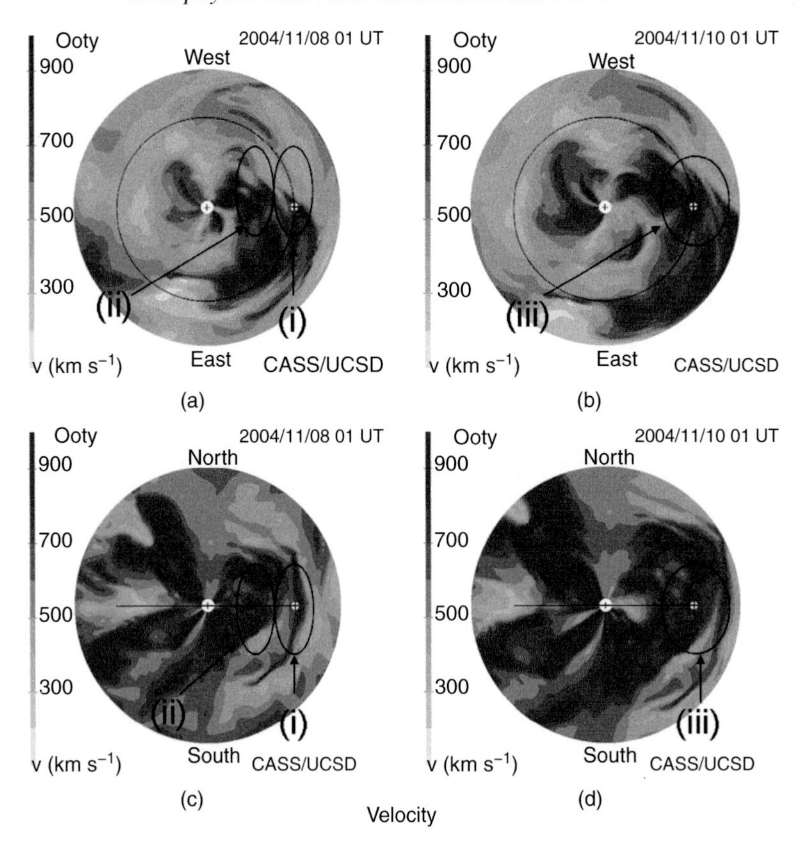

Fig. 13.11 Summary figure of the Ooty ecliptic, (a) and (b), and meridional, (c) and (d), cuts through the 3D velocity (this page) and density (next page) reconstruction out to 1.5 AU at the date and times shown. Various features are circled in the images which relate to various CME features from interacting CMEs which ultimately result in multiple geomagnetic storms at Earth; the in-situ time series from such are shown in Fig. 13.12. Earth's orbit is shown as a near-circle or line with the Earth, $\oplus$, indicated on each panel. Velocity contours are shown to the left of each of the four velocity images and similarly those are also given for the density cuts. These are prime examples of how the IPS CAT can work for interacting CMEs through the inner heliosphere. (From Bisi *et al.*, 2009b.)

case, $20° \times 20°$ digital resolution in latitude and longitude with 0.1 AU distance increments from the Sun on a daily cadence broken into six-hourly interpolations – typical resolutions used for STELab or EISCAT IPS data), or at a higher resolution showing many more detailed features as in Fig. 13.11 (taken from Bisi *et al.*, 2009b, where a more-in-depth discussion and explanation of these labeled features are given) using IPS data from Ooty (at a higher $10° \times 10°$ digital resolution in latitude and longitude with 0.1 AU distance increments from the Sun on a half-daily cadence broken into three-hourly interpolations). In addition, Fig. 13.12 shows an

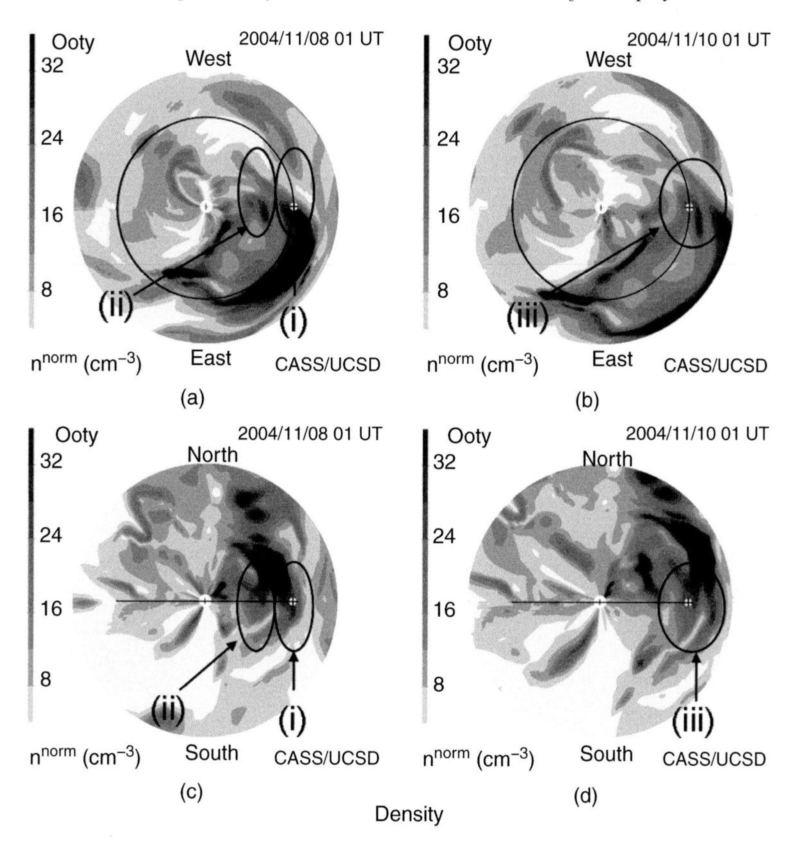

Fig. 13.11 (cont.)

in-situ time series extraction from both the velocity (top) and density (bottom) as compared with the hourly-averaged in-situ measurements recorded by the Wind spacecraft. As with Fig. 13.11, a detailed description of these labeled features in Fig. 13.12 is given in Bisi *et al.* (2009b).

13.6.4 The future of space weather science and forecasting: the IPS contribution to heliophysics

As a reminder, one of the strongest points of the IPS technique is that it enables us to make global observations of the inner heliosphere including a high coverage of the polar regions that are inaccessible to in-situ measurements until another out-of-ecliptic spacecraft mission becomes available, as well as within a large range of distances out from the Sun (from as near as a few $R_\odot$ to around 3 AU from Sun center – which includes the near-Sun regions inside the orbits of Mercury and Venus). The global observations of the solar wind in this current solar cycle (Cycle 24) are particularly valuable from the viewpoint of response to peculiar

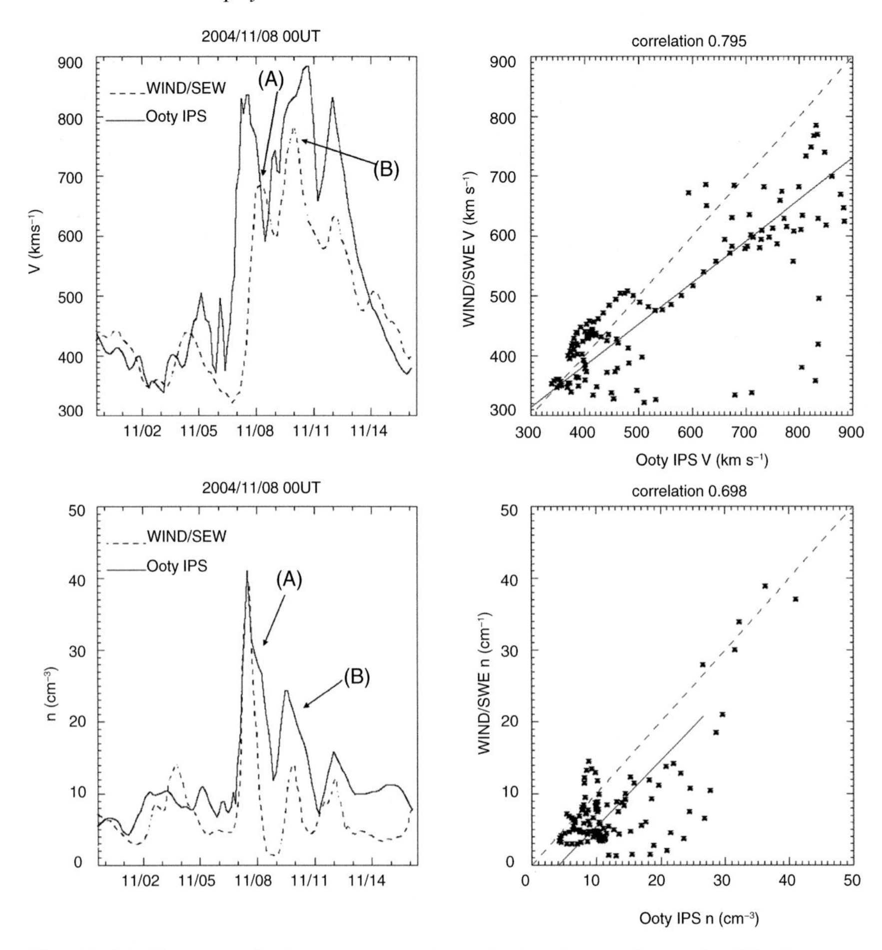

Fig. 13.12 The top-left plot compares the velocity time series at the Wind spacecraft extracted from the Ooty IPS reconstruction (solid line) with Wind solar wind velocity measurements (dashed line), and similarly the bottom-left plot for Wind density measurements. Both the Wind spacecraft velocity and density data are hourly-averaged data that were further averaged with a half-daily cadence to match that of the 3D reconstruction cadence. The top-right plot shows a scatter diagram for the two data sets for velocity, and similarly the bottom-right plot for density; the dashed line on each correlation plot is for a 100% correlation while the solid line shows the best-fit of the data here. (A) and (B) on the left-hand plots relate to features also highlighted earlier in Fig. 13.11. (From Bisi *et al.*, 2009b.)

solar activity (perhaps signifying an exit from a period of grand maxima) and also from the viewpoint of crucial information to interpret Voyager measurements at the outer boundary region of the heliosphere (M. Tokumaru, private communication, 2015).

IPS has been shown to be able to detect Earth-directed CMEs, CMEs with a glancing blow on the Earth, and CMEs elsewhere throughout the inner heliosphere (e.g., Manoharan *et al.*, 2000; Fallows, 2001; Breen *et al.*, 2006; Jones, 2007; Jackson *et al.*, 2007; Jones *et al.*, 2007; Breen *et al.*, 2008; Bisi *et al.*, 2008b; Bisi *et al.*, 2010a, d; Jackson, 2011; Jackson *et al.*, 2011b; and references therein) making it a technique with improved proven potential for space-weather studies and for studying Earth-directed events with its forecasting capabilities continuously being advanced through new analyses and modeling techniques (e.g., Jackson *et al.*, 2013; Bisi *et al.*, 2014; Aguilar-Rodriguez *et al.*, 2014; Jackson *et al.*, 2015a, b; Aguilar-Rodriguez *et al.*, 2015; Jian *et al.*, 2015; and references therein).

It may be possible for multi-site observations of IPS to provide a way of directly estimating the magnetic-field direction of the plasma outflow from the Sun, at least when observing certain solar-wind/transient conditions. Under some circumstances, usually during an observation of a CME, the CCF displays a negative lobe near to or at the zero time lag (e.g., Klinglesmith *et al.*, 1996; Klinglesmith, 1997; Jones *et al.*, 2007; Bisi *et al.*, 2014). Most, if not all, CMEs are now considered to contain magnetic flux ropes that are ejected from the Sun (N. Gopalswamy, private communication, 2013). The density irregularities giving rise to IPS are therefore expected to be anisotropic, usually assumed to be "cigar-shaped" with the long axis parallel to the interplanetary magnetic field. In the regular solar wind, this elongation in the irregularities is usually parallel to the solar wind flow along the outflow direction (approximately radial in direction from the Sun). Modeling of the spatial correlation function by Klinglesmith *et al.* (1996) relates the negative lobe seen in IPS CCFs to the rotation of density irregularities such that their long axis is now more perpendicular to the solar-wind outflow, thus indicating rotation of the magnetic field relative to this flow. Whilst this has often been used as a signature that a CME is likely to be present crossing the LOS, no attempt (thus far) has been made to relate this more precisely to the magnetic-field rotation itself, or to use observations of IPS to sample the spatial correlation function in order to more-directly assess the irregularity orientation in detail.

In order to estimate the orientation of density irregularities in the solar wind, multi-site observations of IPS are required to form different CCFs with different baseline lengths and B_{Par} orientations relative to the plasma outflow. LOFAR has the distinct advantage for such an investigation of the CCF due to the large number of widely spaced stations that allow the spatial correlation function to be sampled in single, short observations, while increasing these numbers further with the inclusion of simultaneous observations of IPS with EISCAT and/or KAIRA much further north in geographic latitude.

Recent LOFAR and KAIRA test observations of both IPS and ionospheric scintillation have enabled the simultaneous observing of well over ~80 MHz in bandwidth. This allows for the creation of dynamic spectra to investigate the scintillation at all observing frequencies and to study the type of scattering and the transition from weak to strong scattering for a particular observation of IPS precisely where this transition occurs for each observation.

In addition, when simultaneous observations of IPS are undertaken at a range of observing frequencies, say f to $6f$ (say, 40 MHz to 240 MHz – a capability of LOFAR now with the exception of the commercial FM band and a small band higher up in frequency), such data would be extremely useful in examining the scale size of the density irregularities responsible for IPS, and the shape of the density and cross-frequency spectra (e.g. Liu and Peng, 2010). It is also very useful for studying the transition between weak and strong scattering of the interplanetary medium and to extract the physical properties of the solar wind (P. K. Manoharan, private communications, 2013, 2015). The correlation between multiple observing frequencies can also provide the effective scale size of IPS in the solar wind and can identify the phase coherence as a function of the size of the irregularities. In the case of a propagating radio burst (e.g., type-III solar radio-burst emission, cf., Chs. 4 and 5 in Vol. III), from low to high coronal heights (generally, higher to lower frequencies as the source region propagates out from the Sun), the scattering by random density fluctuations plays an important role in increasing the angular width of visibility of radio emission with respect to the center (or direction) of the electron beam causing the emission. In other words, this scattering distorts the characteristics of the frequency of solar radio bursts. The smallest scale size of density irregularities in the case of IPS, the cut-off scale or inner-scale size (e.g., Coles and Harmon, 1978) present in the solar wind will determine the typical scale size of type-III radio emission.

The IPS of individual radio sources gives information on the small-scale density turbulence level and the solar-wind speed crossing the LOS in the direction of the radio source. The nature of the turbulence changes depending on the features crossing over the LOS and this is reflected in the spectrum (e.g., Coles and Harmon, 1978, 1989; Manoharan *et al.*, 2000). The study of the scintillation turbulence aspect of IPS has also had relevance much further afield, such as in the detection of submarines using sonar buoy arrays, and in providing laser-guided bombs (B. V. Jackson, private communication, 2015).

The IPS information is also important for the problem of coronal sources of the solar wind streams with different properties as well as for solar-wind acceleration. The IPS information, in particular about the solar-wind-speeds spatial distribution, can be used as initial conditions in the problem of the outer-heliosphere modeling. Further investigations into the solar-wind acceleration regions are open to higher-frequency ($\sim$1420 MHz and higher) observations of IPS probing close-in to the Sun (the inner-most part of the heliosphere right down to the low corona well inside coronagraph fields of view). This, coupled with recent work on velocities obtained from coronal imaging (e.g., Jackson *et al.*, 2014; and references therein), paves a way to investigate precisely where the solar-wind acceleration takes place and to help pin down methods by which the solar wind undergoes this acceleration. This also allows another method of studying the possible bi-modal nature of polar fast solar wind as a result (noted earlier) of IPS and Ulysses investigations carried out by Bisi *et al.* (2007a).

The comparison between the data on small-scale density turbulence, from one side of IPS capabilities, and the large-scale density and other plasma parameters measured by in-situ and spacecraft radio occultation methods, from the other side, show that IPS is important for the problem of the physical nature of the solar-wind plasma turbulence and for the problem of connection between the mean solar-wind parameters and the statistical turbulence parameters throughout the inner heliosphere (I. Chashey, private communication, 2015).

More recently, the UCSD IPS 3D CAT has been shown to be able to successfully combine the current-sheet source surface (CSSS) field modeling (Zhao and Hoeksema, 1995a,b) with the velocity information gleaned from the IPS to provide a system that allows the extrapolation upward of closed fields from near the solar surface (Jackson *et al.*, 2015a). This method allows for the comparison of the north–south component of magnetic fields to be investigated and compared with near-Earth in-situ measurements for solar-minimum conditions. It has successfully been tested for two recent CME case studies with much success (3–5 January 2015 stream interaction region (SIR) and the 7–8 January 2015 CME, as well as the St. Patrick's Day 2015 CME events), and provides a potentially viable near-future method of forecasting the southward component of the magnetic field benefiting greatly from the incorporation of IPS data.

Finally, Jian *et al.* (2015), conducted a survey of the current coronal and heliospheric models that are hosted and run at NASA's Community Coordinated Modeling Center (CCMC). A total of 10 model runs were carried out for the coronal/heliospheric modeling with results from the UCSD 3D CAT matching the best overall with observed time series of velocity and density. These modeled values also work best for SIRs (including the recurring co-rotating interaction regions or CIRs) throughout seven solar (Carrington) rotations modeled in 2007 (around the time of solar minimum). This matches well with the prospects and recent results discussed by Bisi *et al.* (2014) and also by Jackson *et al.* (2015a) for the IPS-driven ENLIL model (ENLIL is a 3D MHD numerical simulation code used for heliospheric modeling and space-weather forecasting around the world; e.g., Odstrcil and Pizzo, 2002; Odstrcil *et al.*, 2005; and references therein), and its potential improvements over other variations in driving ENLIL.

In summary, observations of IPS will play a unique role complementing (and leading) investigations of the inner heliosphere. Space-weather forecasting requires a complex combination of solar, interplanetary, magnetospheric, ionospheric, and geomagnetic observations and measurements. As noted, the strongest geomagnetic storms are produced by the passing of CMEs on the Earth's bow shock. IPS can be used in combination with other remote-sensing techniques to study the "evolution" of CMEs from their onset at the Sun and as they propagate through the inner heliosphere (e.g., Manoharan, 2006; Hardwick *et al.*, 2013; and references therein). Observations of IPS can provide unique information on the shape, size, and velocity of such interplanetary disturbances out to say 3 AU or more distance from the Sun. Therefore, coordinated observations of IPS using the IPS-capable instruments located at different geographic longitudes, in principle, should give the possibility for continuous monitoring of the same plasma features flowing out in the solar wind or one and the same spatial region (A. Gonzalez-Esparza, private communication, 2015). However, there is not currently a unified methodology to infer solar-wind parameters from the observations of IPS and so it is required that the community needs to carry out a survey and testing of the different IPS methodologies in order to unify their results and ensure IPS results are consistent from system to system. This also provides a large potential for continuous CME tracking, for example, as well as much-improved inputs to, say, the UCSD 3D CAT reconstructions, which would allow for higher-resolution results and should further improve the space-weather forecasting capabilities of such a system as well as when using outputs from the UCSD 3D CAT to drive 3D MHD code such as the aforementioned ENLIL or for driving other forms of 3D MHD modeling (e.g., Kim *et al.*, 2014; Jackson *et al.*, 2015b; Yu *et al.*, 2015; and references therein).

13.7 Faraday rotation in heliophysics

An almost-unique means of estimating the true magnetic field of inner-heliospheric structure over wide regions is by Faraday rotation (FR), particularly for large-scale disturbances such as CMEs (Jensen and Russell, 2009; Spangler and Whiting, 2008; Jensen *et al.*, 2010; You *et al.*, 2012). Faraday rotation is the rotation that occurs as an EM wave traverses a birefringent medium such as the plasma of the solar corona and solar wind. It is typically an astrophysical technique that uses pulsars and extragalactic radio sources to study the galactic magnetic field. It is the integrated product of the electron density and the component of the solar magnetic field parallel to the wave vector of the EM wave:

$$FR = \frac{e^3 \lambda^2}{8\pi^2 \epsilon_0 m^2 c^3} \int_{Obs}^{Src} N_e(\mathbf{R}) \, \mathbf{B}(\mathbf{R}) \cdot d\mathbf{s} \tag{13.17}$$

for electron density and **B** distributions, $N_e(\mathbf{R})$ and $\mathbf{B}(\mathbf{R})$ (where the parameters are as previously described for the IPS equations earlier in the chapter).

Faraday rotation is the only readily available (and most-obvious) method to measure the magnetic field of the corona on large spatial scales outside of active regions on the solar surface. As a polarized EM wave from a spacecraft, a bright pulsar, the galactic background synchrotron radio emission, a radio galaxy, or other natural radio source propagates through the birefringent plasma of the solar corona or inner heliosphere, the plane of polarization of the EM wave rotates as a function of the magnetic field that is carried by the plasma.

When a CME, propagating outward from the Sun (in an assumed radial motion), crosses the LOS over a period of time, the FR changes due to changes in the electron density and the parallel component of the magnetic field within the CME structure crossing the LOS. The measurements of FR through the corona and inner solar wind have thus far only been carried out at relatively high observing frequencies ($\sim$1400 MHz and upwards) and only close-in to the Sun (inside of $\sim 8^\circ$ elongation/$\sim 32\, R_\odot$). The magnitude of FR is proportional to the LOS integral of the plasma density (something which can be obtained through IPS and 3D reconstructions of the inner heliosphere) and the LOS component of the magnetic field.

Determining the intrinsic magnetic field within CMEs (and SIRs/CIRs) as they propagate through the inner heliosphere is thus of utmost importance for the prediction and forecast of incoming events with space-weather potential at the Earth, as already highlighted in the IPS discussion at the end of Sect. 13.3. In all, space-weather prediction is key on the determination of the parameters of velocity, density, and magnetic-field orientation combined, because these are the three main contributing factors determining how geo-effective a CME/SIR/CIR will be.

13.7.1 The potential ionospheric dominance in FR observations

Observations of FR in the inner heliosphere have been carried out before using spacecraft beacons at periods of close conjunction to the Sun (e.g., Jensen and Russell, 2008a, b). However, for a global picture, natural radio sources have to be used – whether they are individual polarized radio sources or the background galactic linearly polarized synchrotron radio emission. In these cases, FR due to both the interstellar medium (ISM) and the Earth's ionosphere need to be taken into account. The magnetospheric contribution is thought to be much smaller and not such a concern. The FR due to the ISM is inherent in the rotation measure (RM) of an observed pulsar and can be assumed to be constant over the proposed duration of these observations. The ionosphere, however, is much more variable and more of an unknown quantity. Indeed, the variation in the ionospheric FR can be greater than the total contribution of the heliospheric FR itself. Recent results using data from LOFAR (Sotomayor-Beltran *et al.*, 2013) have used total electron content (TEC) maps obtained from Global Positioning Satellite (GPS) data to calculate the expected RM due to the ionosphere and compare this with the measured RM from a number of well-known and well-categorized pulsars. The results are encouraging enough to expect that adequately correcting for the ionosphere is possible via this method and a successful preliminary heliospheric FR observation and slow-solar-wind result has been reported by, e.g., Bisi *et al.* (2014). Using observations from different receiver stations will also result in different RMs with the differences due only to the ionosphere, which could then be used to estimate its overall effect. Any such differences due to the heliosphere should not be noticeable over the range of baselines covered by LOFAR.

13.7.2 Overview of coronal investigations with FR

Previous comparisons have revealed magnetic flux ropes within CMEs where the flux-rope structures measured in situ are often well aligned with the flux cylinder reconstructed in three dimensions using white-light imagery (Jensen *et al.*, 2010). FR observations and **B**-field determination are available from spacecraft transmission signals, non-CME ionospheric backgrounds, and modeling (e.g., Jensen *et al.*, 2010; Sotomayor-Beltran *et al.*, 2013; Jensen *et al.*, 2013a, b; Xiong *et al.*, 2013). Pioneering FR measurements performed by Stelzried *et al.* (1970) using spacecraft-based radio signals not only demonstrated the strong effect of the quiescent solar magnetic field on FR signals, but they also revealed the presence of the magnetic flux ropes (Jensen and Russell, 2008a,b). They also found that a particular advantage of using FR to identify CMEs is that the flux-rope orientation defines its signature in FR with a range of shapes varying from paraboloid dips and hills or via forward and reverse sigmoids. However, this solution was degenerate between the handedness of the helical magnetic field and the orientation of its main axis. The degeneracy could be overcome using a radio telescope capable of simultaneously/near-simultaneously imaging, such as that of LOFAR, or indeed eventually the SKA. Furthermore, Jensen *et al.* (2010) found that the structure varies on the order of hours creating a boundary condition through which to separate out the structure from the background. Results so far established using LOFAR suggest a strong viability of heliospheric FR for both science and as a real potential for future use in space-weather forecasting, but much work is still needed on refining the experiment as well as incorporating other data sources (e.g., You *et al.*, 2012; Sotomayor-Beltran *et al.*, 2013; Bisi *et al.*, 2014).

13.7.3 An introduction to heliospheric FR

A single observation of FR allows the parallel component of the magnetic field to be determined. Therefore, to establish the parallel component in the first instance, the electron density along the LOS needs to be estimated. The column density along the LOS is directly related to the dispersion measure (DM) of the pulsar (i.e., the number of free electrons between us and the pulsar per unit area). The DM (normally determined in units of pc cm^{-3}, where pc is the unit of parsec) is seen as a broadening of the sharp pulse normally observed from the pulsar when the pulse is observed over a finite bandwidth. The density contributions from the ISM are accounted for in the inherent DM of the radio-source signal, leaving only the heliospheric and ionospheric components remaining. The electron density is simply obtained by dividing the DM by the distance to the pulsar (in pc) to provide the electron density, N_e cm^{-3}. Because ionospheric TEC has been demonstrated as being treated sufficiently, leaving only the heliospheric component remaining (e.g., Sotomayor-Beltran *et al.*, 2013), then heliospheric FR is a well-developing scientific and space-weather capability. Ancillary data from coronal and heliospheric white-light imagers and from IPS can also be used to establish general solar-wind conditions along the LOS to model contributions of these conditions to the overall heliospheric DMs and RMs.

For self-consistent modeling of the observed FR, especially when exploiting the continuities along time and space axes, any time-evolving CME model will yield a very-tightly constrained model for the CME flux rope. In fact, using the time series of FR observed along a single LOS provided by satellite beacons, Jensen and Russell (2008a) have successfully constrained the flux-rope orientation, position, size, velocity, rate of change of rope radius and pitch angle. The λ^2 dependence of FR makes the observable effect grow larger with increasing wavelength λ, making it easier to discern at lower frequencies – another reason why LOFAR (and the low-frequency antennas of the SKA) are (will be) perfectly suited receivers for such observations of FR. MWA is another potential radio instrument for observations of FR, and its potential for such is under investigation.

A practical way to observe potential space-weather events will be to initiate FR observations in response to an external trigger, most likely say from a space-based observatory/monitor, which will provide information of the launch time, direction, and transit velocity of a CME. Using this information, and perhaps near-real-time information from IPS or some form of white-light heliospheric imagers, the patch of the sky to be monitored at a given time will be chosen. However, continuous observations from a dedicated system would be preferred both for scientific and forecasting perspectives. It should be noted, though, that the ionospheric FR signal is expected to possibly dominate over the CME FR signal in the bulk of region of interest, and will have its own time and direction dependence. This could be circumvented using multiple simultaneous observations from different receivers.

For a complete unwinding of the FR signal and determination of the southward-component of the magnetic field, multiple observations of FR are necessary. For example, using the background linearly polarized synchrotron radio signal or the polarized signal from multiple individual radio sources as signals pass through a CME's plasma from multiple lines of sight. If a large number of such polarized radio sources are observed along, say, the projected path of a CME on the sky plane, then the results of these FR observations can provide multiple independent values of FR along different lines of sight passing through different parts of the CME. As noted, FR observations are sensitive only to the LOS component of the magnetic field, but like all propagation effects, the observed FR corresponds to the integral along the entire LOS. However, the observations of hundreds (or thousands) of lines of sight through the CME (and the rest of the inner heliosphere) as material travels through the inner heliosphere provide a large number of independent constraints. It is likely that such observations can also be incorporated into tomographic techniques such as those already described for IPS, and these could allow for full three-component magnetic fields to be reconstructed (Jensen *et al.*, 2010; Bisi *et al.*, 2014) as well as for background magnetic fields as can already be achieved with the UCSD CAT (Dunn *et al.*, 2005; Jackson *et al.*, 2015a, b; Yu *et al.*, 2015).

The determination of heliospheric FR, combined with observations of IPS, will have a high potential to provide essential information on the Sun's extended magnetic-field structure out into the inner heliosphere, especially when also combined with other forms of remote-sensing/heliospheric imaging data, and in-situ measurements.

Appendix I Authors and editors

Frances Bagenal (editor)
Laboratory for Atmospheric and Space Physics
UCB 600 University of Colorado
3665 Discovery Drive
Boulder, CO 80303, USA
email: Frances.Bagenal@colorado.edu

Mario M. Bisi
Science and Technology Facilities Council
Rutherford Appleton Laboratory
Harwell, Oxford OX11 OQX, UK
email: Mario.Bisi@stfc.ac.uk

Stephen W. Bougher
Atmospheric, Oceanic, and Space Science Department
2455 Hayward Avenue
University of Michigan
Ann Arbor, MI 48109, USA
email: bougher@umich.edu

David Brain
Laboratory for Atmospheric and Space Physics
University of Colorado
3665 Discovery Drive
Boulder, CO 80303, USA
email: David.Brain@Colorado.EDU

Ofer Cohen
Harvard-Smithsonian Center for Astrophysics
60 Garden St.
Cambridge, MA 02138, USA
email: ocohen@cfa.harvard.edu

Debra Fischer
Department of Astronomy,
Yale University,
New Haven, CT 06520, USA
email: debra.fischer@yale.edu

Marina Galand
Department of Physics
Imperial College London
Prince Consort Road
London SW7 2AZ, UK
email: m.galand@imperial.ac.uk

Mihály Horányi
Laboratory for Atmospheric and Space Physics, and
Department of Physics
University of Colorado, Boulder, CO 80303, USA
email: Mihaly.Horanyi@lasp.colorado.edu

Margaret G. Kivelson
Department of Earth, Planetary, and Space Sciences
University of California, Los Angeles
Los Angeles, CA 90095-1567, USA
and
Department of Atmospheric, Oceanic and Space Sciences
University of Michigan
Ann Arbor, MI 48109-2143, USA
email: mkivelso@igpp.ucla.edu

Norbert Krupp
Max-Planck-Institut für Sonnensystem-forschung
Justus-von-Liebig-Weg 3
37077 Göttingen
Germany
email: krupp@mps.mpg.de

Jeffrey L. Linsky
JILA, University of Colorado and NIST
Boulder, CO 80309, USA
email: jlinsky@jila.colorado.edu

Luke Moore
Center for Space Physics
Boston University
Boston, MA 02215, USA
email: moore@bu.edu

Rachel Osten
Space Telescope Science Institute
3700 San Martin Drive
Baltimore, MD 21218, USA
email: osten@stsci.edu

Carolus J. Schrijver (editor)
Solar and Astrophysics Laboratory
Lockheed Martin Avanced Technology Center
3251 Hanover Street, Bldg. 252
Palo Alto, CA 94304-1191, USA
email: schryver@lmsal.com

David E. Siskind
Space Science Division
Naval Research Laboratory
4555 Overlook Ave. SW
Washington DC, 20375, USA
email: david.siskind@nrl.navy.mil

Jan J. Sojka (editor)
Center for Atmospheric and Space Sciences
Utah State University
4405 Old Main Hill
Logan, UT 84322-4405, USA
email: sojka@cc.usu.edu

Tom Stallard
Department of Physics and Astronomy
University of Leicester
University Road
Leicester, LE2 3AF, UK
email: t.stallard@ion.le.ac.uk

Sabine Stanley
Rm 516B, Department of Physics
University of Toronto
60 St. George St., Toronto, ON, M5S1A7, Canada
email: stanley@physics.utoronto.ca

Ji Wang
Department of Astronomy,
Yale University,
New Haven, CT 06520, USA

Brian E. Wood
Naval Research Laboratory
Space Science Division
Washington, DC 20375, USA
email: brian.wood@nrl.navy.mil

List of illustrations

List of Tables

References

Aarnio, A. N., Stassun, K. G., Hughes, W. J., & McGregor, S. L.: 2011, *Solar Phys.* **268**, 195, doi:10.1007/s11207-010-9672-7

Aarnio, A. N., Matt, S. P., & Stassun, K. G.: 2012, *ApJ* **760**, 9, doi:10.1088/0004-637X/760/1/9

Abel, B. & Thorne, R. M.: 2003, *Icarus* **166**(2), 311–319, doi:10.1016/j.icarus.2003.08.017

Acuña, M. H., Connerney, J. E. P., Wasilewski, P., *et al.*: 1998, *Science* **279**, 1676

Aguilar-Rodriguez, E., Mejia-Ambriz, J. C., Jackson, B. V., *et al.*: 2015, *Solar Phys.*, in press

Aguilar-Rodriguez, E., Rodriguez-Martinez, M., Romero-Hernandez, E., *et al.*: 2014, *Geophys. Res. Lett.* **41**, 3331, doi:10.1002/2014GL060047

Alboussiere, T., Deguen, R., & Melzani, M.: 2010, *Nature* **466**, 744

Albrecht, S., Winn, J. N., Johnson, J. A., *et al.*: 2012, *Appl. Phys. J.* **757**, 18, doi:10.1088/0004-637X/757/1/18

Albrecht, S., Winn, J. N., Marcy, G. W., *et al.*: 2013, *Appl. Phys. J.* **771**, 11, doi:10.1088/0004-637X/771/1/11

Allred, J. C., Hawley, S. L., Abbett, W. P., & Carlsson, M.: 2006, *ApJ* **644**, 484, doi:10.1086/503314

Amit, H., Christensen, U. R., & Langlais, B.: 2011, *Phys. Earth Planet. Int.* **189**, 63

Ananthakrishnan, S., Tokumaru, M., Kojima, M., *et al.*: 1999, in S. R. Habbal, R. Esser, J. V. Hollweg, and P. A. Isenberg (Eds.), *American Institute of Physics Conference Series*, Vol. 471 of *American Institute of Physics Conference Series*, 321

Anderson, B. J., Johnson, C. L., Korth, H., *et al.*: 2012, *J. Geophys. Res.* **117**, E00L12

Andriopoulou, M., Roussos, E., Krupp, N., *et al.*: 2012, *Icarus* **220**, 503, doi:10.1016/j.icarus.2012.05.010

Arkani-Hamed, J. & Olson, P.: 2010, *Geophys. Res. Lett.* **37**, L02201

Armstrong, D. J., Osborn, H. P., Brown, D. J. A., *et al.*: 2014, *Mon. Not. R. Astr. Soc.* **444**, 1873, doi:10.1093/mnras/stu1570

Armstrong, J. W. & Coles, W. A.: 1972, *J. Geophys. Res.* **77**, 4602

Armstrong, J. W., Coles, W. A., & Rickett, B. J.: 1972, *J. Geophys. Res.* **77**(16), 2739

Atreya, S., Donahue, T., & Festou, M.: 1981, *ApJ* **247**, L43L4

Atreya, S. K., Trainer, M. G., Franz, H. B., *et al.*: 2013, *Geophys. Res. Lett.* **40**, 5605, doi:10.1002/2013GL057763

Aubert, J., Finla, C. C., & Fournier, A.: 2013, *Nature* **502**, 219

Audard, M., Güdel, M., Drake, J. J., & Kashyap, V. L.: 2000, *ApJ* **541**, 396, doi:10.1086/309426

Ayres, T. & France, K.: 2010, *ApJL* **723**, L38, doi:10.1088/2041-8205/723/1/L38
Ayres, T. R.: 1997, *J. Geophys. Research* **102**, 1641, doi:10.1029/96JE03306
Ayres, T. R., Fleming, T. A., Simon, T., *et al.*: 1995, *ApJSS* **96**, 223, doi:10.1086/192118
Ayres, T. R., Osten, R. A., & Brown, A.: 1999, *ApJ* **526**, 445, doi:10.1086/308001
Ayres, T. R., Osten, R. A., & Brown, A.: 2001, *ApJL* **562**, L83, doi:10.1086/337971
Bagenal, F.: 2009, in C. J. Schrijver and G. Siscoe (Eds.), *Heliophysics: Plasma Physics of the Local Cosmos*, Cambridge University Press, Cambridge
Bagenal, F., Cravens, T. E., Luhmann, J. G., McNutt, R. L., & Cheng, A. F.: 1997, in Pluto's interaction with the solar wind, in *Pluto and Charon*, University of Arizona Press, Tucson, 523–555
Bagenal, F. *et al.*: 2014, *Space Sci. Rev.*, doi:10.1007/s11214-014-0036-8
Bakos, G. Á., Noyes, R. W., Kovács, G., *et al.*: 2007, *Appl. Phys. J.* **656**, 552, doi:10.1086 /509874
Balbus, S. A. & Hawley, J. F.: 1991, *ApJ* **376**, 214, doi:10.1086/170270
Bally, J., Licht, D., Smith, N., & Walawender, J.: 2006, *Astron. J.* **131**, 473, doi:10. 1086/498265
Balona, L. A.: 2012, *Mon. Not. R. Astron. Soc.* **423**, 3420, doi:10.1111/j.1365-2966.2012. 21135.x
Banks, P. M. & Kockarts, G.: 1973, *Aeronomy, B*, Academic Press
Barabash, S., Fedorov, A., Lundin, R., & Sauvaud, J.-A.: 2007, *Science* **315**(5811), 501–503, doi:10.1126/science.1134358
Baranov, V. B. & Malama, Y. G.: 1993, *J. Geophys. Res.* **98**, 15157, doi:10. 1029/93JA01171
Baranov, V. B. & Malama, Y. G.: 1995, *J. Geophys. Res.* **100**, 14755, doi:10. 1029/95JA00655
Barkan, A., Merlino, R. L., & D'Angelo, N.: 1995, *Physics of Plasmas* **2**, 3563, doi:10. 1063/1.871121
Barkan, A., D'Angelo, N., & Merlino, R. L.: 1996, *Planetary and Space Science* **44**, 239, doi:10.1016/0032-0633(95)00109-3
Barman, T. S.: 2008, *ApJL* **676**, L61, doi:10.1086/587056
Barnes, J. R.: 1990, *J. Geophys. Res.* **95**, 1401, doi:10.1029/JB095iB02p01401
Barrow, D. & Matcheva, K. I.: 2011, *Icarus* **211**(1), 609622, doi:10.1016/j.icarus.2010.10. 017
Barth, C. A., Tobiska, W. K., Siskind, D. E., & Cleary, D. D.: 1988, *Geophys. Res. Lett.* **15**, 92
Batygin, K.: 2012, *Nature* **491**, 418, doi:10.1038/nature11560
Batygin, K. & Stevenson, D. J.: 2010, *ApJL* **714**, L238
Batygin, K., Stanley, S., & Stevenson, D.: 2013, *ApJ* **776**, 53
Behannon, K. W., Acuna, M. H., Burlaga, L. F., *et al.*: 1977, *Space Sci. Rev.* **21**, 235, doi:10.1007/BF00211541
Bekker, A., Holland, H. D., Wang, P. L., *et al.*: 2004, *Nature* **427**(6), 117120
Bellan, P. M.: 2006, *Fundamentals of Plasma Physics*, Cambridge University Press
Benz, A. O. & Güdel, M.: 2010, *Ann. Rev. Astron. Astrophys.* **48**, 241, doi:10.1146/annurev -astro-082708-101757
Berghöfer, T. W. & Breitschwerdt, D.: 2002, *A&A* **390**, 299, doi:10.1051/0004- 6361:20020627
Beuermann, K., Hessman, F. V., Dreizler, S., *et al.*: 2010, *Astron. Astrophys.* **521**, L60, doi :10.1051/0004-6361/201015472
Bhardwaj, A. & Gladstone, G. R.: 2000, *Adv. Sp. Res.* **26**(10), 1551–1558

Bisi, M. M.: 2006, PhD Thesis (http://cadair.aber.ac.uk/dspace/handle/2160/4064), The University of Wales, Aberystwyth

Bisi, M. M., Breen, A. R., Fallows, R. A., *et al.*: 2005, in *ESA SP-592: Solar Wind 11/SOHO 16, Connecting Sun and Heliosphere*, Vol. 16, European Space Agency, 593

Bisi, M. M., Fallows, R. A., Breen, A. R., Habbal, S. R., & Jones, R. A.: 2007a, *J. Geophys. Res.* **112**, A06101, doi:10.1029/2006JA012166

Bisi, M. M., Jackson, B. V., Fallows, R. A., *et al.*: 2007b, in *Society of Photo-Optical Instrumentation Engineers (SPIE) Conference Series*, Vol. 6689 of *Society of Photo-Optical Instrumentation Engineers (SPIE) Conference Series*

Bisi, M. M., Jackson, B. V., Hick, P. P., Buffington, A., & Clover, J. M.: 2007c, *Proceedings of the AOGS 2007 Meeting, Advances in Geosciences*, Vol. 14, Chapt. 12, World Scientific Publishing Company

Bisi, M. M., Jackson, B. V., Buffington, A., *et al.*: 2008a, (invited oral presentation) *AOGS 2008 Meeting*

Bisi, M. M., Jackson, B. V., Fallows, R. A., *et al.*: 2008b, *Adv. Geophys.* **21**

Bisi, M. M., Jackson, B. V., Hick, P. P., *et al.*: 2008c, *J. Geophys. Res.* **113**(A12), A00A11, doi:10.1029/2008JA013222

Bisi, M. M., Jackson, B. V., Buffington, A., *et al.*: 2009a, *Solar Phys.* **256**, 201, doi:10.1007/s11207-009-9350-9

Bisi, M. M., Jackson, B. V., Clover, J. M., *et al.*: 2009b, *Ann. Geophys.* **27**, 4479, doi:10.5194/angeo-27-4479-2009

Bisi, M. M., Breen, A. R., Jackson, B. V., *et al.*: 2010a, *Solar Phys.* **265**, 49, doi:10.1007/s11207-010-9602-8

Bisi, M. M., Fallows, R. A., Breen, A. R., & O'Neill, I. J.: 2010b, *Solar Phys.* **261**, 149, doi:10.1007/s11207-009-9471-1

Bisi, M. M., Jackson, B. V., Breen, A. R., *et al.*: 2010c, *Solar Phys.* **265**, 233, doi:10.1007/s11207-010-9594-4

Bisi, M. M., Jackson, B. V., Hick, P. P., *et al.*: 2010d, *ApJL* **715**, L104, doi:10.1088/2041-8205/715/2/L104

Bisi, M. M., Fallows, R. A., Eftekhari, T., *et al.*: 2014, (invited oral presentation) *AOGS 2014 Meeting*

Blandford, R. D. & Payne, D. G.: 1982, *Mon. Not. R. Astron. Soc.* **199**, 883

Bliokh, P., Sinitsin, V., & Yaroshenko, V.: 1995, *Dusty and Self-Gravitational Plasmas in Space*, Kluwer (Astrophysics and Space Science Library)

Bodenheimer, P.: 1995, *Ann. Rev. Astron. Astrophys.* **33**, 199, doi:10.1146/annurev.aa.33.090195.001215

Bonfond, B., Grodent, D., Gérard, J.-C., *et al.*: 2008, *Geophys. Res. Lett.* **35**(5), L05107, doi:10.1029/2007GL032418

Borovikov, S. N. & Pogorelov, N. V.: 2014, *ApJL* **783**, L16, doi:10.1088/2041-8205/783/1/L16

Borucki, W. J., Koch, D. G., Lissauer, J. J., *et al.*: 2003, in J. C. Blades and O. H. W. Siegmund (Eds.), *Future EUV/UV and Visible Space Astrophysics Missions and Instrumentation.*, Vol. 4854 of *Society of Photo-Optical Instrumentation Engineers (SPIE) Conference Series*, 129

Bougher, S. W. & Roble, R. G.: 1991, *J. Geophys. Res.* **96**, 11045

Bougher, S. W., Engel, S., Roble, R. G., & Foster, B.: 1999, *J. Geophys. Res.* **104**, 16591, doi:10.1029/1998JE001019

Bougher, S. W., Engel, S., Roble, R. G., & Foster, B.: 2000, *J. Geophys. Res.* **105**, 17669, doi:10.1029/1999JE001232

Bougher, S. W., Engel, S., Hinson, D. P., & Forbes, J. M.: 2001, *Geophys. Res. Lett.* **28**, 3091, doi:10.1029/2001GL012884

Bougher, S. W., Engel, S., Hinson, D. P., & Murphy, J. R.: 2004, *J. Geophys. Res.* **109**, E03010, doi:10.1029/2003JE002154

Bougher, S. W., Jr., Waite, J. H., Majeed, T., & Gladstone, G. R.: 2005, *J. Geophys. Res.* **110**(E4), E04008, doi:10.1029/2003JE002230

Bougher, S. W., McDunn, T. M., Zoldak, K. A., & Forbes, J. M.: 2009, *Geophys. Res. Lett.* **36**, L05201, doi:10.1029/2008GL036376

Bougher, S. W., Brain, D. A., Fox, J. L., *et al.*: 2014, in D. P. Drob (Ed.), *Mars Book II*, Cambridge University Press

Bourgois, G., Coles, W. A., Daigne, G., *et al.*: 1985, *A&A* **144**, 452

Bouvier, A. & Wadhwa, M.: 2010, *Nature Geoscience* **3**, 637, doi:10.1038/ngeo941

Bower, G. C., Plambeck, R. L., Bolatto, A., *et al.*: 2003, *ApJ* **598**, 1140, doi:10.1086/379101

Brain, D. A: 2006, *Space Sci. Rev.* **126**(1), 77112, doi:10.1007/s11214-006-9122-x

Brain, D. A., Baker, A. H., Briggs, J., *et al.*: 2010a, *Geophys. Res. Lett.* **37**(1), 14108, doi:10.1029/2010GL043916

Brain, D., Barabash, S., Boesswetter, A., *et al.*: 2010b, *Icarus* **206**, 139, doi:10.1016/j.icarus.2009.06.030

Brain, D. A., Dong, Y., Fortier, K., *et al.*: 2015, *Lunar and Planetary Inst. Technical Report* **46**, 2663

Brandt, J. C.: 1970, in *Introduction to the Solar Wind*, W. H. Freeman and Co., 103–117

Branduardi-Raymont, G., Bhardwaj, A., Elsner, R. F., *et al.*: 2007, *Astron. Astrophys.* **463**, 761–774

Branduardi-Raymont, G., Elsner, R. F., Galand, M., *et al.*: 2008, *J. Geophys. Res.* **113**(A2), A02202, doi:10.1029/2007JA012600

Brasseur, G. P., Marsh, D., & Schmidt, H.: 2010, in *Heliophysics: Evolving Solar Activity and the Climates of Space and Earth*, Cambridge University Press

Breen, A. R., Coles, W. A., Grall, R., *et al.*: 1996a, *Journal of Atmospheric and Terrestrial Physics* **58**, 507

Breen, A. R., Coles, W. A., Grall, R. R., *et al.*: 1996b, *Ann. Geophys.* **14**, 1235

Breen, A. R., Fallows, R. A., Bisi, M. M., *et al.*: 2006, *J. Geophys. Res.* **111**(A10), 8104, doi:10.1029/2005JA011485

Breen, A. R., Fallows, R. A., Bisi, M. M., *et al.*: 2008, *ApJL* **683**, L79, doi:10.1086/591520

Brinkman, A. C., Behar, E., Güdel, M., *et al.*: 2001, *A&A* **365**, L324, doi:10.1051/0004-6361:20000047

Broadfoot, A. L. *et al.*: 1981, *Science* **212**(4491), 206211

Broadfoot, A. L. *et al.*: 1989, *Science* **246**(4936), 145966, doi:10.1126/science.246.4936.1459

Brown, T. M., Charbonneau, D., Gilliland, R. L., Noyes, R. W., & Burrows, A.: 2001, *Appl. Phys. J.* **552**, 699, doi:10.1086/320580

Brown, T. M., Latham, D. W., Everett, M. E., & Esquerdo, G. A.: 2011, *ApJ* **142**, 112, doi:10.1088/0004-6256/142/4/112

Browne, S. E., Welsh, B. Y., & Wheatley, J.: 2009, *PASPs* **121**, 450, doi:10.1086/599365

Brownsberger, S. & Romani, R. W.: 2014, *ApJ* **784**, 154, doi:10.1088/0004-637X/784/2/154

Buchhave, L. A., Latham, D. W., Johansen, A. *et al.*: 2012, *Nature* **486**, 375, doi:10.1038/nature11121

Buchhave, L. A., Bizzarro, M., Latham, D. W., *et al.*: 2014, *Nature* **509**, 593, doi:10.1038/nature13254

Buffett, B.: 2009, *Geophys. J. Int.* **179**, 711
Buffett, B.: 2014, *Nature* **507**, 484
Bullard, E. C. & Gellman, H.: 1954, *Proc. R. Soc. Lond., Ser. A* **A 247**, 213
Bullock, M. A. & Grinspoon, D. H.: 2013, in S. J. Mackwell, A. A. Simon-Miller, J. W. Harder, & M. A. Bullock (Eds.), *Comparative Climatology of Terrestrial Planets*, 19–54, doi:10.2458/a2u_uapress_9780816530595-ch002
Burns, A. G., Killeen, T. L., Crowley, G., Emery, B. A., & Roble, R. G.: 1989, *J. Geophys. Res.* **94**, 16961
Butler, C. J., Rodono, M., & Foing, B. H.: 1988, *A&A* **206**, L1
Cahoy, K. L., Hinson, D.P., & Tyler, G. L.: 2006, *J. Geophys. Res.* **111**, E05003
Caldwell, J., Tokunaga, A. T., & Gillett, F. C.: 1980, *Icarus* **44**(3), 667–675, doi:10.1016/0019-1035(80)90135-9
Canals, A.: 2002, PhD Thesis, The University of Wales, Aberystwyth
Cao, H., Russell, C. T., Christensen, U. R., Wicht, J., & Dougherty, M. K.: 2012, *Icarus* **221**, 388
Cao, H., Aurnou, J. M., Wicht, J., & Dietrich, W.: 2014, *Geophys. Res. Lett.* **41**, 4127, doi:10.1002/2014GL060196
Capone, L. A., Whitten, R. C., Prasad, S. S., & Dubach, J.: 1977, *ApJ* **215**, 977983
Caramazza, M., Flaccomio, E., Micela, G., *et al.*: 2007, *A&A* **471**, 645, doi:10.1051/0004-6361:20077195
Carporzen, L., Weiss, B. P., Elkins-Tanton, L. T., *et al.*: 2011, *Proc. Natl Acad. Sci.* **108**, 6386
Carruthers, G. R. & Page, T.: 1976, *J. Geophys. Res.* **81**, 483, doi:10.1029/JA081i004p00483
Carslaw, K. S., Harrison, R. G., & Kirkby, J.: 2002, *Science* **298**, 1732, doi:10.1126/science.1076964
Chamberlain, J. W & Hunten, D. M.: 1987, *Theory of Planetary Atmospheres*, Academic Press
Chandler, M. & Waite, Jr., J. H.: 1986, *Geophys. Res. Lett.* **13**(1), 69
Chang, L. C., Polo, S. E., & Liu, H.-L.: 2011, *J. Geophys. Res.* **116**, doi:10.1029/2010JD014996
Charbonneau, D., Berta, Z. K., Irwin, J., *et al.*: 2009, *Nature* **462**, 891, doi:10.1038/nature08679
Charbonneau, D., Brown, T. M., Latham, D. W., & Mayor, M.: 2000, *ApJL* **529**, L45, doi:10.1086/312457
Charbonneau, P.: 2010, *Living Reviews in Solar Physics* **7**, 3, doi:10.12942/lrsp-2010-3
Chau, J. L., Fejer, B. G., & Goncharenko, L. P.: 2009, *Geophys. Res. Lett.* **36**, L05101, doi:10.1029/2008GL036785
Chaufray, J. Y., Leblanc, F., Quémerais, E., & Bertaux, J.-L.: 2009, *J. Geophys. Res.* **114**, E02006
Chen, F. F.: 1974, *Introduction to Plasma Physics*, New York: Plenum Press
Chen, P. R.: 1992, *J. Geophys. Res.* **97**, 6343–6357.
Chiang, E. & Laughlin, G.: 2013, *Mon. Not. R. Astr. Soc.* **431**, 3444, doi:10.1093/mnras/stt424
Christensen, U. R.: 2010, in C. J. Schrijver and G. Siscoe (Eds.), *Heliophysics: Evolving Solar Activity and the Climates of Space and Earth*, Cambridge University Press, Cambridge
Christensen, U. R. & Wicht, J.: 2008, *Icarus* **196**, 16
Clarke, J. T. *et al.*: 2002, *Nature* **415**(6875), 997–1000, doi:10.1038/415997a

Clarke, J. T., Grodent, D., Cowley, S. W. H., *et al.*: 2004, in W. B. McKinnon, F. Bagenal, T. E. Dowling (Eds.), *Jupiter: The Planet, Satellites and Magnetosphere*, Cambridge University Press
Clarke, J. T., Wannawichian, S., Hernandez, N., *et al.*: 2011, in *EPSC-DPS Joint Meeting, vol. 114*, Nantes, 1468
Cleeves, L. I., Adams, F. C., & Bergin, E. A.: 2013, *ApJ* **772**, 5, doi:10.1088/0004-637X/772/1/5
Cochran, W. D., Hatzes, A. P., Butler, R. P., & Marcy, G. W.: 1997, *Appl. Phys. J.* **483**, 457
Cohen, M. H., Gundermann, E. J., Hardebeck, H. E., & Sharp, L. E.: 1967, *ApJ* **147**(2), 449
Cohen, O.: 2011, *Mon. Not. R. Astron. Soc.* **417**, 2592, doi:10.1111/j.1365-2966.2011.19428.x
Cohen, O. & Drake, J. J.: 2014, *ApJ* **783**, 55, doi:10.1088/0004-637X/783/1/55
Cohen, O., Drake, J. J., Kashyap, V. L., *et al.*: 2009, *ApJ* **704**, L85
Cohen, O., Drake, J. J., Kashyap, V. L., Hussain, G. A. J., & Gombosi, T. I.: 2010a, *ApJ* **721**, 80, doi:10.1088/0004-637X/721/1/80
Cohen, O., Drake, J. J., Kashyap, V. L., *et al.*: 2010b, *ApJ* **719**, 299, doi:10.1088/0004-637X/719/1/299
Cohen, O., Kashyap, V. L., Drake, J. J., Sokolov, I. V., & Gombosi, T. I.: 2011, *ApJ* **738**, 166, doi:10.1088/0004-637X/738/2/166
Cohen, O., Drake, J. J., & Kóta, J.: 2012, *ApJ* **760**, 85, doi:10.1088/0004-637X/760/1/85
Colegrove, F. D., Hanson, W. B., & Johnson, F. S.: 1965, *J. Geophys. Res.* **70**, 4931
Colegrove, F. D., Johnson, F. S., & Hanson, W. B.: 1966, *J. Geophys. Res.* **71**, 2227
Coles, W. A.: 1978, *Space Sci. Rev.* **21**, 411, doi:10.1007/BF00173067
Coles, W. A.: 1996, *Ap&SS* **243**, 87, doi:10.1007/BF00644037
Coles, W. A. & Harmon, J. K.: 1978, *J. Geophys. Res.* **83**, 1413, doi:10.1029/JA083iA04p01413
Coles, W. A. & Harmon, J. K.: 1989, *ApJ* **337**, 1023
Coles, W. A. & Rickett, B. J.: 1976, *J. Geophys. Res.* **81**, 4797
Coles, W. A., Rickett, B. J., Rumsey, V. H., *et al.*: 1980, *Nature* **286**, 239, doi:10.1038/286239a0
Collier, M. R., Moore, T. E., Ogilvie, K. W., *et al.*: 2001, *J. Geophys. Research* **106**, 24893, doi:10.1029/2000JA000382
Connerney, J.: 2013, *Nature* **496**(7444), 178179
Connerney, J. & Waite, J.: 1984, *Nature* **312**, 136138
Connerney, J. E. P., Baron, R., Satoh, T., & Owen, T.: 1993, *Science* **262**(5136), 1035–1038
Court, R. W., Sephton, M. A., Parnell, J., & Gilmour, I.: 2006, *Geochim. Cosmochim. Acta* **70**, 1020, doi:10.1016/j.gca.2005.10.017
Cowley, S. W. H.: 2013, *J. Geophys. Res. Space Physics* **118**(6), 2897–2902, doi:10.1002/jgra.50323
Cowley, S. W. H. & Bunce, E. J.: 2001, *Planet. Space Sci.* **49**(10–11), 1067–1088, doi:10.1016/S0032-0633(00)00167-7
Cowley, S. W. H. & Bunce, E. J.: 2003, *Ann. Geophys.* **21**(8), 1691–1707, doi:10.5194/angeo-21-1691-2003
Cowley, S. W. H., Bunce, E., Stallard, T., & Miller, S.: 2003, *Geophys. Res. Lett.* **30**(5), 1220, doi:10.1029/2002GL016030
Cowley, S. W. H., Bunce, E. J., & O'Rourke, J. M.: 2004, *J. Geophys. Res.* **109**(A5), A05212, doi:10.1029/2003JA010375

Cowley, S. W. H., Arridge, C. S., Bunce, E. J., *et al.*: 2008, *Ann. Geophys.* **26**(9), 2613–2630, doi:10.5194/angeo-26-2613-2008
Cowling, T. G.: 1933, *Monthly Notices RAS* **94**, 0039
Cranmer, S. R. & Saar, S. H.: 2011, *ApJ* **741**, 54, doi:10.1088/0004-637X/741/1/54
Cravens, T. E.: 1987, *J. Geophys. Res.* **92**(5), 11 083–11 100
Cravens, T. E.: 2000, *ApJL* **532**, L153, doi:10.1086/312574
Crouzet, N., McCullough, P. R., Burke, C., & Long, D.: 2012, *Appl. Phys. J.* **761**, 7, doi:10.1088/0004-637X/761/1/7
Crowley, G., Hackert, C. L., Meier, R. R., *et al.*: 2006, *J. Geophys. Res.* **111**, A10S18, doi :10.1029/2005JA011518
Crutcher, R. M.: 1982, *ApJ* **254**, 82, doi:10.1086/159707
Dartnell, L. R.: 2011, *Astrobiology* **11**, 551, doi:10.1089/ast.2010.0528
Deeg, H. J., Ocaña, B., Kozhevnikov, V. P., *et al.*: 2008, *Astron. Astrophys.* **480**, 563, doi :10.1051/0004-6361:20079000
Delamere, P. A.: 2009, *J. Geophys. Res.* **114**, A03220, doi:10.1029/2008JA013756
Delamere, P. A. & Bagenal, F.: 2010, *J. Geophys. Res.* **115**(A10), A10201, doi:10.1029 /2010JA015347
Deleuil, M., Barge, P., Defay, C., *et al.*: 2000, in G. Garzón, C. Eiroa, D. de Winter, and T. J. Mahoney (Eds.), *Disks, Planetesimals, and Planets*, Vol. 219 of *Astronomical Society of the Pacific Conference Series*, 656
Dennison, P. A. & Hewish, A.: 1967, *Nature* **213**, 343
Dere, K. P., Landi, E., Mason, H. E., Monsignori Fossi, B. C., & Young, P. R.: 1997, *Astronomy and Astrophysics Supplement* **125**, 149, doi:10.1051/aas:1997368
Desidera, S. & Barbieri, M.: 2007, *Astron. Astrophys.* **462**, 345, doi:10.1051/0004-6361 :20066319
Dessler, A. J.: 1980, *Icarus* **44**(2), 291–295, doi:10.1016/0019-1035(80)90024-X
Dietrich, W. & Wicht, J.: 2011, *Phys. Earth Planet. Int.* **217**, 10
Djuth, F. T., Zhang, L. D., Livneh, D. J., *et al.*: 2010, *J. Geophys. Res.* **115**, A08305, doi:10.1029/2009JA014799
Dohlen, K., Beuzit, J.-L., Feldt, M., *et al.*: 2006, in *Society of Photo-Optical Instrumentation Engineers (SPIE) Conference Series*, Vol. 6269 of *Society of Photo-Optical Instrumentation Engineers (SPIE) Conference Series*
Donahue, T. M., Hoffman, J. H., Hodges, R. R., & Watson, A. J.: 1982, *Science* **216**, 630633, doi:10.1126/science.216.4546.630
Donati, J.-F. & Collier Cameron, A.: 1997, *Mon. Not. R. Astron. Soc.* **291**, 1
Donati, J.-F. & Landstreet, J. D.: 2009, *Ann. Rev. Astron. Astrophys.* **47**, 333, doi:10.1146 /annurev-astro-082708-101833
Dong, C., Bougher, S. W., Ma, Y., *et al.*: 2014, *Geophys. Res. Lett.* **41**(8), 27082715, doi :10.1002/2014GL059515
Dorrian, G. D., Breen, A. R., Fallows, R. A., & Bisi, M. M.: 2013, *Solar Phys.* **285**, 97, doi:10.1007/s11207-012-0081-y
Dougherty, M. K., Kellock, S., Southwood, D. J., *et al.*: 2004, *Space Sci. Rev.* **114**, 331, doi:10.1007/s11214-004-1432-2
Doyle, L. R. & Grun, E.: 1990, *Icarus* **85**, 168, doi:10.1016/0019-1035(90)90109-M
Doyle, L. R., Carter, J. A., Fabrycky, D. C., *et al.*: 2011, *Science* **333**, 1602, doi:10.1126 /science.1210923
Drake, J. J., Cohen, O., Yashiro, S., & Gopalswamy, N.: 2013, *ApJ* **764**, 170, doi:10.1088 /0004-637X/764/2/170
Driscoll, P. & Olson, P.: 2011, *Icarus* **213**, 12
Drossart, P. *et al.*: 1989, *Nature* **340**, 539541

Drossart, P., Bézard, B., Atreya, S., *et al.*: 1993, *J. Geophys. Res.* **98**(E10), 18 803–18 811
Duarte, L. D. V., Gastine, T., & Wicht, J.: 2013, *Phys. Earth Planet. Int.* **222**, 22
Duchêne, G. & Kraus, A.: 2013, *Ann. Rev. Astron. Astrophys.* **51**, 269, doi:10.1146/annurev-astro-081710-102602
Dumusque, X., Pepe, F., Lovis, C., *et al.*: 2012, *Nature* **491**, 207, doi:10.1038/nature11572
Dungey, J.: 1961, *Phys. Rev. Lett.* **6**(2), 47–48
Dunn, T., Jackson, B. V., Hick, P. P., Buffington, A., & Zhao, X. P.: 2005, *Solar Phys.* **227**, 339, doi:10.1007/s11207-005-2759-x
Duquennoy, A. & Mayor, M.: 1991, *Astron. Astrophys.* **248**, 485
Duru, F., Gurnett, D. A., Morgan, D. D., *et al.*: 2008, *J. Geophys. Res.* **113**, A07302, doi:10.1029/2008JA013073
Dwyer, C. A., Stevenson, D. J., & Nimmo, F.: 2011, *Nature* **479**, 212
Dwyer, J. R., Smith, D. M., & Cummer, S. A.: 2012, *Space Sci. Rev.* **173**, 133, doi:10.1007/s11214-012-9894-0
Dyudina, U., Ingersoll, A. P., Wellington, D., Ewald, S. P., & Porco, C.: 2011, in *EPSC-DPS Joint Meeting, vol. 6*, Nantes, France, 604
Eckermann, S. D., Hopped, K. W., Coy, L., *et al.*: 2009, *Journal of Atmospheric and Solar-Terrestrial Physics* **71**, 531–551
Eggenberger, A., Udry, S., & Mayor, M.: 2004, *Astron. Astrophys.* **417**, 353, doi:10.1051/0004-6361:20034164
Einstein, A.: 1905, *Annalen der Physik* **322**, 891, doi:10.1002/andp.19053221004
Ekers, R. D. & Little, L. T.: 1971, *A&A* **10**, 310
Elkins-Tanton, L. T., Weiss, B. P., & Zuber, M. T.: 2011, *Earth Planet. Sci. Lett.* **305**, 1
Elphic, R. C. & Ershkovich, A. I.: 1984, *J. Geophys. Res. Space* **89**, 997–1002
Emmert, J. T. & Picone, J. M.: 2010, *J. Geophys. Res.* **115**, A09326, doi:10.1029/2010JA015298
Emmert, J. T., Picone, J. M., & Meier, R. R.: 2008, *Geophys. Res. Lett.* **35**, L05101, doi:10.1029/2007GL032809
Emslie, A. G., Dennis, B. R., Shih, A. Y., *et al.*: 2012, *ApJ* **759**, 71, doi:10.1088/0004-637X/759/1/71
Eplee, R. E. & Smith, B. A.: 1984, *Icarus* **59**, 188, doi:10.1016/0019-1035(84)90022-8
Eshleman, V. R.: 1973, *Planet. Space Sci.* **21**, 1521, doi:10.1016/0032-0633(73)90059-7
Eshleman, V., Tyler, G., Wood, G., *et al.*: 1979, *Science* **204**(4396), 976978
Evans, A. J., Zuber, M. T., Weiss, B. P., & Tikoo, S. M.: 2014, *J. Geophys. Res.* **119**, 1061
Evans, N., Calvet, N., Cieza, L., *et al.*: 2009, ArXiv e-prints
Everett, M. E., Howell, S. B., Silva, D. R., & Szkody, P.: 2013, *Appl. Phys. J.* **771**, 107, doi:10.1088/0004-637X/771/2/107
Eyles, C. J., Simnett, G. M., Cooke, M. P., *et al.*: 2003, *Solar Phys.* **217**, 319
Fabrycky, D. & Tremaine, S.: 2007, *Appl. Phys. J.* **669**, 1298, doi:10.1086/521702
Fallows, R. A.: 2001, PhD Thesis, The University of Wales, Aberystwyth
Fallows, R. A., Breen, A. R., Bisi, M. M., Jones, R. A., & Wannberg, G.: 2006, *Geophys. Res. Lett.* **33**, 11106, doi:10.1029/2006GL025804
Fallows, R. A., Breen, A. R., & Dorrian, G. D.: 2008, *Ann. Geophys.* **26**, 2229
Fallows, R. A., Asgekar, A., Bisi, M. M., Breen, A. R., & ter-Veen, S.: 2013, *Solar Phys.* **285**, 127, doi:10.1007/s11207-012-9989-5
Fallows, R. A., Coles, W. A., McKay-Bukowski, D., *et al.*: 2014, *J. Geophys. Res. Space Physics* **119**(A18), 10 544, doi:10.1002/2014JA020406
Farmer, A. J. & Goldreich, P.: 2005, *Icarus* **179**, 535, doi:10.1016/j.icarus.2005.07.025
Farrell, W. M., Kaiser, M. L., & Desch, M. D.: 1999, *Geophys. Res. Lett.* **26**(16), 26012604

Farrell, W. M., Desch, M. D., Kaiser, M. L., Kurth, W. S., & Gurnett, D. A.: 2006, *Geophys. Res. Lett.* **33**, 7203, doi:10.1029/2005GL024922

Favata, F., Flaccomio, E., Reale, F., *et al.*: 2005, *ApJSS* **160**, 469, doi:10.1086/432542

Feigelson, E. D. & Montmerle, T.: 1999, *Ann. Rev. Astron. Astrophys.* **37**, 363, doi:10.1146/annurev.astro.37.1.363

Feigelson, E. D., Garmire, G. P., & Pravdo, S. H.: 2002, *ApJ* **572**, 335, doi:10.1086/340340

Feigelson, E. D., Welty, A. D., Imhoff, C., *et al.*: 1994, *ApJ* **432**, 373, doi:10.1086/174575

Festou, M., Atreya, S., Donahue, T., *et al.*: 1981, *J. Geophys. Res.* **86**(A7), 57155725

Festou, M. C. & Atreya, S. K.: 1982, *Geophys. Res. Lett.* **9**(10), 11471150

Feuchtgruber, H., Lellouch, E., & De Graauw, T.: 1997, *Nature* **389** (September), 159162

Fiege, K., Trieloff, M., Hillier, J. K., *et al.*: 2014, *Icarus* **241**, 336, doi:10.1016/j.icarus.2014.07.015

Fields, B. D., Athanassiadou, T., & Johnson, S. R.: 2008, *ApJ* **678**, 549, doi:10.1086/523622

Fischer, D. A. & Marcy, G. W.: 1992, *Appl. Phys. J.* **396**, 178, doi:10.1086/171708

Fischer, D. A. & Valenti, J.: 2005, *Appl. Phys. J.* **622**, 1102, doi:10.1086/428383

Fischer, D. A., Marcy, G. W., & Spronck, J. F. P.: 2014, *ApJSS* **210**, 5, doi:10.1088/0067-0049/210/1/5

Fischer, G., Gurnett, D. A., Zarka, P., Moore, L., & Dyudina, U. A.: 2011, *J. Geophys. Res.* **116**(A4), A04315, doi:10.1029/2010JA016187

Fisk, L. A. & Gloeckler, G.: 2008, *ApJ* **686**, 1466, doi:10.1086/591543

Fleshman, B. L., Delamere, P. A., Bagenal, F., & Cassidy, T.: 2012, *J. Geophys. Res.* **117**(E5), E05007, doi:10.1029/2011JE003996

Fogg, M. J. & Nelson, R. P.: 2007, *Astron. Astrophys.* **461**(3), 11951208, doi:10.1051/0004-6361:20066171

Forbes, J. M., Bridger, A. F. C., Hagan, M. E., *et al.*: 2002, *J. Geophys. Res.* **107**, 5113, doi:10.1029/2001JE001582

Forbes, J. M., Lemoine, F. G., Bruinsma, S. L., Smith, M. D., & Zhang, X.: 2008, *Geophys. Res. Lett.* **35**, L01201, doi:10.1029/2007GL031904

Forbush, S. E.: 1946, *Physical Review* **70**, 771, doi:10.1103/PhysRev.70.771

Ford, E. B. & Rasio, F. A.: 2008, *Appl. Phys. J.* **686**, 621, doi:10.1086/590926

Franciosini, E., Pallavicini, R., & Tagliaferri, G.: 2001, *A&A* **375**, 196, doi:10.1051/0004-6361:20010830

Frisch, P. C.: 1994, *Science* **265**, 1423, doi:10.1126/science.265.5177.1423

Frisch, P. C., Dorschner, J. M., Geiss, J., *et al.*: 1999, *ApJ* **525**, 492, doi:10.1086/307869

Frisch, P. C., Grodnicki, L., & Welty, D. E.: 2002, *ApJ* **574**, 834, doi:10.1086/341001

Frisch, P. C., Redfield, S., & Slavin, J. D.: 2011, *Ann. Rev. Astron. Astrophys.* **49**, 237, doi:10.1146/annurev-astro-081710-102613

Fritts, D. C., Wang, L., & Tolson, R. H.: 2006, *J. Geophys. Res.* **11**, A12304, doi:10.1029/2006JA011897

Fromang, S., Terquem, C., & Balbus, S. A.: 2002, *Mon. Not. R. Astron. Soc.* **329**, 18, doi:10.1046/j.1365-8711.2002.04940.x

Fu, R. R., Weiss, B. P., Shuster, D. L., *et al.*: 2012, *Science* **338**, 238

Fuller-Rowell, T. J.: 1998, *J. Geophys. Res.* **103**, 3951

Fuller-Rowell, T. J., Codrescu, M. V., Rishbeth, H., Moffett, R. J., & Quegan, S.: 1996, *J. Geophys. Res.* **101**, 2343

Futaana, Y. *et al.*: 2008, *Planetary and Space Science* **56**(6), 873880, doi:10.1016/j.pss.2007.10.014

Gagne, M., Valenti, J., Johns-Krull, C., *et al.*: 1998, in R. A. Donahue and J. A. Bookbinder (Eds.), *Cool Stars, Stellar Systems, and the Sun*, Vol. 154 of *Astronomical Society of the Pacific Conference Series*, 1484

Gaidos, E. J., Güdel, M., & Blake, G. A.: 2000, *Geophys. Res. Lett.* **27**, 501, doi:10.1029/1999GL010740

Gail, H.-P. & Sedlmayr, E.: 1975, *A&A* **41**, 359

Galand, M., Moore, L., Charnay, B., Müller-Wodarg, I., & Mendillo, M.: 2009, *J. Geophys. Res.* **114**(A6), A06313, doi:10.1029/2008JA013981

Galand, M., Moore, L., Müller-Wodarg, I., Mendillo, M., & Miller, S.: 2011, *J. Geophys. Res.* **116**(A9), A09306, doi:10.1029/2010JA016412

Gammie, C. F.: 1996, *ApJ* **457**, 355, doi:10.1086/176735

Garcia, R.R.: 1991, *J. Atmos. Sci.* **48**, 1405

Garcia, R. R. & Solomon, S.: 1985, *J. Geophys. Res.* **90**, 3850

Garcia, R. R., Puertas, M. L., Funke, B., *et al.*: 2014, *J. Geophys. Res.* **119**, doi:10.1002/2013JD021208

Garrick-Bethell, I., Weiss, B. P., Shuster, D. L., & Buz, Z.: 2009, *Science* **323**, 356

Gayley, K. G., Zank, G. P., Pauls, H. L., Frisch, P. C., & Welty, D. E.: 1997, *ApJ* **487**, 259

Gérard, J.-C., Bonfond, B., Gustin, J., *et al.*: 2009, *Geophys. Res. Lett.* **36**(2), doi:10.1029/2008GL036554

Getman, K. V., Broos, P. S., Salter, D. M., Garmire, G. P., & Hogerheijde, M. R.: 2011, *ApJ* **730**, 6, doi:10.1088/0004-637X/730/1/6

Gladstone, G. R. *et al.*: 2002, *Nature* **415**(6875), 1000–3, doi:10.1038/4151000a

Glassmeier, K.-H., Boehnhardt, H., Koschny, D., Kührt, E., & Richter, I.: 2007, *Space Sci. Rev.* **128**, 1, doi:10.1007/s11214-006-9140-8

Gloeckler, G.: 2010, in C. J. Schrijver and G. L. Siscoe (Eds.), *Heliophysics: Space Storms and Radiation: Causes and Effects*, Cambridge University Press, 43

Gloeckler, G., Fisk, L. A., & Geiss, J.: 2010, *Twelfth International Solar Wind Conference 1210*, 514, doi:10.1063/1.3395915

Glyantsev, A. V., Tyul'bashev, S. A., Chashei, I. V., & Shishov, V. I.: 2015, *Astronomy Reports* **59**, 40, doi:10.1134/S1063772915010047

Goertz, C. K. & Ip, W.-H.: 1984, *Geophys. Res. Lett.* **11**, 349, doi:10.1029/GL011i004p00349

Goertz, C. K. & Morfill, G. E.: 1983, *Icarus* **53**, 219

Gombosi, T. I., Armstrong, T. P., Arridge, C. S., *et al.*: 2009, in *Saturn from Cassini–Huygens*, 209–255

Gómez-Pérez, N. & Heimpel, M.: 2007, *Geophys. Astrophys. Fluid Dyn.* **101**, 371

Goncharenko, L., Chau, J., Liu, H.-L., & Coster, A. J.: 2010, *Geophys. Res. Lett.* **37**, L10101, doi:10.1029/2010GL043125

Gonzalez, G.: 1997, *Mon. Not. R. Astr. Soc.* **285**, 403

Gopalswamy, N., Akiyama, S., Yashiro, S., Michalek, G., & Lepping, R. P.: 2008, *Journal of Atmospheric and Solar-Terrestrial Physics* **70**, 245, doi:10.1016/j.jastp.2007.08.070

Gothoskar, P. & Pramesh Rao, A.: 1996, *Ap&SS* **243**, 225, doi:10.1007/BF00644060

Graedel, T. E., Sackmann, I.-J., & Boothroyd, A. I.: 1991, *Geophys. Res. Lett.* **18**, 1881, doi:10.1029/91GL02314

Grall, R. R.: 1995, PhD Thesis, University of California, San Diego

Griessmeier, J.-M., Zarka, P., & Girard, J. N.: 2011, *Radio Sci.* **46**, RS0F09

Grillmair, C. J., Burrows, A., Charbonneau, D., *et al.*: 2008, *Nature* **456**, 767, doi:10.1038/nature07574

Grodent, D., Clarke, J., Jr., Waite, J. H. *et al.*: 2003, *J. Geophys. Res.* **108**(A10), 1366, doi :10.1029/2003JA010017

Grodent, D., Gérard, J.-C., Gustin, J., *et al.*: 2006, *Geophys. Res. Lett.* **33**(6), L06201, doi :10.1029/2005GL025487

Grodent, D., Radioti, A., Bonfond, B., & Gérard, J.-C.: 2010, *J. Geophys. Res.* **115**(A8), A08219, doi:10.1029/2009JA014901

Gronoff, G., Simon-Wedlund, C., Barthelemy, C. J., *et al.*: 2012, *J. Geophys. Res.* **117**, A05309, doi:10.1029/2011JA017308

Grun, E., Morfill, G. E., Terrile, R. J., Johnson, T. V., & Schwehm, G.: 1983, *Icarus* **54**, 227, doi:10.1016/0019-1035(83)90194-X

Grün, E., Fechtig, H., Hanner, M. S., *et al.*: 1992, *Space Sci. Rev.* **60**, 317, doi:10.1007 /BF00216860

Grün, E., Zook, H. A., Baguhl, M., *et al.*: 1993, *Nature* **362**, 428

Grün, E., Gustafson, B., Mann, I., *et al.*: 1994, *A&A* **286**, 915

Grün, E., Staubach, P., Baguhl, M., *et al.*: 1997, *Icarus* **129**, 270

Grün, E., Kempf, S., Krüger, H., Moragas-Klostermeyer, G., & Srama, R.: 2003, *Meteoritics & Planetary Science* **38**, 5208

Gry, C. & Jenkins, E. B.: 2014, *A&A* **567**, A58, doi:10.1051/0004-6361/201323342

Güdel, M.: 2007, *Living Reviews in Solar Physics* **4**, 3, doi:10.12942/lrsp-2007-3

Güdel, M., Audard, M., Reale, F., Skinner, S. L., & Linsky, J. L.: 2004, *A&A* **416**, 713, doi:10.1051/0004-6361:20031471

Güdel, M. & Nazé, Y.: 2009, *Astron. Astrophys. Rev.* **17**, 309, doi:10.1007/s00159-009-0022-4

Guedel, M., Schmitt, J. H. M. M., Benz, A. O., & Elias, II, N. M.: 1995, *A&A* **301**, 201

Guenther, E. W. & Ball, M.: 1999, *A&A* **347**, 508

Gurevich, A. V., Zybin, K. P., & Roussel-Dupre, R. A.: 1999, *Physics Letters A* **254**, 79, doi:10.1016/S0375-9601(99)00091-2

Gurnett, D. A., Grun, E., Gallagher, D., Kurth, W. S., & Scarf, F. L.: 1983, *Icarus* **53**, 236, doi:10.1016/0019-1035(83)90145-8

Gurnett, D. A., Kurth, W. S., Burlaga, L. F., & Ness, N. F.: 2013, *Science* **341**, 1489, doi :10.1126/science.1241681

Hagan, M. E., Maute, A., & Roble, R. G.: 2009, *J. Geophys. Res.* **114**, A01302, doi:10 .1029/2008JA013637

Haider, S. A., Mahajan, K. K., & Kallio, E.: 2011, *Rev. Geophys.* 49

Hall, C. M., Meek, C. E., Manson, A. H., & Nozawa, S.: 2008, *J. Geophys. Res.* **113**, D13104, doi:10.1029/2008JD009938

Hallett, J. T., Shemansky, D. E., & Liu, X.: 2005, *Geophys. Res. Lett.* **32**(2), L02204, doi :10.1029/2004GL021327

Hamilton, D. P.: 2006, *Bull. Amer. Astron. Soc.* **38**, 578

Hansen, B. M. S. & Murray, N.: 2013, *Appl. Phys. J.* **775**, 53, doi:10.1088/0004-637X/775 /1/53

Hansen, J., Sato, M., Russell, G., & Kharecha, P.: 2013, *Phil. Trans. R. Soc. Lond. A: Mathematical* **371**, 20294, doi:10.1098/rsta.2012.0294

Hansen, K. C. *et al.*: 2007, *Space Sci. Rev.* **128**, 133–166

Hanson, W. B., Sanatani, S., & Zuccaro, D. R.: 1977, *J. Geophys. Res.* **82**, 4351

Hardwick, S. A., Bisi, M. M., Davies, J. A., *et al.*: 2013, *Solar Phys.* **285**, 111, doi:10.1007 /s11207-013-0223-x

Haro, G. & Parsamian, E.: 1969, *Boletin de los Observatorios Tonantzintla y Tacubaya* **5**, 41

Hauber, E., Brož, P., Jagert, F., Jodłowski, P., & Platz, T.: 2011, *Geophys. Res. Lett.* **38**(1), L10201, doi:10.1029/2011GL047310

Hauck, S. A., Aurnou, J. M., & Dombard, A. J.: 2006, *J. Geophys. Res.* **111**, E09008

Havnes, O.: 1984, *Advances in Space Research* **4**(9), 75, doi:10.1016/0273-1177(84)90010-3

Havnes, O., Aanesen, T. K., & Melandso, F.: 1990, *J. Geophys. Research* **95**, 6581, doi:10.1029/JA095iA05p06581

Hawley, S. L., Fisher, G. H., Simon, T., *et al.*: 1995, *ApJ* **453**, 464, doi:10.1086/176408

Hawley, S. L., Allred, J. C., Johns-Krull, C. M., *et al.*: 2003, *ApJ* **597**, 535, doi:10.1086/378351

Hays, P. B., Jones, R. A., & Rees, M. H.: 1973, *Planet. Space Sci.* **21**, 559

Heath, C. & Brain, D. A.: 2014, *AGU Fall Meeting Abstracts*, c4031

Heerikhuisen, J., Pogorelov, N. V., Zank, G. P., *et al.*: 2010, *ApJL* **708**, L126, doi:10.1088/2041-8205/708/2/L126

Henning T. & Semenov, D.: 2013, *Chem. Rev.* **113**, 9016, doi:10.1021/cr 400128P

Henry, G. W., Marcy, G. W., Butler, R. P., & Vogt, S. S.: 2000, *ApJL* **529**, L41, doi:10.1086/312458

Herbert, F., Sandel, B. R., Yelle, R. V., *et al.*: 1987, *J. Geophys. Res.* **92**(A13), 15 093–15 109, doi:10.1029/JA092iA13p15093

Hewish, A.: 1955, *Proc. R. Soc. Lond. A* **238**, 238

Hewish, A.: 1989, *Journal of Atmospheric and Terrestrial Physics* **51**(9/10), 743

Hewish, A. & Okoye, S. E.: 1965, *Nature* **207**, 59

Hewish, A., Scott, P. F., & Wills, D.: 1964, *Nature* **203**, 1214

Hey, J. S.: 1983, *The Radio Universe*, Pergamon Press, third edition (First Edition 1971, Second Edition 1975, Third Edition 1983)

Hick, P. P. & Jackson, B. V.: 2004, in S. Fineschi and M. A. Gummin (Eds.), *Telescopes and Instrumentation for Solar Astrophysics. Proc. SPIE*, Vol. 5171, 287

Hickey, M., Waltersheid, R., & Schubert, G.: 2000, *Icarus* **148**(1), 266–281, doi:10.1006/icar.2000.6472

Hill, J. R. & Mendis, D. A.: 1982, *J. Geophys. Research* **87**, 7413, doi:10.1029/JA087iA09p07413

Hilton, E. J.: 2011, PhD thesis, University of Washington

Hinkley, S., Oppenheimer, B. R., Brenner, D., *et al.*: 2008, in *Society of Photo-Optical Instrumentation Engineers (SPIE) Conference Series*, Vol. 7015 of *Society of Photo-Optical Instrumentation Engineers (SPIE) Conference Series*

Hinson, D., Flasar, F., Kliore, A., *et al.*: 1997, *Geophys. Res. Lett.* **24**(17), 2107–2110

Hinson, D., Twicken, J., & Karayel, E.: 1998, *J. Geophys. Res.* **103**(A5), 9505–9520

Hiraki, Y. & Tao, C.: 2008, *Annales Geophysicae* 77-86

Holzer, T. E.: 1989, *Ann. Rev. Astron. Astrophys.* **27**, 199, doi:10.1146/annurev.aa.27.090189.001215

Horányi, M.: 1996, *Ann. Rev. Astron. Astrophys.* **34**, 383, doi:10.1146/annurev.astro.34.1.383

Horányi, M.: 2000, *Phys. of Plasmas* **7**(10), 3847

Horányi, M. & Juhász, A.: 2000, *AAS/Division for Planetary Sciences Meeting 32*, 1088

Horányi, M., Morfill, G., & Grün, E.: 1993a, *J. Geophys. Res.* **98**, 21245

Horányi, M., Morfill, G., & Grün, E.: 1993b, *Nature* **363**, 144

Horányi, M., Hartquist, T. W., Havnes, O., Mendis, D. A., & Morfill, G. E.: 2004, *Reviews of Geophysics* **42**, 4002, doi:10.1029/2004RG000151

Houdebine, E. R.: 1996, in Y. Uchida, T. Kosugi, and H. S. Hudson (Eds.), *IAU Colloq. 153: Magnetodynamic Phenomena in the Solar Atmosphere – Prototypes of Stellar Magnetic Activity*, 147

Houghton, J.: 2002, *The Physics of Atmospheres*, 3rd edition, Cambridge University Press

Houminer, Z.: 1971, *Nature Physical Science* **231**, 165, doi:10.1038/physci231165a0

Howard, A. W., Sanchis-Ojeda, R., Marcy, G. W., *et al.*: 2013, *Nature* **503**, 381, doi:10.1038/nature12767

Hsu, H. W., Kempf, S., & Jackman, C. M.: 2010, *Icarus*, **206**, 653–661

Huba, J. D., Ossakow, S. L., Joyce, G., Krall, J., & England, S. L.: 2009, *Geophys. Res. Lett.* **36**, L19106, doi:10.1029/2009GL040284

Hubbard, W. B., Dougherty, M. K., Gautier, D., & Jacobson, R.: 2009, in *Saturn from Cassini-Huygens*, 7581

Hubbard, W., Haemmerle, V., Porco, C., Rieke, G., & Rieke, M.: 1995, *Icarus* **113**, 103–109

Hubbard, W. B. *et al.*: 1997, *Icarus* **130**, 404425

Hunter, D. M.: 1992, *American Geophysical Union Geophysical Monograph Series*, **66**, 1–5

Hussain, G. A. J.: 2012, *Astronomische Nachrichten* **333**, 4, doi:10.1002/asna.201111627

Hussain, G. A. J., Jardine, M., Donati, J.-F., *et al.*: 2007, *Mon. Not. R. Astron. Soc.* **377**, 1488, doi:10.1111/j.1365-2966.2007.11692.x

Ilgner, M. & Nelson, R. P.: 2006, *A&A* **445**, 205, doi:10.1051/0004-6361:20053678

Imamura, T., Toda, T., Tomiki, A., *et al.*: 2011, *Earth, Planets, and Space* **63**, 493, doi:10.5047/eps.2011.03.009

Imamura, T., Tokumaru, M., Isobe, H., *et al.*: 2014, *ApJ* **788**, 117, doi:10.1088/0004-637X/788/2/117

Immel, T. J., Crowley, G., Craven, J. D., & Roble, R. G.: 2001, *J. Geophys. Res.* **106**, A8, 15471

Immel, T. J., Sagawa, E., England, S. L., *et al.*: 2006, *Geophys. Res. Lett.* **33**, L15108, doi:10.1029/2006GL02616

Izmodenov, V., Wood, B. E., & Lallement, R.: 2002, *J. Geophys. Res. Space Physics* **107**, 1308, doi:10.1029/2002JA009394

Jackson, B. V.: 1984, in M. A. Shea, D. F. Smart, and S. M. P. McKenna-Lawlor (Eds.), *Solar/Interplanetary Intervals*, 169

Jackson, B. V.: 2011, *Adv. Geophys.* **30**, 69

Jackson, B. V. & Hick, P. P.: 2005, in D. E. Gary and C. U. Keller (Eds.), *Solar and Space Weather Radiophysics: Current status and future developments*, Vol. 314 of *Astrophys. and Space Sci. Lib.*, Chapt. 17, 355–386, Kluwer Academic Publ., Dordrecht

Jackson, B. V., Hick, P. P., Kojima, M., & Yokobe, A.: 1998, *J. Geophys. Res.* **103**(A6), 12 049

Jackson, B. V., Hick, P. P., Buffington, A., *et al.*: 2003, in M. Velli, R. Bruno, F. Malara, and B. Bucci (Eds.), *Solar Wind Ten*, Vol. 679 of *American Institute of Physics Conference Series*, 75

Jackson, B. V., Buffington, A., Hick, P. P., *et al.*: 2004, *Solar Phys.* **225**, 177, doi:10.1007/s11207-004-2766-3

Jackson, B. V., Boyer, J. A., Hick, P. P., *et al.*: 2007, *Solar Phys.* **241**(2), 385, doi:10.1007/s11207-007-0276-9

Jackson, B. V., Bisi, M. M., Hick, P. P., *et al.*: 2008, *J. Geophys. Res.* **113**(A00A15), doi:10.1029/2008JA013224

Jackson, B. V., Buffington, A., Hick, P. P., Bisi, M. M., & Clover, J. M.: 2010a, *Solar Phys.* **265**, 257, doi:10.1007/s11207-010-9579-3

Jackson, B. V., Hick, P. P., Bisi, M. M., Clover, J. M., & Buffington, A.: 2010b, *Solar Phys.* **265**, 245, doi:10.1007/s11207-010-9529-0

Jackson, B. V., Hamilton, M. S., Hick, P. P., *et al.*: 2011a, *Journal of Atmospheric and Solar-Terrestrial Physics* **73**, 1317, doi:10.1016/j.jastp.2010.11.023

Jackson, B. V., Hick, P. P., Buffington, A., *et al.*: 2011b, *Journal of Atmospheric and Solar-Terrestrial Physics* **73**, 1214, doi:10.1016/j.jastp.2010.10.007
Jackson, B. V., Clover, J. M., Hick, P. P., *et al.*: 2013, *Solar Phys.* **285**, 151, doi:10.1007/s11207-012-0102-x
Jackson, B. V., Yu, H.-S., Buffington, A., & Hick, P. P.: 2014, *ApJ* **793**, 54, doi:10.1088/0004-637X/793/1/54
Jackson, B. V., Hick, P. P., Buffington, A., *et al.*: 2015a, *ApJL* **803**, L1, doi:10.1088/2041-8205/803/1/L1
Jackson, B. V., Odstrcil, D., Yu, H.-S., *et al.*: 2015b, *Space Weather* **13**, 104, doi:10.1002/2014SW001130
Jaffe, L. D. & Herrell, L. M.: 1997, *J. Spacecr. Rockets* **34**(4), 509–521
Jakosky, B. M. & Phillips, R. J.: 2001, *Nature* **412**(6), 237244
Jakosky, B. M., Pepin, R. O., Johnson, R. E., & Fox, J. L.: 1994, *Icarus* **111**, 271, doi:10.1006/icar.1994.1145
Jakosky, B. M., Lin, R. P., Grebowsky, J. M., *et al.*: 2015 *Lunar and Planetary Inst. Technical Report* **46**, 1370
Jensen, E. A. & Russell, C. T.: 2008a, *Geophys. Res. Lett.* **35**, 2103, doi:10.1029/2007GL031038
Jensen, E. A. & Russell, C. T.: 2008b, *Planet. Space Sci.* **56**, 1562, doi:10.1016/j.pss.2008.07.010
Jensen, E. A. & Russell, C. T.: 2009, *Geophys. Res. Lett.* **36**, 5104, doi:10.1029/2008GL036257
Jensen, E. A., Hick, P. P., Bisi, M. M., *et al.*: 2010, *Solar Phys.* **265**, 31, doi:10.1007/s11207-010-9543-2
Jensen, E. A., Bisi, M. M., Breen, A. R., *et al.*: 2013a, *Solar Phys.* **285**, 83, doi:10.1007/s11207-012-0213-4
Jensen, E. A., Nolan, M., Bisi, M. M., Chashei, I., & Vilas, F.: 2013b, *Solar Phys.* **285**, 71, doi:10.1007/s11207-012-0162-y
Jia, X., Walker, R. J., Kivelson, M. G., Khurana, K. K., & Linker, J. A.: 2008, *J. Geophys. Res.* **113**, A06212, doi:10.1029/2007JA012748
Jia, X., Kivelson, M. G., Khurana, K. K., & Walker, R. J.: 2010, *Space Sci. Rev.* **152**, 271
Jian, L. K., MacNeice, P. J., Taktakishvili, A., *et al.*: 2015, *Space Weather*, doi:10.1002/2015SW001174
Jin, H., Miyoshi, Y., Fujiwara, H., *et al.*: 2011, *J. Geophys. Res.* **116**, A01316, doi:1029/2010JA015925
Johnson, J. A., Aller, K. M., Howard, A. W., & Crepp, J. R.: 2010, *PASP* **122**, 905, doi:10.1086/655775
Johnson, R. E.: 1994, *Space Sci. Rev.* **69**(3), 215–253, doi:10.1007/BF02101697
Jones, C. A.: 2007, in P. Olson (Ed.), *Treatise of Geophysics, Volume 8: Core dynamics*, 131-186, Elsevier, Amsterdam
Jones, C.A.: 2008, in P. Cardin and L. F. Cugliandolo (Eds.), *Les Houches, Session LXXXVIII, 2007, Dynamos*, Elsevier, Amsterdam
Jones, R. A., Breen, A. R., Fallows, R. A., *et al.*: 2007, *J. Geophys. Res.* **112**(A11), 8107, doi:10.1029/2006JA011875
Judge, P. G., Solomon, S. C., & Ayres, T. R.: 2003, *ApJ* **593**, 534, doi:10.1086/376405
Juhász, A., Horányi, M., & Morfill, G. E.: 2007, *Geophys. Res. Lett.* **34**, 9104, doi:10.1029/2006GL029120
Kahler, S. W.: 2013a, *Appl. Phys. J.* **769**, 110, doi:10.1088/0004-637X/769/2/110
Kahler, S. W.: 2013b, *Appl. Phys. J.* **769**, 35, doi:10.1088/0004-637X/769/1/35
Kahler, S. W. & Vourlidas, A.: 2013, *Appl. Phys. J.* **769**, 143, doi:10.1088/0004-637X/769/2/143

Kaiser, M. L., Desch, M. D., & Connerney, J. E. P.: 1984, *J. Geophys. Res.* **89**(A4), 2371, doi:10.1029/JA089iA04p02371

Kalas, P., Graham, J. R., Chiang, E., *et al.*: 2008, *Science* **322**, 1345, doi:10.1126/science.1166609

Kallenbach, R., Czechowski, A., Hilchenbach, M., & Wurz, P.: 2006, in V. V. Izmodenov & R. Kallenbach (Eds.), *The Physics of the Heliospheric Boundaries*, ISSI Scientific Report, 203

Kallenrode, M.-B.: 2003, *Journal of Physics G Nuclear Physics* **29**, 965

Kallio, E. & Barabash, S.: 2012, *Earth* **64**(2), 149156, doi:10.5047/eps.2011.07.008

Kane, S. R., Kreplin, R. W., Martres, M.-J., Pick, M., & Soru-Escaut, I.: 1974, *Sol. Phys.* **38**, 483, doi:10.1007/BF00155083

Kashyap, V. L., Drake, J. J., Güdel, M., & Audard, M.: 2002, *ApJ* **580**, 1118, doi:10.1086/343869

Kashyap, V. L., Drake, J. J., & Saar, S. H.: 2008, *ApJ* **687**, 1339, doi:10.1086/591922

Kasprzak, W. T., Keating, G. M., Hsu, N. C., *et al.*: 1997, in *Venus II*, University of Arizona Press, 225

Kass, D. M., Kleinboehl, A., McCleese, D. J., Schofield, J. T., & Smith, M. D.: 2013, *AGU Fall Meeting Abstracts* A7

Kass, D. M., Kleinböhl, A., McCleese, D. J., Schofield, J. T., & Smith, M. D.: 2014, in F. Forget and M. Millour (Eds.), *Mars Atmosphere: Modelling and Observation, 5th International Workshop, 2014*

Kasting, J. F. & Ackerman, T. P.: 1986, *Science* **234**, 1383, doi:10.1126/science.234.4782.1383

Kaufman, A. J., Johnston, D. T., Farquhar, J., *et al.*: 2007, *Science* **317**(5), 1900, doi:10.1126/science.1138700

Keating, G. M., Bougher, S. W., Zurek, R. W., *et al.*: 1998, *Science* **279**, 1672, doi:10.1126/science.279.5357.1672

Kellermann, K. I. & Pauliny-Toth, I. I. K.: 1969, *ApJL* **155**, L71, doi:10.1086/180305

Kelley, M. C.: 2009, *The Earth's Ionosphere: Plasma Physics and Electrodynamics*, Academic Press

Kelly, M. C., Larsen, M. F., & Hoz, C. La: 1981, *J. Geophys. Res.* **86**, 9087

Kempf, S., Srama, R., Horányi, M., *et al.*: 2005, *Nature* **433**, 289

Khodachenko, M. L., Ribas, I., Lammer, H., *et al.*: 2007, *Astrobiology* **7**, 167, doi:10.1089/ast.2006.0127

Khodachenko, M. L., Alexeev, I., Belenkaya, E., *et al.*: 2012, *ApJ* **744**, 70, doi:10.1088/0004-637X/744/1/70

Kim, S., Drossart, P., Caldwell, J., & Maillard, J.-P.: 1990, *Icarus* **84**, 54–61

Kim, T. K., Pogorelov, N. V., Borovikov, S. N., *et al.*: 2014, *J. Geophys. Res. Space Physics* **119**, 7981, doi:10.1002/2013JA019755

Kim, Y. & Fox, J.: 1994, *Icarus* **112**, 310–324

Kim, Y., Pesnell, D., Grebowsky, J., & Fox, J.: 2001, *Icarus* **150**(2), 261–278, doi:10.1006/icar.2001.6590

Kim, Y. H., Fox, J. L., Black, J. H., & Moses, J. I.: 2014, *J. Geophys. Res. Sp. Phys.* **119**, 112, doi:10.1002/2013JA019022

Kirkby, J., Curtius, J., Almeida, J., *et al.*: 2011, *Nature* **476**, 429, doi:10.1038/nature10343

Kivelson, M.: 1995, in *Introduction to Space Physics*, Cambridge University Press, 330

Kivelson, M. G. & Bagenal, F.: 2007, in P. Weissman, L.-A. McFadden, and T. Johnson (Eds.), *Planetary Magnetospheres, The Encyclopedia of the Solar System*, 2nd Edition, Academic Press, 519

Kivelson, M. G., Bagenal, F., Kurth, W. S., *et al.*: 2004, in *Jupiter: The Planet, Satellites and Magnetosphere*, 513–536

Kley, W. & Nelson, R. P.: 2008, *Astron. Astrophys.* **486**, 617, doi:10.1051/0004-6361:20079324

Klinglesmith, M.: 1997, PhD Thesis, University of California, San Diego (UCSD)

Klinglesmith, M. T., Grall, R. R., & Coles, W. A.: 1996, in *Solar Wind 8*, Vol. 382, 180

Kliore, A., Cain, D. L., Levy, G. S., *et al.*: 1965, *Science* **149**, 1243, doi:10.1126/science.149.3689.1243

Kliore, A. J., Nagy, A. F., Marouf, E. A., *et al.*: 2009, *J. Geophys. Res.* **114**(A4), A04315, doi:10.1029/2008JA013900

Knie, K., Korschinek, G., Faestermann, T., *et al.*: 2004, *Physical Review Letters* **93**(17), 171103, doi:10.1103/PhysRevLett.93.171103

Knight, S.: 1973, *Planet. Space Sci.* **21**(5), 741–750, doi:10.1016/0032-0633(73)90093-7

Knutson, H. A., Charbonneau, D., Allen, L. E., *et al.*: 2007a, *Nature* **447**, 183, doi:10.1038/nature05782

Knutson, H. A., Charbonneau, D., Noyes, R. W., Brown, T. M., & Gilliland, R. L.: 2007b, *Appl. Phys. J.* **655**, 564, doi:10.1086/510111

Kobulnicky, H. A., Gilbert, I. J., & Kiminki, D. C.: 2010, *ApJ* **710**, 549, doi:10.1088/0004-637X/710/1/549

Kojima, M. & Kakinuma, T.: 1987, *J. Geophys. Res.* **92**, 7269

Kopp, R. A. & Poletto, G.: 1984, *MmSAI* **55**, 737

Koskinen, T. T., Sandel, B. R., Yelle, R. V., *et al.*: 2013, *Icarus* **226**(2), 1318–1330, doi:10.1016/j.icarus.2013.07.037

Kóspál, Á., Salter, D. M., Hogerheijde, M. R., Moór, A., & Blake, G. A.: 2011, *A&A* **527**, A96, doi:10.1051/0004-6361/201015917

Kostiuk, T., Romani, P., Espenak, F., Livengood, T. A., & Goldstein, J. J.: 1993, *J. Geophys. Res.* **98**(E10), 18 823–18 830, doi:10.1029/93JE01332

Koutroumpa, D., Lallement, R., Raymond, J. C., & Kharchenko, V.: 2009, *ApJ* **696**, 1517, doi:10.1088/0004-637X/696/2/1517

Kouyama, T., Imamura, T., Nakamura, M., Satoh, T., & Futaana, Y.: 2013, *J. Geophys. Res. Planet* **118**(1), 3746, doi:10.1029/2011JE004013

Kowalski, A. F., Hawley, S. L., Holtzman, J. A., Wisniewski, J. P., & Hilton, E. J.: 2010, *ApJL* **714**, L98, doi:10.1088/2041-8205/714/1/L98

Kowalski, A. F., Hawley, S. L., Wisniewski, J. P., *et al.*: 2013, *ApJSS* **207**, 15, doi:10.1088/0067-0049/207/1/15

Kraus, A. L., Ireland, M. J., Hillenbrand, L. A., & Martinache, F.: 2012, *Appl. Phys. J.* **745**, 19, doi:10.1088/0004-637X/745/1/19

Kretzschmar, M.: 2011, *A&A* **530**, A84, doi:10.1051/0004-6361/201015930

Krimigis, S. M., Decker, R. B., Hill, M. E., *et al.*: 2003, *Nature* **426**, 45

Krucker, S., Giménez de Castro, C. G., Hudson, H. S., *et al.*: 2013, *Astron. Astrophys. Rev.* **21**, 58, doi:10.1007/s00159-013-0058-3

Krüger, H. & Grün, E.: 2009, *Space Science Reviews* **143**, 347, doi:10.1007/s11214-008-9431-3

Krupp, N., Lagg, A., Livi, S., *et al.*: 2001, *J. Geophys. Res.* **106**, 26017, doi:10.1029/2000JA900138

Krymskii, A. M., Breus, T. K., Ness, N. F., Hinson, D. P., & Bojkov, D. I: 2003, *J. Geophys. Res. Space Physics* **108**, 1431

Kurth, W. S. *et al.*: 2009, in *Saturn from Cassini–Huygens*, 333–374

Lada, C. J. & Wilking, B. A.: 1984, *Appl. Phys. J.* **287**, 610, doi:10.1086/162719

Lagg, A., Krupp, N., Woch, J., & Williams, D. J.: 2003, *Geophys. Res. Lett.* **30**(11), 1556, doi:10.1029/2003GL017214

Lagrange, A.-M., Bonnefoy, M., Chauvin, G., *et al.*: 2010, *Science* **329**, 57, doi:10.1126/science.1187187

Lallement, R. & Bertin, P.: 1992, *A&A* **266**, 479
Lallement, R., Quémerais, E., Bertaux, J. L., *et al.*: 2005, *Science* **307**, 1447, doi:10.1126/science.1107953
Lam, H. A., Achilleos, N., Miller, S., *et al.*: 1997, *Icarus* **393**, 379–393
Lammer, H., Lichtenegger, H. I. M., Kolb, C., *et al.*: 2003, *Icarus* **165**, 9, doi:10.1016/S0019-1035(03)00170-2
Lammer, H., Lichtenegger, H. I. M., Kulikov, Y. N., *et al.*: 2007, *Astrobiology* **7**, 185, doi:10.1089/ast.2006.0128
Lammer, H., Kasting, J. F., Chassefière, E., *et al.*: 2009, *Comparative Aeronomy* **2**, 399, doi:10.1007/978-0-387-87825-6-11
Lamy, L., Cecconi, B., Prangé, R., *et al.*: 2009, *J. Geophys. Res.* **114**(A10), A10212, doi:10.1029/2009JA014401
Lamy, L. *et al.*: 2012, *Geophys. Res. Lett.* **39**(7), doi:10.1029/2012GL051312
Landgraf, M.: 2000, *J. Geophys. Research* **105**, 10303, doi:10.1029/1999JA900243
Lang, K. R.: 2009, *The Sun from Space*, Springer
Langmuir, I.: 1923, *Science* **58**, 290, doi:10.1126/science.58.1502.290
Lanza, A. F.: 2010, *Astron. Astrophys.* **512**, A77, doi:10.1051/0004-6361/200912789
Laskar, J., Correia, A. C. M., Gastineau, M., *et al.*: 2004, *Icarus* **170**(2), 343364, doi:10.1016/j.icarus.2004.04.005
Latham, D. W., Stefanik, R. P., Mazeh, T., Mayor, M., & Burki, G.: 1989, *Nature* **339**, 38, doi:10.1038/339038a0
Le Bars, M., Wieczorek, M. A., Karatekin, O., Cebron, D., & Laneuville, M.: 2011, *Nature* **479**, 215
Leitzinger, M., Odert, P., Ribas, I., *et al.*: 2011, *A&A* **536**, A62, doi:10.1051/0004-6361/201015985
Lemaire, J.: 1990, *ApJ* **360**, 288, doi:10.1086/169119
Levrard, B., Forget, F., Montmessin, F., & Laskar, J.: 2004, *Nature* **431**(7), 1072–1075, doi:10.1038/nature03055
Lillis, R. J., Frey, H. V., & Manga, M.: 2008, *Geophys. Res. Lett.* **35**(1), 14203, doi:10.1029/2008GL034338
Lin, D. N. C., Bodenheimer, P., & Richardson, D. C.: 1996, *Nature* **380**, 606, doi:10.1038/380606a0
Lindal, G.: 1992, *Astron. J.* **103**(3), 967982
Lindal, G. F., Sweetnam, D. N., & Eshleman, V. R.: 1985, *Astron. J.* **90**(6), 1136–1146
Lindal, G. F., Lyons, J. R., Sweetnam, D. N., *et al.*: 1987, *J. Geophys. Res.* **92**(A13), 14987, doi:10.1029/JA092iA13p14987
Lindzen, R. S.: 1970, *J. Atm. Sci.* **27**, 536
Linker, J. A., Khurana, K. K., Kivelson, M. G., & Walker, R. J.: 1998, *Geophys. Res.* **103**, 19 867
Linsky, J. L. & Redfield, S.: 2014, *Ap&SS*, doi:10.1007/s10509-014-1943-6
Linsky, J. L. & Wood, B. E.: 1996, *ApJ* **463**, 254, doi:10.1086/177238
Liu, H.-L. & Richmond, A. D.: 2013, *J. Geophys. Res.* **118**, 2452–2465
Liu, H.-L., Wang, W., Richmond, A. D., & Roble, R. C.: 2010, *J. Geophys. Res.* **115**, A00G01, doi:10.1029/2009JA015188
Liu, L. & Peng, B.: 2010, *Science China Physics, Mechanics, and Astronomy* **53**, 187, doi:10.1007/s11433-010-0076-3
Livengood, T. A., Strobel, D. F., & Moos, H. W.: 1990, *J. Geophys. Res.* **95**(A7), 10 375–10 388
Lopez, E. D. & Fortney, J. J.: 2014, *Appl. Phys. J.* **792**, 1, doi:10.1088/0004-637X/792/1/1
Luhmann, J. G., Johnson, R. E., & Zhang, M. H. G.: 1992, *Geophys. Res. Lett.* **19**, 2151, doi:10.1029/92GL02485

Lundin, R., Zakharov, A., Pellinen, R., *et al.*: 1990, *Geophys. Res. Lett.* **17**, 873876, doi:10.1029/GL017i006p00873

Lyons, J. R.: 1995, *Science* **267**(5198), 648651

Lystrup, M. B., Miller, S., Russo, N. Dello, Vervack, J. R. J., & Stallard, T.: 2008, *ApJ* **677**(1), 790–797, doi:10.1086/529509

Machin, K. E. & Smith, F. G.: 1952, *Nature* **170**, 319, doi:10.1038/170319b0

Macintosh, B., Graham, J., Palmer, D., *et al.*: 2006, in *Society of Photo-Optical Instrumentation Engineers (SPIE) Conference Series*, Vol. 6272 of *Society of Photo-Optical Instrumentation Engineers (SPIE) Conference Series*

Madhusudhan, N. & Seager, S.: 2009, *Appl. Phys. J.* **707**, 24, doi:10.1088/0004-637X/707/1/24

Maehara, H., Shibayama, T., Notsu, S., *et al.*: 2012, *Nature* **485**, 478, doi:10.1038/nature11063

Maggio, A., Pallavicini, R., Reale, F., & Tagliaferri, G.: 2000, *A&A* **356**, 627

Majeed, T. & McConnell, J.: 1996, *J. Geophys. Res.* **101**, 75897598

Majeed, T., McConnell, J., & Yelle, R.: 1991, *Planet. Space Sci.* **39**(11), 15911606

Majeed, T., Waite, J. H., Bougher, S. W., & Gladstone, G. R.: 2005. *J. Geophys. Res.* **110**, 12007, doi:10.1029/2004JE002351

Mamajek, E. E.: 2009, in T. Usuda, M. Tamura, and M. Ishii (Eds.), *American Institute of Physics Conference Series*, Vol. 1158 of *American Institute of Physics Conference Series*, 3

Mamajek, E. E. & Hillenbrand, L. A.: 2008, *ApJ* **687**, 1264, doi:10.1086/591785

Manchester, W. B., Gombosi, T. I., Roussev, I. *et al.*: 2004, *J. Geophys. Res. Space Physics* **109**, 2107, doi:10.1029/2003JA010150

Mann, I.: 2010, *Ann. Rev. Astron. Astrophys.* **48**, 173, doi:10.1146/annurev-astro-081309-130846

Mann, I. & Czechowski, A.: 2005, *ApJL* **621**, L73, doi:10.1086/429129

Manoharan, P. K.: 2006, *Solar Phys.* **235**, 345, doi:10.1007/s11207-006-0100-y

Manoharan, P. K. & Ananthakrishnan, S.: 1990, *Mon. Not. R. Astron. Soc.* **244**, 691

Manoharan, P. K., Kojima, M., Gopalswamy, N., Kondo, T., & Smith, Z.: 2000, *ApJ* **530**, 1061, doi:10.1086/308378

Marcq, E., Bertaux, J.-L., Montmessin, F., & Belyaev, D.: 2013, *Nature Geoscience* **6**(1), 2528, doi:10.1038/ngeo1650

Marois, C., Macintosh, B., Barman, T., *et al.*: 2008, *Science* **322**, 1348, doi:10.1126/science.1166585

Martin, D. C., Seibert, M., Neill, J. D., *et al.*: 2007, *Nature* **448**, 780, doi:10.1038/nature06003

Massi, M., Forbrich, J., Menten, K. M., *et al.*: 2006, *A&A* **453**, 959, doi:10.1051/0004-6361:20053535

Matcheva, K. I. & Strobel, D. F.: 1999, *Icarus* **140**, 328, doi:10.1006/icar.1999.6151

Matranga, M., Mathioudakis, M., Kay, H. R. M., & Keenan, F. P.: 2005, *ApJL* **621**, L125, doi:10.1086/429288

Matsui, H., Iwagami, N., Hosouchi, M., Ohtsuki, S., & Hashimoto, G. L.: 2012, *Icarus* **217**(2), 610614, doi:10.1016/j.icarus.2011.07.026

Matt, S. & Pudritz, R. E.: 2007, in J. Bouvier and I. Appenzeller (Eds.), *IAU Symposium*, Vol. 243 of *IAU Symposium*, 299

Matt, S. & Pudritz, R. E.: 2008, *ApJ* **678**, 1109, doi:10.1086/533428

Matt, S. P., MacGregor, K. B., Pinsonneault, M. H., & Greene, T. P.: 2012, *ApJL* **754**, L26, doi:10.1088/2041-8205/754/2/L26

Maxwell, James Clerk: 1873, *Treatise on Electricity and Magnetism*, Oxford

Mayor, M. & Queloz, D.: 1995, *Nature* **378**, 355, doi:10.1038/378355a0

McCleary, J. E. & Wolk, S. J.: 2011, *Astron. J.* **141**, 201, doi:10.1088/0004-6256/141/6/201

McComas, D. J. *et al.*: 2007, *Reviews of Geophysics* **45**, 1004, doi:10.1029/2006RG000195

McComas, D., Allegrini, F., Bagenal, F., *et al.*: 2008, *Space Sci. Rev.* **140**, 261–313

McComas, D. J., Allegrini, F., Bochsler, P., *et al.*: 2009, *Science* **326**, 959, doi:10.1126/science.1180906

McComas, D. J., Alexashov, D., Bzowski, M., *et al.*: 2012, *Science* **336**, 1291, doi:10.1126/science.1221054

McConnell, J. C., Holberg, J. B., Smith, G. R., *et al.*: 1982, *Planet. Space Sci.* **30**(2), 151167

McCullough, P. R., Stys, J. E., Valenti, J. A., *et al.*: 2005, *PASP* 117, 783, doi:10.1086/432024

McElroy, M.: 1973, *Space Sci. Rev.* **14**, 460473

McGhee, C. A., French, R. G., Dones, L., *et al.*: 2005, *Icarus* **173**, 508, doi:10.1016/j.icarus.2004.09.001

McKee, C. F. & Ostriker, J. P.: 1977, *ApJ* **218**, 148, doi:10.1086/155667

McLaughlin, D. B.: 1924, *Appl. Phys. J.* **60**, 22, doi:10.1086/142826

McNeil, D. S. & Nelson, R. P.: 2010, *Mon. Not. R. Astron. Soc.* **401**(3), 16911708, doi:10.1111/j.1365-2966.2009.15805.x

McNutt, R. L., Andrews, G. B., Gold, R. E., *et al.*: 2004, *Advances in Space Research* **34**, 192, doi:10.1016/j.asr.2003.03.053

McNutt Jr., R. L., Livi, S. A., Gurnee, R. S., *et al.*: 2008, *Space Sci. Rev.* **140**, 315–385

Medvedev, M. V. & Melott, A. L.: 2007, *ApJ* **664**, 879, doi:10.1086/518757

Melin, H., Stallard, T., Miller, S., *et al.*: 2011a, *ApJ* **729**(2), 134, doi:10.1088/0004-637X/729/2/134

Melin, H., Stallard, T., Miller, S., *et al.*: 2011b, *Geophys. Res. Lett.* **38**(15), doi:10.1029/2011GL048457

Melin, H., Stallard, T. S., Miller, S., *et al.*: 2013, *Icarus* **223**(2), 741–748, doi:10.1016/j.icarus.2013.01.012

Mendillo, M., Nagy, A., & Waite, J. H.: 2002, in *Introduction*, in *Atmospheres in the Solar System: Comparative Aeronomy*, American Geophysical Union, Washington, D.C., 13

Mendis, D. A.: 2002, *Plasma Sources Science Technology* **11**(26), A219

Mendis, D. A. & Horányi, M.: 2013a, *Reviews of Geophysics* **51**, 53, doi:10.1002/rog.20005

Mendis, D. A. & Horányi, M.: 2013b, *J. Plasma Phys.* **79**, 1067

Mendis, D. A. & Rosenberg, M.: 1994, *Ann. Rev. Astron. Astrophys.* **32**, 419, doi:10.1146/annurev.aa.32.090194.002223

Menou, K. & Rauscher, E.: 2010, *ApJ* **713**, 1174

Metzger, A. E., Gilman, D. A., Luthey, J. L., *et al.*: 1983, *J. Geophys. Res.* **88**(A10), 7731–7741

Mewe, R.: 1999, in J. van Paradijs and J. A. M. Bleeker (Eds.), *X-Ray Spectroscopy in Astrophysics*, Vol. 520 of *Lecture Notes in Physics*, Berlin: Springer Verlag, 109

Meyer-Vernet, N., Maksimovic, M., Czechowski, A., *et al.*: 2009, *Solar Phys.* **256**, 463, doi:10.1007/s11207-009-9349-2

Miller, B. P., Gallo, E., Wright, J. T., & Pearson, E. G.: 2015, *ApJ* **799**, 163, doi:10.1088/0004-637X/799/2/163

Miller, S., Achilleos, N., Ballester, G. E., *et al.*: 1997, *Icarus* **130**, 5767

Miller, S., Aylward, A., & Millward, G.: 2005, *Space Sci. Rev.* **116**(1–2), 319–343, doi:10.1007/s11214-005-1960-4

Miller, S., Stallard, T., Melin, H., & Tennyson, J.: 2010, *Faraday Discuss.* **147**, 283, doi:10.1039/c004152c

Millward, G., Miller, S., Stallard, T., & Aylward, A.: 2002, *Icarus* **160**(1), 95–107, doi:10.1006/icar.2002.6951

Mitchell, C. J., Horányi, M., Havnes, O., & Porco, C. C.: 2006, *Science* **311**, 1587, doi:10.1126/science.1123783

Mitchell, C. J., Porco, C. C., Dones, H. L., & Spitale, J. N.: 2013, *Icarus* **225**, 446, doi:10.1016/j.icarus.2013.02.011

Mitchell, D. L., Lin, R. P., Mazelle, C., *et al.*: 2001, *J. Geophys. Res.* **106**, 23419

Mitchell, D. L., Halekas, J. S., Lin, R. P., *et al.*: 2008, *Icarus* **194**, 401, doi:10.1016/j.icarus.2007.10.027

Mlynczak, M. G., Hunt, L. A., Marshall, B. T., *et al.*: 2010, *J. Geophys. Res.* **115**, A03309, doi:10.1029/2009JA014713

Monnereau, M., Calvet, M., Margerin, L., & Souriau, A.: 2010, *Science* **502**, 1014

Moore, L., Galand, M., Müller-Wodarg, I., Yelle, R., & Mendillo, M.: 2008, *J. Geophys. Res.* **113**(A10), A10306, doi:10.1029/2008JA013373

Moore, L., Müller-Wodarg, I., Galand, M., Kliore, A., & Mendillo, M.: 2010, *J. Geophys. Res.* **115**(A11), A11317, doi:10.1029/2010JA015692

Moore, L., Fischer, G., Müller-Wodarg, I., Galand, M., & Mendillo, M.: 2012, *Icarus* **221**(2), 508–516, doi:10.1016/j.icarus.2012.08.010

Moore, L., O'Donoghue, J., Müller-Wodarg, I., Galand, M., & Mendillo, M.: 2015, *Icarus*, **245**, 355–366, doi:10.1016/j.icarus.2014.08.041

Moore, T. E. & Khazanov, G. V.: 2010, *J. Geophys. Res.* **115**, doi:10.1029/2009JA014905

Moran, P. J.: 1998, PhD Thesis, The University of Wales, Aberystwyth

Moran, P. J., Breen, A. R., Varley, C. A., *et al.*: 1998, *Ann. Geophys.* **16**, 1259

Morfill, G. E. & Thomas, H. M.: 2005, *Icarus* **179**, 539, doi:10.1016/j.icarus.2005.08.008

Morooka, M. W., Wahlund, J.-E., Eriksson, A. I., *et al.*: 2011, *J. Geophys. Res. Space Physics* **116**(A15), 12221, doi:10.1029/2011JA017038

Moses, J. & Bass, S.: 2000, *J. Geophys. Res.* **105**(1999), 70137052

Moses, J., Bézard, B., Lellouch, E., *et al.*: 2000, *Icarus* **145**(1), 166202, doi:10.1006/icar.1999.6320

Moudden, Y. & Forbes, J. M.: 2010, *J. Geophys. Res.* **115**, 9005

Müller, H.-R. & Zank, G. P.: 2004, *J. Geophy. Res. Space Physics* **109**, 7104, doi:10.1029/2003JA010269

Müller, H.-R., Frisch, P. C., Florinski, V., & Zank, G. P.: 2006, *ApJ* **647**, 1491, doi:10.1086/505588

Müller, H.-R., Florinski, V., Heerikhuisen, J., *et al.*: 2008, *A&A* **491**, 43, doi:10.1051/0004-6361:20078708

Müller, H.-R., Frisch, P. C., Fields, B. D., & Zank, G. P.: 2009, *Space Sci. Rev.* **143**, 415, doi:10.1007/s11214-008-9448-7

Müller-Wodarg, I. C. F., Mendillo, M., Yelle, R., & Aylward, A.: 2006, *Icarus* **180**(1), 147–160, doi:10.1016/j.icarus.2005.09.002

Müller-Wodarg, I. C. F., Moore, L., Galand, M., Miller, S., & Mendillo, M.: 2012, *Icarus* **221**(2), 481–494, doi:10.1016/j.icarus.2012.08.034

Mutel, R. L., Molnar, L. A., Waltman, E. B., & Ghigo, F. D.: 1998, *ApJ* **507**, 371, doi:10.1086/306311

Nagasawa, M., Ida, S., & Bessho, T.: 2008, *Appl. Phys. J.* **678**, 498, doi:10.1086/529369

Nagy, A. F., Kliore, A. J., Marouf, E., *et al.*: 2006, *J. Geophys. Res. Space Physics* **111**, 6310, doi:10.1029/2005JA011519

Nagy, A. F., Kliore, A. J., Mendillo, M., *et al.*: 2009, in *Saturn from Cassini–Huygens*, Springer Netherlands, Dordrecht, 181

Nair, H., Allen, M., Anbar, A., Yung, Y. L., & Clancy, R. T.: 1994, *Icarus* **111**, 124

Naoz, S., Farr, W. M., Lithwick, Y., Rasio, F. A., & Teyssandier, J.: 2011, *Nature* **473**, 187, doi:10.1038/nature10076

Nellis, W. J.: 2010, *Phys. Rev. B* **82**, 092101

Nelson, G. A.: 2002, *Gravit. Space Biol. Bull.* **16**, 29

Neves, V., Bonfils, X., Santos, N. C., *et al.*: 2013, *VizieR Online Data Catalog* **355**, 19036

Newkirk, Jr., G.: 1980, in R. O. Pepei, J. A. Eddy, & R. B. Merrill (Eds.), *The Ancient Sun: Fossil Record in the Earth, Moon and Meteorites*, 293–320

Nichols, J. D., Burleigh, M. R., Casewell, S. L., *et al.*: 2012, *ApJ* **760**(59), 19, doi:10.1088/0004-637X/760/1/59

Niciejewski, R., Skinner, W., Cooper, M., *et al.*: 2011, *J. Geophys. Res.* **116**, doi:10.1029/2010JA16277

Nielsen, E., Fraenz, M., Zou, H., *et al.*: 2007, *Planet. Space Sci.* **55**, 2164

Nier, A. O. & McElroy, M. B.: 1977, *J. Geophys. Res.* **82**, 4341

Nilsson, H., Edberg, N., Stenberg, G., & Barabash, S.: 2011, *Icarus*, doi:10.1016/j.icarus.2011.08.003

Nimmo, F. & Gilmore, M. S.: 2001, *J. Geophys. Res.* **106**, 12315

Nogami, D., Notsu, Y., Honda, S., *et al.*: 2014, *Publ. Astron. Soc. Japan* **66**, L4, doi:10.1093/pasj/psu012

Northrop, T. G.: 1963, *Rev. Geophys. Space Phys.* **1**, 283, doi:10.1029/RG001i003p00283

Northrop, T. G. & Birmingham, T. J.: 1982, *J. Geophys. Res.* **87**, 661, doi:10.1029/JA087iA02p00661

Northrop, T. G. & Hill, J. R.: 1983, *J. Geophys. Research* **88**, 1, doi:10.1029/JA088iA01p00001

Noyes, R. W., Hartmann, L. W., Baliunas, S. L., Duncan, D. K., & Vaughan, A. H.: 1984, *ApJ* **279**, 763, doi:10.1086/161945

Nutzman, P. A., Fabrycky, D. C., & Fortney, J. J.: 2011, *ApJL* 740, L10, doi:10.1088/2041-8205/740/1/L10

O'dell, C. R. & Wen, Z.: 1994, *Appl. Phys. J.* **436**, 194, doi:10.1086/174892

O'Donoghue, J., Stallard, T. S., Melin, H., *et al.*: 2013, *Nature* **496**(7444), 193–195, doi:10.1038/nature12049

Odstrcil, D. & Pizzo, V. J.: 2002, in H. Sawaya-Lacoste (Ed.), *Solspa 2001, Proceedings of the Second Solar Cycle and Space Weather Euroconference*, Vol. 477 of *ESA Special Publication*, 281

Odstrcil, D., Pizzo, V. J., & Arge, C. N.: 2005, *J. Geophys. Res.* **110**(A9), 2106, doi:10.1029/2004JA010745

Ohta, K., Cohen, R. E., Hirose, K., *et al.*: 2012, *Phys. Rev. Lett.* **108**, 026403

Opher, M., Bibi, F. A., Toth, G., *et al.*: 2009, *Nature* **462**, 1036, doi:10.1038/nature08567

Oppenheimer, B. R.: 1999, PhD thesis, California Institute of Technology

Oppenheimer, B. R. & Hinkley, S.: 2009, *Ann. Rev. Astron. Astrophys.* **47**, 253, doi:10.1146/annurev-astro-082708-101717

Oppenheimer, B. R., Golimowski, D. A., Kulkarni, S. R., *et al.*: 2001, *ApJ* **121**, 2189, doi:10.1086/319941

Osten, R. A.: 2002, Multiwavelength Observations of flares and variability in the coronae of active binary systems, PhD thesis, University of Colorado at Boulder

Osten, R., Livio, M., Lubow, S., *et al.*: 2013, *ApJL* **765**, L44, doi:10.1088/2041-8205/765/2/L44

Osten, R. A. & Bastian, T. S.: 2008, *ApJ* **674**, 1078, doi:10.1086/525013

Osten, R. A. & Brown, A.: 1999, *ApJ* **515**, 746, doi:10.1086/307034

Osten, R. A., Brown, A., Ayres, T. R., *et al.*: 2000, *ApJ* **544**, 953, doi:10.1086/317249

Osten, R. A., Brown, A., Ayres, T. R., *et al.*: 2004, *ApJSS* **153**, 317, doi:10.1086/420770

Osten, R. A., Hawley, S. L., Allred, J. C., Johns-Krull. C. M., & Roark, C.: 2005, *ApJ* **621**, 398, doi:10.1086/427275

Osten, R. A., Hawley, S. L., Allred, J., *et al.*: 2006, *ApJ* **647**, 1349, doi:10.1086/504889

Osten, R. A., Drake, S., Tueller, J., *et al.*: 2007, *ApJ* **654**, 1052, doi:10.1086/509252

Osten, R. A., Godet, O., Drake, S., *et al.*: 2010, *ApJ* **721**, 785, doi:10.1088/0004-637X/721/1/785

Osten, R. A., Kowalski, A., Sahu, K., & Hawley, S. L.: 2012, *ApJ* **754**, 4, doi:10.1088/0004-637X/754/1/4

Owens, M. J. & Forsyth, R. J.: 2013, *Living Reviews in Solar Physics* **10**(5), doi:10.12942/lrsp-2013-5

Owens, M. J., Crooker, N. U., Schwadron, N. A., *et al.*: 2008, *Geophys. Res. Lett.* **35**, 20108, doi:10.1029/2008GL035813

Pace, G.: 2013, *A&A* **551**, L8, doi:10.1051/0004-6361/201220364

Pace, G. & Pasquini, L.: 2004, *A&A* **426**, 1021, doi:10.1051/0004-6361:20040568

Paranicas, C. & Cheng, A. F.: 1997, *Icarus* **125**, 380, doi:10.1006/icar.1996.5635

Paranicas, C., Cooper, J. F., Garrett, H. B., Johnson, R. E., & Sturner, S. J.: 2009, in R. T. Pappalardo, W. B. McKinnon, and K. K. Khurana (Eds.), *Europa*, University of Arizona Press, Tucson, 529 ISBN: 9780816528448

Parker, E. N.: 1958, *ApJ* **128**, 664

Parker, E. N.: 1961, *ApJ* **134**, 20, doi:10.1086/147124

Parker, E. N.: 1988, *ApJ* **330**, 474, doi:10.1086/166485

Paulson, D. B., Allred, J. C., Anderson, R. B., *et al.*: 2006, *PASPs* **118**, 227, doi:10.1086/499497

Pavlov, A. A., Toon, O. B., Pavlov, A. K., Bally, J., & Pollard, D.: 2005, *Geophys. Res. Lett.* **32**, 3705, doi:10.1029/2004GL021890

Pedatella, N, M., Hagan, M. E., & Maute, A.: 2012, *Geophys. Res. Lett.*, **39** doi:10.1029/2012GL053643

Pedatella, N., Fuller-Rowell, T., Wang, H., *et al.*: 2014a, *J. Geophys. Res.* **119**, 1306, doi:10.1002/2013JA019421

Pedatella, N., Liu, H.-L, Richmond, A. D., Maute, A., & Fang, T.-W.: 2014b, *J. Geophys. Res.* **117**, A08326, doi:10.1029/2012JA017858

Peek, J. E. G., Heiles, C., Peek, K. M. G., Meyer, D. M., & Lauroesch, J. T.: 2011, *ApJ* **735**, 129, doi:10.1088/0004-637X/735/2/129

Penz, T. *et al.*: 2004, *Planetary and Space Science* **52**(13), 1157–1167, doi:10.1016/j.pss.2004.06.001

Perna, R., Menou, K., & Rauscher, E.: 2010, *ApJ* **719**, 1421

Perry, J., Kim, Y., Fox, J., & Porter, H.: 1999, *J. Geophys. Res.* **104**(E7), 16 541–16 565

Petigura, E. A., Howard, A. W., & Marcy, G. W.: 2013, *Proc. Natl Acad. Sci. USA* **110**(48), 1927319278, doi:10.1073/pnas.1319909110

Petit, P., Dintrans, B., Solanki, S. K., *et al.*: 2008, *Mon. Not. R. Astron. Soc.* **388**, 80, doi:10.1111/j.1365-2966.2008.13411.x

Phillips, J. L., Balogh, A., Bame, S. J., *et al.*: 1994, *Geophys. Res. Lett.* **21**, 1105

Phillips, J. L., Bame, S. J., Feldman, W. C., *et al.*: 1995, *Science* **268**, 1030

Phillips, R. J., Raubertas, R. F., Arvidson, R. E., *et al.*: 1992, *J. Geophys. Res.* **97**, 15 923, doi:10.1029/92JE01696

Piel, A.: 2010, *Plasma Physics*, Springer Verlag

Pilinski, M. D., Argrow, B. M., Palo, S. E., & Bowman, B. R.: 2013, *J. Spacecraft and Rockets* **50**, 556

Piters, A. J. M., Schrijver, C. J., Schmitt, J. H. M. M., *et al.*: 1997, *A&A* **325**, 1115

Pogorelov, N. V., Heerikhuisen, J., & Zank, G. P.: 2008, *ApJL* **675**, L41, doi:10.1086/529547

Pollack, J. B., Hubickyj, O., Bodenheimer, P., Lissauer, J. J., Podolak, M., & Greenzweig, Y.: 1996, *Icarus* **124**, 62, doi:10.1006/icar.1996.0190

Pont, F.: 2009, *Mon. Not. R. Astron. Soc.* **396**, 1789, doi:10.1111/j.1365-2966.2009.14868.x

Porco, C. A. & Danielson, G. E.: 1982, *Astron. J.* **87**, 826, doi:10.1086/113162

Porco, C. C. *et al.*: 2006, *Science* **311**(5766), 1393401, doi:10.1126/science.1123013

Pozzo, M., Davies, C., Gubbins, D., & Alfe, D.: 2012, *Nature* **485**, 355

Prolss, G. W.: 2012, *Surv. Geophys.* **32**, 101

Pryor, W. R. *et al.*: 2011, *Nature* **472**(7343), 3313, doi:10.1038/nature09928

Purucker, M. E. & Nicholas, J. B.: 2010, *J. Geophys. Res.* **115**, E12007

Qian, L., Solomon, S. C., & Kane, T. J.: 2009, *J. Geophys. Res.* **114**, A01312, doi:10.1029/2008JA013643

Qian, L., Burns, A. G., Solomon, S. C., & Wang, W.: 2013, *Geophys. Res. Lett.*, doi:10.1002/grl.50448

Radioti, A., Lystrup, M., Bonfond, B., Grodent, D., & Gérard, J.-C.: 2013, *J. Geophys. Res. Sp. Phys.* **118**(5), 2286–2295, doi:10.1002/jgra.50245

Rafkin, S. C. R., Hollingsworth, J. L., Mischna, M. A., Newmen, C. E., & Richardson, M I.: 2013, in S. J. Mackwell, A. A. Simon-Miller, J. W. Harder, & M. A. Bullock (Eds.), *Comparative Climatology of Terrestrial Planets*, 55–89, doi:10.2458/azu_uapress_9780816530595-ch003

Raghavan, D., McAlister, H. A., Henry, T. J., *et al.*: 2010, *ApJSS* **190**, 1, doi:10.1088/0067-0049/190/1/1

Rao, N. N., Shukla, P. K., & Yu, M. Y.: 1990, *Planet. Space Sci.* **38**, 543, doi:10.1016/0032-0633(90)90147

Rauscher, E. & Menou, K.: 2013, *ApJ* **764**, 103

Ray, L. C., Su, Y.-J., Ergun, R. E., Delamere, P. A., & Bagenal, F.: 2009, *J. Geophys. Res.* **114**(A4), A04214, doi:10.1029/2008JA013969

Read, P. L., Dowling, T. E., & Schubert, G.: 2009, *Nature* **460**(7255), 608–610, doi:10.1038/nature08194

Reale, F., Betta, R., Peres, G., Serio, S., & McTiernan, J.: 1997, *A&A* **325**, 782

Reames, D. V.: 1995, *Reviews of Geophysics* **33**, 585, doi:10.1029/95RG00188

Reames, D. V.: 2013, *Space Sci. Rev.* **175**, 53, doi:10.1007/s11214-013-9958-9

Reames, D. V., Ng, C. K., & Tylka, A. J.: 2013, *Sol. Phys.* **285**, 233, doi:10.1007/s11207-012-0038-1

Redfield, S. & Falcon, R. E.: 2008, *ApJ* **683**, 207–225, doi:10.1086/589230

Redfield, S. & Linsky, J. L.: 2008, *ApJ* **673**, 283, doi:10.1086/524002

Redfield, S. & Linsky, J. L.: 2015, *ApJ* **812**, 125, doi:10.1088/0004-637X/812/2/125

Redmer, R., Mattsson, T. R., Netelmann, N., & French, M.: 2011, *Icarus* **211**, 798

Rego, D., Miller, S., Achilleos, N., Prangé, R., & Joseph, R. D.: 2000, *Icarus* **147**(2), 366–385, doi:10.1006/icar.2000.6444

Reiners, A.: 2012, *Living Reviews in Solar Physics* **9**, 1, doi:10.12942/lrsp-2012-1

Rempel, M.: 2009, in C. J. Schrijver and G. Siscoe (Eds.), *Heliophysics: Plasma Physics of the Local Cosmos*, Cambridge University Press, Cambridge

Ribas, I., Guinan, E. F., Güdel, M., & Audard, M.: 2005, *ApJ* **622**, 680, doi:10.1086/427977

Richardson, J. D. & Sittler, Jr., E. C.: 1990, *J. Geophys. Research* **95**, 12019, doi:10.1029/JA095iA08p12019

Richardson, J. D., Kasper, J. C., Wang, C., Belcher, J. W., & Lazarus, A. J.: 2008, *Nature* **454**, 63, doi:10.1038/nature07024

Rickett, B.: 1992, in *Solar Wind Seven Colloquium*, 255

Rickett, B. J. & Coles, W. A.: 1991, *J. Geophys. Res.* **96**, 1717
Rishbeth, H. & Williams, P. J. S.: 1985, *Mon. Not. R. Astron. Soc.* **26**, 478
Robinson, L. J.: 1980, *Sky and Telescope* **60**, 481
Robinson, R. D., Carpenter, K. G., & Brown, A.: 1998, *ApJ* **503**, 396, doi:10.1086/305971
Roble, R. G.: 1987, *J. Geophys. Res.* **92**, A8, 8745
Roble, R. G. & Dickinson, R. E.: 1989, *Geophys. Res. Lett.* **16**, 1441
Rocha-Pinto, H. J., Castilho, B. V., & Maciel, W. J.: 2002, *A&A* **384**, 912, doi:10.1051/0004-6361:20011815
Rodono, M.: 1974, *A&A* **32**, 337
Rodriguez, L.: 2005, PhD thesis, Technical University Braunschweig, Germany
Roederer, J. G.: 1970, *Dynamics of Geomagnetically Trapped Radiation*, Springer, Berlin
Rogers, L. A.: 2014, ArXiv e-prints
Rogers, L. A. & Seager, S.: 2010, *Appl. Phys. J.* **712**, 974, doi:10.1088/0004-637X/712/2/974
Rogers, T. M. & Showman, A. P.: 2014, *ApJL* **782**, L4
Rosenberg, M.: 1993, *Planet. Space Sci.* **41**, 229, doi:10.1016/0032-0633(93)90062-7
Rossiter, R. A.: 1924, *Appl. Phys. J.* **60**, 15, doi:10.1086/142825
Roussos, E.: 2008, PhD thesis, Universität Braunschweig, Germany
Roussos, E., Jones, G. H., Krupp, N., *et al.*: 2007, *J. Geophys. Res.* **112**(A11), 6214, doi:10.1029/2006JA012027
Roussos, E., Jones, G. H., Krupp, N., *et al.*: 2008, *Icarus* **193**, 455, doi:10.1016/j.icarus.2007.03.034
Rubin, M., Koenders, C., Atlwegg, K., *et al.*: 2014, *Icarus* 38–49
Russell, C. T., Aroian, R., Arghavani, M., & Nock, K.: 1984, *Science* **226**, 43, doi:10.1126/science.226.4670.43
Russell, C. T., Raymond, C. A., Coradini, A., *et al.*: 2012, *Science* **336**, 684
Sackmann, I.-J. & Boothroyd, A. I.: 2003, *ApJ* **583**, 1024, doi:10.1086/345408
Sagan, C. & Mullen, G.: 1972, *Science* **177**(4), 5256, doi:10.1126/science.177.4043.52
Salpeter, E. E.: 1967, *ApJ* **147**, 433
Sanchis-Ojeda, R., Fabrycky, D. C., Winn, J. N., *et al.*: 2012, *Nature* **487**, 449, doi:10.1038/nature11301
Sano, T., Miyama, S. M., Umebayashi, T., & Nakano, T.: 2000, *ApJ* **543**, 486, doi:10.1086/317075
Santos, N. C., Israelian, G., & Mayor, M.: 2004, *Astron. Astrophys.* **415**, 1153, doi:10.1051/0004-6361:20034469
Schaber, G. G., Strom, R. G., Moore, H. J., *et al.*: 1992, *J. Geophys. Res.* **97**, 13257, doi:10.1029/92JE01246
Schaefer, B. E., King, J. R., & Deliyannis, C. P.: 2000, *ApJ* **529**, 1026, doi:10.1086/308325
Scherer, K. & Fichtner, H.: 2014, *ApJ* **782**, 25, doi:10.1088/0004-637X/782/1/25
Schlichting, H. E.: 2014, ArXiv e-prints
Schmidt, S. J., Prieto, J. L., Stanek, K. Z., *et al.*: 2014, *ApJL* **781**, L24, doi:10.1088/2041-8205/781/2/L24
Schrijver, C. J.: 2009, *ApJL* **699**, L148, doi:10.1088/0004-637X/699/2/L148
Schrijver, C. J. & Beer, J.: 2014, *EOS*, doi:10.1002/2014EO240001
Schrijver, C. J. & Title, A. M.: 2001, *ApJ* **551**, 1099, doi:10.1086/320237
Schrijver, C. J. & Zwaan, C.: 2000, *Solar and Stellar Magnetic Activity*, Cambridge University Press
Schrijver, C. J., De Rosa, M. L., & Title, A. M.: 2003, *ApJ* **590**, 493, doi:10.1086/374982

Schrijver, C. J., Beer, J., Baltensperger, U., *et al.*: 2012, *Journal of Geophysical Research Space Physics* **117**, 8103, doi:10.1029/2012JA017706
Schuessler, M. & Solanki, S. K.: 1992, *A&A* **264**, L13
Schunk, R. W., & Nagy, A. F.: 2009, *Ionospheres second edition: Physics, Plasma Physics, and Chemistry*, Cambridge University Press
Schwadron, N. A. & McComas, D. J.: 2013, *ApJL* **778**, L33, doi:10.1088/2041-8205/778/2/L33
Schwadron, N. A., Allegrini, F., Bzowski, M., *et al.*: 2011, *ApJ* **731**, 56, doi:10.1088/0004-637X/731/1/56
Schwamb, M. E., Orosz, J. A., Carter, J. A., *et al.*: 2013, *Appl. Phys. J.* **768**, 127, doi:10.1088/0004-637X/768/2/127
Scott, S. L., Coles, W. A., & Bourgois, G.: 1983, *A&A* **123**, 207
Seager, S., Kuchner, M., Hier-Majumder, C. A., & Militzer, B.: 2007, *Appl. Phys. J.* **669**, 1279, doi:10.1086/521346
Segura, A., Walkowicz, L. M., Meadows, V., Kasting, J., & Hawley, S.: 2010, *Astrobiology* **10**, 751, doi:10.1089/ast.2009.0376
Seiff, A., Kirk, D. B., Knight, T. C. D., *et al.*: 1998, *J. Geophys. Res.* **103**, 22 857–22 890
Shapiro, V. D., Bingham, R., Kellett, B. J., *et al.*: 2005, *Physica Scripta Volume T* **116**, 83, doi:10.1238/Physica.Topical.116a00083
Shaviv, N. J.: 2003, *Journal of Geophysical Research Space Physics* **108**, 1437, doi:10.1029/2003JA009997
Shaviv, N. J.: 2005a, *J. Geophys. Research* **110**, 8105, doi:10.1029/2004JA010866
Shaviv, N. J.: 2005b, *International Journal of Modern Physics A* **20**, 6662, doi:10.1142/S0217751X05029733
Shea, E. K., Weiss, B. P., Cassata, W. S., *et al.*: 2012, *Science* **335**, 453
Shemansky, D. E., Ajello, J. M., & Hall, D. T.: 1985, *ApJ* **296**, 765–773
Shibata, K. & Magara, T.: 2011, *Living Reviews in Solar Physics* **8**, 6, doi:10.12942/lrsp-2011-6
Shkolnik, E. L.: 2013, *ApJ* **766**, 9, doi:10.1088/0004-637X/766/1/9
Shkolnik, E., Walker, G. A. H., & Bohlender, D. A.: 2003, *ApJ* **597**, 1092, doi:10.1086/378583
Shkolnik, E., Walker, G. A. H., Bohlender, D. A., Gu, P.-G., & Kurster, M.: 2005a, *ApJ* **622**, 1075, doi:10.1086/428037
Shkolnik, E., Walker, G. A. H., Rucinski, S. M., Bohlender, D. A., & Davidge, T. J.: 2005b, *Astronom. J.* **130**, 799, doi:10.1086/431364
Shkolnik, E., Bohlender, D. A., Walker, G. A. H., & Collier Cameron, A.: 2008, *ApJ* **676**, 628, doi:10.1086/527351
Showman, A. P., Cho, J. Y.-K., & Menou, K.: 2011, in S. Seager (Ed.), *Exoplanets*, 471–516, University of Arizona Press, Tucson
Shukla, P. K. & Mamun, A. A.: 2002, *Introduction to the Physics of Dusty Plasmas*, IOP Series in Plasma Physics
Shukla, P. K. & Silin, V. P.: 1992, *Physica Scripta* **45**, 508, doi:10.1088/0031-8949/45/5/015
Silvestri, N. M., Hawley, S. L., & Oswalt, T. D.: 2005, *Astron. J.* **129**, 2428, doi:10.1086/429593
Simakov, M. B., Kuzicheva, E. A., Antropov, A. E., & Dodonova, N. Y.: 2002, *Advances in Space Research* **30**, 1489, doi:10.1016/S0273-1177(02)00510-0
Siskind, D. E. & McCormack, J. P.: 2014, *Geophys. Res. Lett.* **41**, doi:10.1002/2013GL058875
Siskind, D. E., Barth, C. A., Evans, D. S., & Roble, R. G.: 1989a, *J. Geophys. Res.* **94**, 16 899

Siskind, D. E., Barth, C. A., & Roble, R. G.: 1989b, *J. Geophys. Res.* **94**, 16 885
Siskind, D. E., Drob, D. P., Dymond, K. F., & McCormack, J. P.: 2014, *J. Geophys. Res.* **119**, doi:10.1002/2013JA019116
Skumanich, A.: 1972, *ApJ* **171**, 565, doi:10.1086/151310
Slavin, J. D. & Frisch, P. C.: 2008, *A&A* **491**, 53, doi:10.1051/0004-6361:20078101
Smart, D. F., Shea, M. A., & Flückiger, E. O.: 2000, *Space Sci. Rev.* **93**, 305, doi:10.1023/A:1026556831199
Smith, A. K.: 2012, *Surv. Geophys.* **33**, 1177, doi:10.1007/S10712-012-9196-9
Smith, B. A., Soderblom, L., Beebe, R. F., *et al.*: 1981, *Science* **212**, 163
Smith, B. A., Soderblom, L., Batson, R. M., *et al.*: 1982, *Science* **215**, 503
Smith, C. G. A., Aylward, A. D., Millward, G. H., Miller, S., & Moore, L. E.: 2007, *Nature* **445**(7126), 399401, doi:10.1038/nature05518
Smith, D. S. & Scalo, J. M.: 2009, *Astrobiology* **9**, 673, doi:10.1089/ast.2009.0337
Smith, G. R. & Hunten, D. M.: 1990, *Rev. Geophys.* **28**(2), 117143
Smith, G. R., Shemansky, D. E., Holberg, J. B., Broadfoot, A. L., & Sandel, B. R.: 1983, *J. Geophys. Res.* **88**(A11), 86678678
Smith, K., Güdel, M., & Audard, M.: 2005, *A&A* **436**, 241, doi:10.1051/0004-6361:20042054
Smrekar, S. E., Stofan, E. R., Mueller, N., *et al.*: 2010, *Science* **328**(5), 605, doi:10.1126/science.1186785
Soderblom, D. R.: 2010, *Ann. Rev. Astron. Astrophys.* **48**, 581, doi:10.1146/annurev-astro-081309-130806
Soderlund, K. M., Heimpel, M. H., King, E. M., & Aurnou, J. M.: 2013, *Icarus* **224**, 97
Solanki, S. K., Motamen, S., & Keppens, R.: 1997, *A&A* **324**, 943
Solomon, S.: 2010, in *Heliophysics: Evolving Solar Activity and the Climates of Space and Earth*, Cambridge University Press
Solomon, S. C., Aharonson, O., Aurnou, J. M., *et al.*: 2005, *Science* **307**, 1214
Som, S. M., Catling, D. C., Harnmeijer, J. P., Polivka, P. M., & Buick, R.: 2012, *Nature* **484**(7), 359362, doi:10.1038/nature10890
Sotomayor-Beltran, C., Sobey, C., Hessels, J. W. T., *et al.*: 2013, *A&A* **552**, A58, doi:10.1051/0004-6361/201220728
Sousa, S. G., Santos, N. C., Mayor, M., *et al.*: 2008, *Astron. Astrophys.* **487**, 373, doi:10.1051/0004-6361:200809698
Spangler, S. R. & Whiting, C. A.: 2008, in *American Astronomical Society Meeting Abstracts #212*, Vol. 40 of *Bulletin of the American Astronomical Society*, 193
Spitzer, L.: 1978, *Physical Processes in the Interstellar Medium*, New York: Wiley-Interscience
Stallard, T. S.: 2003, *Geophys. Res. Lett.* **30**(5), 1221, doi:10.1029/2002GL016031
Stallard, T., Miller, S., Millward, G., & Joseph, R. D.: 2001, *Icarus* **154**(2), 475–491, doi:10.1006/icar.2001.6681
Stallard, T., Miller, S., Millward, G., & Joseph, R. D.: 2002, *Icarus* **156**(2), 498–514, doi:10.1006/icar.2001.6793
Stallard, T., Miller, S., Melin, H., *et al.*: 2008, *Nature* **453**(7198), 10835, doi:10.1038/nature07077
Stallard, T., Melin, H., Cowley, S. W. H., Miller, S., & Lystrup, M. B.: 2010, *ApJ* **722**(1), L85–L89, doi:10.1088/2041-8205/722/1/L85
Stallard, T. S., Melin, H., Miller, S., *et al.*: 2012a, *Geophys. Res. Lett.* **39**(15), 15, doi:10.1029/2012GL052806
Stallard, T. S., Melin, H., Miller, S., *et al.*: 2012b, *Philos. Trans. A. Math. Phys. Eng. Sci.* **370**(1978), 521324, doi:10.1098/rsta.2012.0028
Stanley, S.: 2010, *Geophys. Res. Lett.* **37**, L05201

Stanley, S. & Bloxham, J.: 2004, *Nature* **428**, 151

Stanley, S. & Bloxham, J.: 2006, *Icarus* **184**, 556

Stanley, S. & Glatzmaier, G. A.: 2010, *Space Sci. Rev.* **152**, 617

Stanley, S., Elkins-Tanton, L. T., Zuber, M. T., & Parmentier, E. M.: 2008, *Science* **321**, 1822

Stauffer, J. R.: 1991, in S. Catalano and J. R. Stauffer (Eds.), *NATO ASIC Proc. 340: Angular Momentum Evolution of Young Stars*, 117

Stelzer, B., Neuhäuser, R., & Hambaryan, V.: 2000, *A&A* **356**, 949

Stelzer, B., Schmitt, J. H. M. M., Micela, G., & Liefke, C.: 2006, *A&A* **460**, L35, doi:10.1051/0004-6361:20066488

Stelzer, B., Flaccomio, E., Briggs, K., *et al.*: 2007, *A&A* **468**, 463, doi:10.1051/0004-6361:20066043

Stelzried, C. T., Levy, G. S., Sato, T., *et al.*: 1970, *Solar Phys.* **14**, 440, doi:10.1007/BF00221330

Sterken, V. J., Altobelli, N., Kempf, S., *et al.*: 2013, *A&A* **552**, A130, doi:10.1051/0004-6361/201219609

Stevenson, D. J.: 1980, *Science* **208**, 746

Stewart, A. I. F.: 1987, *Revised time dependent model of the martian atmosphere for use in orbit lifetime and sustenance studies, LASP-JPL Internal Report, NQ-802429*, Jet Propulsion Lab, Pasadena, California

Stewart, A. I. F., Alexander, M. J., Meier, R. R., *et al.*: 1992, *J. Geophys. Res.* **97**, 91

Stone, E. C., Cummings, A. C., McDonald, F. B., *et al.*: 2005, *Science* **309**, 2017, doi:10.1126/science.1117684

Stone, E. C., Cummings, A. C., McDonald, F. B., *et al.*: 2008, *Nature* **454**, 71, doi:10.1038/nature07022

Stone, E. C., Cummings, A. C., McDonald, F. B., *et al.*: 2013, *Science* **341**, 150, doi:10.1126/science.1236408

Strangeway, R. J., Ergun, R. E., Su, Y.-J., Carlson, C. W., & Elphic, R. C.: 2005, *J. Geophys. Res.* **110**(A), 03221, doi:10.1029/2004JA010829

Strangeway, R. J., Rusell, C. T., Luhmann, J. G., *et al.*: 2010, *AGU Fall Meeting Abstracts*, B1893

Strassmeier, K. G.: 1996, in K. G. Strassmeier and J. L. Linsky (Eds.), *Stellar Surface Structure*, IAU Symp. 176, Kluwer Academic Publishers, Dordrecht, The Netherlands, 289

Strassmeier, K. G.: 2001, in R. J. Garcia Lopez, R. Rebolo, and M. R. Zapaterio Osorio (Eds.), *11th Cambridge Workshop on Cool Stars, Stellar Systems and the Sun*, Vol. 223 of *Astronomical Society of the Pacific Conference Series*, 271

Strassmeier, K. G.: 2002, *Astronomische Nachrichten* **323**, 309, doi:10.1002/1521-3994(200208)323:3/4<309::AID-ASNA309>3.0.CO;2-U

Strickland, D. J., Stewart, A. I., Barth, C. A., Hord, C. W., & Lane, A. L.: 1973, *J. Geophys. Res.* **78**, 4547

Strickland, D. J., Thomas, G. E., & Sparks, P. R.: 1972, *J. Geophys. Res.* **77**, 4052

Strobel, D. F.: 2002, in *Atmospheres in the Solar System: Comparative Aeronomy, vol. 130*, AGU, Washington, D.C., 722

Strobel, D. F.: 2009, *Icarus* **202**(2), 632641, doi:10.1016/j.icarus.2009.03.007

Suavet, C., Weiss, B. P., Cassata, W. S., *et al.*: 2013, *Proc. Natl. Acad. Sci.* **110**, 8453

Svensmark, H. & Friis-Christensen, E.: 1997, *Journal of Atmospheric and Solar-Terrestrial Physics* **59**, 1225, doi:10.1016/S1364-6826(97)00001-1

Svensmark, H., Bondo, T., & Svensmark, J.: 2009, *Geophys. Res. Lett.* **36**, 15101, doi:10.1029/2009GL038429

Swift, J. J., Johnson, J. A., Morton, T. D., *et al.*: 2013, *Appl. Phys. J.* **764**, 105, doi:10.1088/0004-637X/764/1/105

Tagger, M., Henriksen, R. N., & Pellat, R.: 1991, *Icarus* **91**, 297, doi:10.1016/0019-1035(91)90026-P

Tappin, S. J. & Howard, T. A.: 2010, *Solar Phys.* **265**, 159, doi:10.1007/s11207-010-9588-2

Tellmann, S., Pätzold, M., Häusler, B., Hinson, D. P., & Tyler, G. L.: 2013, *J. Geophys. Res. Planets* **118**, 306, doi:10.1002/jgre.20058

Thebault, P.: 2011, *Celestial Mechanics and Dynamical Astronomy* **111**, 29, doi:10.1007/s10569-011-9346-2

Thébault, P., Marzari, F., & Scholl, H.: 2006, *Icarus* **183**, 193, doi:10.1016/j.icarus.2006.01.022

Thomasson, P.: 1986, *Q. J. R. Astron. Soc.* **27**, 413

Thompson, C., Barkan, A., Merlino, R. L., & D'Angelo, N.: 1999, *IEEE Transactions on Plasma Science* **27**, 146, doi:10.1109/27.763096

Tian, B. Y. & Stanley, S.: 2013, *ApJ* **768**, 156

Tian, B. Y., Stanley, S., Tikoo, S. M., & Weiss, B. P.: 2014, submitted

Tian, F. & Toon, O. B.: 2005, *Geophys. Res. Lett.* **32**(1), L18201, doi:10.1029/2005GL023510

Tian, F., Toon, O. B., Pavlov, A. A., & Sterck, H. De: 2005, *ApJ* **621**(2), 10491060, doi:10.1086/427204

Tian, F., Chassefière, E., Leblanc, F. & Brain, D.: 2013, in S. J. Mackwell, A. A. Simon-Miller, J. W. Harder & M. A. Bullock (Eds.), *Comparative Climatology of Terrestrial Planets*, 567–581, doi:10.2458/azu_uapress_9780816530595-ch23

Tinetti, G., Deroo, P., Swain, M. R., *et al.*: 2010, *ApJL* **712**, L139, doi:10.1088/2041-8205/712/2/L139

Tokumaru, M., Kojima, M., Fujiki, K., Yamashita, M., & Jackson, B. V.: 2007, *J. Geophys. Res. Space Physics* **112**, 5106, doi:10.1029/2006JA012043

Tóth, G., Kovács, D., Hansen, K. C., & Gombosi, T. I.: 2004, *J. Geophys. Res.* **109**, A11210, doi:10.1029/2004JA010406

Tseng, W.-L., Ip, W.-H., Johnson, R. E., Cassidy, T. A., & Elrod, M. K.: 2010, *Icarus* **206**(2), 382389, doi:10.1016/j.icarus.2009.05.019

Tsuchiya, T. & Tsuchiya, J.: 2011, *Proc. Natl. Acad. Sci.* **108**, 1252

Tsytovich, V., Morfill, G., Vladimirov, S., & Thomas, H.: 2010, *Elementary Physics of Complex Plasmas*, Springer (Lecture Notes in Physics)

Turner, N. J. & Drake, J. F.: 2009, *ApJ* **703**, 2152, doi:10.1088/0004-637X/703/2/2152

Turner, N. J. & Sano, T.: 2008, *ApJL* **679**, L131, doi:10.1086/589540

Uscinski, B. J.: 1977, *The Elements of Wave Propagation in Radom Media*, New York: McGraw-Hill

Vadas, S. L. & Liu, H.-L.: 2009, *J. Geophys. Res.* **14**, A10310, doi:10.1029/2009JA014108

Vallerga, J. V. & Welsh, B. Y.: 1995, *ApJ* **444**, 702, doi:10.1086/175643

van Allen, J. A., Thomsen, M. F., Randall, B. A., Rairden, R. L., & Grosskreutz, C. L.: 1980, *Science* **207**, 415

van den Oord, G. H. J., Mewe, R., & Brinkman, A. C.: 1988, *A&A* **205**, 181

van Marle, A. J., Decin, L., & Meliani, Z.: 2014, *A&A* **561**, A152, doi:10.1051/0004-6361/201321968

Vasavada, R., Bouchez, H., Ingersoll, A., Little, B., & Anger, C. D.: 1999, *J. Geophys. Res.* **104**(E11), 27 133–27 142

Vasyliūnas, V. M.: 1983, in *Physics of the Jovian Magnetosphere*, Cambridge University Press, Cambridge and New York, 395–453

Vasyliūnas, V. M.: 2008, *Ann. Geophys.* **26**(6), 1341–1343, doi:10.5194/angeo-26-1341-2008
Vasyliūnas, V. M., & Song, P.: 2005, *J. Geophys. Res.* **110**(A2), A02301, doi:10.1029/2004JA010615
Vidal-Madjar, A., Lecavelier des Etangs, A., Desert, J.-M., *et al.*: 2003, *Nature* **422**(6), 143146, doi:10.1038/nature01448.
Vidotto, A. A., Jardine, M., Cameron, A. C., *et al.*: 2014, ArXiv e-prints
Vilim, R., Stanley, S., & Elkins-Tanton, L.: 2013, *ApJL* **768**, L30
von Zahn, U., Fricke, K. H., Hunten, D. M., *et al.*: 1980, *J. Geophys. Res.* **85**, 78290
Vourlidas, A., Howard, R. A., Esfandiari, E., *et al.*: 2010, *ApJ* **722**, 1522, doi:10.1088/0004-637X/722/2/1522
Waite, J. H., Cravens, T. E., Ip, W.-H., *et al.*: 2005, *Science* **307**, 1260, doi:10.1126/science.1105734
Wallmann, K.: 2004, *Geochemistry, Geophysics, Geosystems* **5**, 6004, doi:10.1029/2003GC000683
Walt, M.: 1994, *Introduction to Geomagnetically Trapped Radiation*, Cambridge University Press
Walterscheid, R.: 2010, in *Heliophysics*, Cambridge University Press
Wang, J. & Fischer, D. A.: 2013, ArXiv e-prints
Wang, J., Fischer, D. A., Xie, J.-W., & Ciardi, D. R.: 2014, *Appl. Phys. J.* **791**, 111, doi:10.1088/0004-637X/791/2/111
Wang, Y.-M. & Sheeley, N. R.: 1990, *ApJ* **355**, 726
Wannberg, G., Wolf, I., Vanhainen, L.-G., *et al.*: 1997, *Radio Science* **32**, 2283
Wannberg, G., Vanhainen, L.-G., Westman, A., Breen, A. R., & Williams, P. J. S.: 2002, in *Conference Proceedings*, Union of Radio Scientists (URSI)
Ward, P. & Brownlee, D.: 2000, *Rare Earth: Why Complex Life is Uncommon in the Universe*, New York: Copernicus
Wargelin, B. J. & Drake, J. J.: 2001, *ApJL* **546**, L57, doi:10.1086/318066
Webb, D. F. & Howard, T. A.: 2012, *Living Reviews in Solar Physics* **9**(3), doi:10.12942/lrsp-2012-3
Weber, E. J. & Davis, Jr., L.: 1967, *ApJ* **148**, 217, doi:10.1086/149138
Weber, R. C., Lin, P. Y., Garnero, E. J., Williams, Q., & Lognonne, P.: 2011, *Science* **331**, 309
Wei, Y. *et al.*: 2012, *J. Geophys. Res.* **117**(A), 03208, doi:10.1029/2011JA017340
Weiss, B. P. & Elkins-Tanton, L. T.: 2013, *Ann. Rev. Earth Planet. Sci.* **41**, 529
Weiss, B. P., Berdahl, J. S., Elkins-Tanton, L., *et al.*: 2008, *Science* **322**, 713
Weiss, B. P., Gattacceca, J., Stanley, S., Rochette, P., & Christensen, U. R.: 2010, *Space Sci. Rev.* **152**, 341
Weiss, L. M. & Marcy, G. W.: 2014, *ApJL* **783**, L6, doi:10.1088/2041-8205/783/1/L6
Welsh, B. Y. & Shelton, R. L.: 2009, *Ap&SS* **323**, 1, doi:10.1007/s10509-009-0053-3
Welsh, B. Y., Lallement, R., Vergely, J.-L., & Raimond, S.: 2010, *A&A* **510**, A54, doi:10.1051/0004-6361/200913202
Welsh, W. F., Orosz, J. A., Carter, J. A., *et al.*: 2012, *Nature* **481**, 475, doi:10.1038/nature10768
Welsh, W. F., Orosz, J. A., Short, D. R., *et al.*: 2014, ArXiv e-prints
Whipple, E. C.: 1981, *Reports on Progress in Physics* **44**, 1197, doi:10.1088/0034-4885/44/11/002
Whipple, E. C., Northrop, T. G., & Mendis, D. A.: 1985, *J. Geophys. Research* **90**, 7405, doi:10.1029/JA090iA08p07405
White, S. M. & Franciosini, E.: 1995, *ApJ* **444**, 342, doi:10.1086/175609

White, S. M., Jackson, P. D., & Kundu, M. R.: 1993, *Astron. J.* **105**, 563, doi:10.1086/116453

Wieczorek, M. A., Jolliff, B. L., Khan, A., *et al.*: 2006, *Rev. Mineral. Geochem.* **60**, 221

Winn, J. N.: 2011, *Exoplanet Transits and Occultations*, 55–77, S. Seager, (Ed.), University of Arizona Press, Tucson, AZ:

Winn, J. N., Fabrycky, D., Albrecht, S., & Johnson, J. A.: 2010, *ApJL* **718**, L145, doi:10.1088/2041-8205/718/2/L145

Winske, D., Gary, S. P., Jones, M. E., *et al.*: 1995, *Geophys. Res. Lett.* **22**, 2069, doi:10.1029/95GL01983

Withers, P. G.: 2009, *Adv. Space Res.* **44**, 277

Withers, P.: 2010, *Advances in Space Research* **46**, 58, doi:10.1016/j.asr.2010.03.004

Withers, P. G. & Pratt, R.: 2013, *Icarus* **225**, 378

Withers, P. G., Bougher, S. W., & Keating, G. M.: 2003, *Icarus* **164**, 14

Witte, M.: 2004, *A&A* **426**, 835, doi:10.1051/0004-6361:20035956

Woch, J., Krupp, N., Lagg, A., & Tomás, A.: 2004, *Adv. Space Res.* **33**, 2030, doi:10.1016/j.asr.2003.04.050

Wolfire, M. G., Hollenbach, D., McKee, C. F., Tielens, A. G. G. M., & Bakes, E. L. O.: 1995, *ApJ* **443**, 152, doi:10.1086/175510

Wolk, S. J., Harnden, Jr., F. R., Flaccomio, E., *et al.*: 2005, *ApJSS* **160**, 423, doi:10.1086/432099

Wolszczan, A. & Frail, D. A.: 1992, *Nature* **355**, 145, doi:10.1038/355145a0

Wood, B. E.: 2004, *Living Reviews in Solar Physics* **1**, 2, doi:10.12942/lrsp-2004-2

Wood, B. E. & Linsky, J. L.: 1997, *ApJL* **474**, L39, doi:10.1086/310428

Wood, B. E., Linsky, J. L., & Ayres, T. R.: 1997, *ApJ* **478**, 745

Wood, B. E., Müller, H.-R., & Zank, G. P.: 2000, *ApJ* **542**, 493, doi:10.1086/309541

Wood, B. E., Müller, H.-R., Zank, G. P., & Linsky, J. L.: 2002, *ApJ* **574**, 412, doi:10.1086/340797

Wood, B. E., Müller, H.-R., Zank, G. P., Linsky, J. L., & Redfield, S.: 2005, *ApJL* **628**, L143, doi:10.1086/432716

Wood, B. E., Harper, G. M., Müller, H.-R., Heerikhuisen, J., & Zank, G. P.: 2007a, *ApJ* **655**, 946, doi:10.1086/510404

Wood, B. E., Izmodenov, V. V., Linsky, J. L., & Malama, Y. G.: 2007b, *ApJ* **657**, 609, doi:10.1086/510844

Wood, B. E., Izmodenov, V. V., Alexashov, D. B., Redfield, S., & Edelman, E.: 2014a, *ApJ* **780**, 108, doi:10.1088/0004-637X/780/1/108

Wood, B. E., Müller, H.-R., Redfield, S., & Edelman, E.: 2014b, *ApJL* **781**, L33, doi:10.1088/2041-8205/781/2/L33

Woodman, R. F.: 2009, *Ann. Geophys.* **27**, 1915

Wu, Y.: 2003, in D. Deming and S. Seager (Eds.), *Scientific Frontiers in Research on Extrasolar Planets*, Vol. 294 of *Astronomical Society of the Pacific Conference Series*, 213

Wu, Y. & Lithwick, Y.: 2011, *Appl. Phys. J.* **735**, 109, doi:10.1088/0004-637X/735/2/109

Wyatt, M. C.: 2008, *Ann. Rev. Astron. Astrophys.* **46**, 339, doi:10.1146/annurev.astro.45.051806.110525

Wyman, K. & Redfield, S.: 2013, *ApJ* **773**, 96, doi:10.1088/0004-637X/773/2/96

Xiong, M., Davies, J. A., Bisi, M. M., *et al.*: 2013, *Solar Phys.* **285**, 369, doi:10.1007/s11207-012-0047-0

Yaroshenko, V. V., Horányi, M., & Morfill, G. E.: 2008, in *Multifacets of Dusty Plasmas, Fifths International Conference on the Physics of Dusty Plasmas*, Vol. 1041, AIP, 215

Yashiro, S. & Gopalswamy, N.: 2009, in N. Gopalswamy and D. F. Webb (Eds.), *IAU Symposium*, Vol. 257 of *IAU Symposium*, 233
Ye, S.-Y., Gurnett, D. A., Kurth, W. S., *et al.*: 2014a, *J. Geophys. Res. Space Physics* **119**, 6294, doi:10.1002/2014JA020024
Ye, S.-Y., Gurnett, D. A., Kurth, W. S., *et al.*: 2014b, *J. Geophys. Res. Space Physics* **119**, 3373, doi:10.1002/2014JA019861
Yelle, R, V. & Miller, S.: 2004, in *Jupiter's Thermosphere and Ionosphere, in Jupiter: The Planet, Satellites and Magnetosphere*, Cambridge University Press, 185–218
Yelle, R. V., McConnell, J. C., Sandel, B. R., & Broadfoot, A. L.: 1987, *J. Geophys. Res.* **92**(A13), 15 110–15 124
Yelle, R., Young, L., Vervack, R. J., Jr., *et al.*: 1996, *J. Geophys. Res.* **101**(E1), 21492161
You, X. P., Coles, W. A., Hobbs, G. B., & Manchester, R. N.: 2012, *Mon. Not. R. Astron. Soc.* **422**, 1160, doi:10.1111/j.1365-2966.2012.20688.x
Young, L. A., Yelle, R., Young, R., Seiff, A., & Kirk, D.: 1997, *Science* **276**(5309), 108111, doi:10.1126/science.276.5309.108
Yu, H.-S., Jackson, B. V., Buffington, A., *et al.*: 2014, *ApJ* **784**, 166, doi:10.1088/0004-637X/784/2/166
Yu, H.-S., Jackson, B. V., Hick, P. P., *et al.*: 2015, *Solar Phys.* in press
Yue, J. & Wang, W.: 2014, *J. Geophys. Res.* **119**, doi:10.1002/2013JA019725
Yue, J., Wang, W., Richmand, A. D., & Liu, H.-L.: 2012, *J. Geophys. Res.* **117**(A7), A07305.
Zahn, J.-P.: 1989, *A&A* **220**, 112
Zalesak, S. T., Ossakow, S. L., & Chaturvedi, P. K.: 1982, *J. Geophys. Res. Space Physics* **87**, A1, 151
Zank, G. P.: 1999, *Space Sci. Rev.* **89**, 413, doi:10.1023/A:1005155601277
Zank, G. P., Pauls, H. L., Williams, L. L., & Hall, D. T.: 1996, *J. Geophys. Research* **101**, 21639, doi:10.1029/96JA02127
Zank, G. P., Heerikhuisen, J., Wood, B. E., *et al.*: 2013, *ApJ* **763**, 20, doi:10.1088/0004-637X/763/1/20
Zarka, P.: 1985, *Astron. Astrophys.* **146**(L), 15–18
Zarka, P.: 1998, *J. Geophys. Res.* **103**(E9), 20 159–20 194
Zhan, X. & Schubert, G.: 2012, *J. Geophys. Res.* **117**, E08011
Zhang, H., Khurana, K. K., Kivelson, M. G., *et al.*: 2014, *J. Geophys. Res.* **119**, 5220, doi:10.1002/2014JA020111
Zhang, K. & Busse, F. H.: 1990, *Phys. Earth Planet. Inter.* **59**, 208
Zhao, X. & Hoeksema, J. T.: 1995a, *Advances in Space Research* **16**, 181, doi:10.1016/0273-1177(95)00331-8
Zhao, X. & Hoeksema, J. T.: 1995b, *J. Geophys. Res.* **100**, 19, doi:10.1029/94JA02266
Zieger, B., Opher, M., Schwadron, N. A., McComas, D. J., & Tóth, G.: 2013, *Geophys. Res. Lett.* **40**, 2923, doi:10.1002/grl.50576
Zuluaga, J. I. & Cuartas, P. A.: 2012, *Icarus* **217**, 88
Zurek, R. W.: 1976, *J. Atmos. Sci.* **33**, 321
Zurek, R. W., Barnes, J. R., Haberle, R. M., *et al.*: 1992, *Dynamics of the Atmosphere of Mars*, 835–933

Index

Printed in the United States
By Bookmasters